WHO
DISCOVERED
WHAT
WHEN

David Ellyard

First published in Australia in 2005 by
Reed New Holland
an imprint of New Holland Publishers (Australia) Pty Ltd
Sydney • Auckland • London • Cape Town
14 Aquatic Drive Frenchs Forest NSW 2086 Australia
218 Lake Road Northcote Auckland New Zealand
86 Edgware Road London W2 2EA United Kingdom
80 McKenzie Street Cape Town 8001 South Africa

National Library of Australia Cataloguing-in-Publication Data:

Ellyard, David, 1942– .

Who discovered what when.

Bibliography.
Includes index.
ISBN 1 877069 22 1.

1. Discoveries in science—Encyclopedias.
2. Science—Encyclopedias. I. Title.

503

Publisher: Louise Egerton
Project Editor: Yani Silvana
Designer: Avril Makula
Production: Linda Bottari
Printer: Tien Wah Press, Singapore

Cover photographs
Top Left: Albert Einstein. Photolibrary/Science Photo Library
Top right: Marie Curie. Photolibrary/Index Stock Imagery
Centre left: DNA spiral. Photolibrary/Photo Researchers, Inc.
Bottom left: Galileo Galilei. Photolibrary
Bottom right: Radio Telescope array (VLA, New Mexico). Dave Finley, NRAO/AUI/NSF

Contents

Acknowledgements

While this book has been brewing at the back of my brain for many years, its appearance in print at this time was made possible only through the generosity of Reed New Holland, who sponsored the Science Book Prize at the 2004 Eureka Awards, staged by the Australian Museum, the prize I won with the proposal for this book. Given that they also undertook to publish the book, which was only 10 per cent written at the time, choosing me as the winner was a bold act of faith on their part, which I hope I have justified. They also provided me with some highly talented people to work with, notably publishing manager Louise Egerton and editor Yani Silvana. I am very grateful for their expertise and insights, their encouragement and their patience. This book is full of their influence.

My family and friends have responded enthusiastically to news of the book's progress. I think particularly of Bill and Jan Crosby, Tony and Lindsay Cotton and Alan and Christine Bishop, who were there to 'hold my hand' on the night the prize was announced.

The book would not have been written without the patient encouragement of my wife Sue through a very busy time. I cannot thank her enough. My children —Rachel, James and Sam—always clamoured to know 'how it was going'.

The ultimate inspiration for this book came from my late father, Samuel Ellyard, who pioneered the teaching of natural science in primary schools in this country, and who opened my eyes to the fascination and human dimension of science at an early age. I also thank my mother, Marjorie Bell Ellyard, and dedicate this book to her memory.

How to Get the Most from this Book

Who Discovered What When (WDWW) will introduce you to the growth of scientific ideas over the past 500 years and to the people who brought them to life. It sets those ideas in a context of time and place and provides a framework for understanding the concepts and theories that have shaped our modern scientific world view.

No scientific idea emerges from a void. WDWW presents science as a cumulative and interactive activity, with researchers' work feeding into that of their contemporaries and successors. Even the greatest scientific minds—Newton, Lavoisier, Darwin, Pasteur, Einstein—drew upon the work of others.

WDWW illuminates what scientists do, highlighting the central role of observation and measurement and the interplay of evidence and theory.

Over the centuries, science has taken many wrong turns. Ideas that have enjoyed much popular support at various times—the Earth-centred universe, phlogiston, caloric, the aether, spontaneous generation of life, neptunism—later proved inadequate or misguided. Yet even these have contributed to a more complete understanding of both the natural order and the enterprise of science, and they therefore have a place in this book.

Each chapter covers 50 years, and is introduced by a brief overview summarising the major developments and trends during that period. Within the chapters, entries are arranged chronologically from around 1500 to almost 2000. This means you can move smoothly through the sequence of discoveries across the broad range of science, appreciating the interrelationships.

Each story highlights a particular discovery or new idea, interwoven with a pen picture of the discoverer in his (and sometimes her) personal circumstances. Most researchers have only a single entry, but the more productive have several, which form in total a summary of their life and work and their contribution to scientific knowledge. Bolded date cross-references point you to an earlier or later part of the story. An arrow symbol and date at the end of a story, e.g. **➻1603**, refer you to the next entry in the life of the particular researcher.

You can find value in this book in a number of ways: look up answers to specific questions that you have, such as about the Michelson–Morley experiment or Mendel's work on genetics or how Joseph Black discovered carbon dioxide; gain a useful overview, say of scientific developments in the first half of the twentieth century; follow through a particular subject area, such as the development of the idea of the atom, or arguments over the age of the Earth or the purpose of chromosomes; or simply dip into it—open a page at random and you will find a self-contained story that will intrigue, inform and entertain.

Whichever way you use it, enjoy!

David Ellyard

Introduction

What We Knew by 1500

In 1500 science as we know it today was almost unknown, even in western Europe where it all began. The notion that observation and experiment might provide a reliable guide to understanding how the world worked was not new, but not widely accepted either. The ideas of the great ancient thinkers were still dominant—Aristotle (384–322 BC) in biology and physics, Ptolemy (c. 85 – c. 165 AD) in astronomy and geography, Galen (131–201 AD) in medicine. Even the wisest often sought answers in those writings rather than in the evidence of their own eyes.

While few doubted that the Earth was round, few challenged Ptolemy's notion that the Earth stood still at the centre of the universe and all the other celestial objects went around it in circular orbits. After all, the great Aristotle had decreed that circular motion was the most perfect and natural.

Just as in ancient times, seven heavenly bodies were known: the Sun, the Moon and the 'wandering stars'—Mercury, Venus, Mars, Jupiter and Saturn. It would be nearly 300 years before that number was added to. The motions of these bodies were still of greatest interest to astrologers, who sought to define human fates from their positions in the sky.

Despite the progress made by experimenters in the Middle East and other areas under the sway of Islam, chemistry was still mostly alchemy: it was preoccupied with the quest for the Philosopher's Stone, which would turn base metals into gold, or with the Elixir of Life. Indeed, it would be a long time before such endeavours were seen as unworthy of science. Even 200 years later, the great Isaac Newton thought that alchemy was at least as important as any of his other work.

To back their quest, the alchemists had a theory that matter was composed of a small number of semi-mystical 'elements'. Aristotle and the other ancients had named these earth, air, fire and water, though more modern thought preferred the triumvirate salt, sulphur and mercury. Seven metals were known: gold, silver, lead, copper, tin, iron and mercury. Each of these had been linked in ancient times with one of the heavenly bodies—for example, gold with the Sun, silver with the Moon, iron with Mars—and the same symbols were used for the metal and the planet. Astrology and alchemy were still closely intertwined.

In physics, Aristotle was still largely dominant, especially since his world view had been harmonised with Christian theology through the writings of Thomas Aquinas. To challenge Aristotle was to challenge the authority of the Church. Most people believed, as he had done, that 'nature abhorred a vacuum'; that atoms were impossible; that an object needed to be continually pushed to keep it in motion; that some substances (those endowed with 'levity') naturally rose while those with 'gravity' sank, each seeking its rightful place in the scheme of nature.

Those who studied living things (the term 'biology' was still 300 years away) had added nothing of significance to the observations of Aristotle, if that were needed. The interest in plants was mostly for medicinal purposes. 'Herbals' full of recipes to treat ailments were popular, though they were not always based on careful observation. It was not until 1500 or after, when explorers in new lands found animals and plants unknown to Aristotle, that his authority began to fade.

Technology, of course, had advanced significantly, even in the absence of any sound scientific knowledge. Trial and error and the ingenuity of craftworkers produced many new ways to do useful things. Some of this technology, including paper, gunpowder and the magnetic compass, reached western Europe from China. Quite accurate clocks had been invented, hundreds of years before Huygens and the pendulum. Firearms and cannons were in regular use, though no one knew why gunpowder exploded. Perhaps most important of all, Gutenberg and others late in the fifteenth century mastered printing using movable type.

In science, too, the new practices of observation and experiment heralded change. In the early 1400s, Nicolaus Cusanus noted that plants appeared to absorb something from the atmosphere. There were references to newly discovered substances such as bismuth. Sailors noted that their magnetic compasses did not always point due north. Over coming decades and centuries findings like those would stimulate major lines of inquiry.

In the late 1400s the great early figures in the development of science were born or became active. Nicholaus Copernicus was born in 1473. Shortly after, Leonardo da Vinci was designing flying machines and experimenting with parachutes and capillary tubes. The monk Regiomontanus viewed what would later be called Halley's Comet from his observatory in Nuremberg, the first in Europe.

Stirrings of this new spirit can be glimpsed much further back, stimulated by the translation of classic writings that made long-suppressed or forgotten ideas accessible. In the thirteenth century, Witelo from Silesia understood much about the workings of the eye and optics in general. In England the Franciscan monk Roger Bacon had known the value of experiment. He explained the laws of reflection and refraction of light. He described an early magnifying glass, so contributing to the invention of spectacles soon after. He knew the composition of gunpowder, even if he did not invent it.

Bacon fought for the teaching of 'natural philosophy' in universities, but was imprisoned for heresy for his pains. The power of ecclesiastical authority over ideas was strong and remained so until Galileo and later. Even when the coercive power of the Church had waned, ideas from theology and the Bible continued to shape much of science. Not until after 1800 did the biblical story of Creation and the Great Flood cease to influence biology and geology.

In the two centuries that followed Bacon, others would promote heretical ideas in physics that contradicted Aristotle and anticipated Galileo and Newton. William of Occam devised his famous 'razor', which prized simplicity in explaining observations. William of Merlee kept records of the weather at Oxford for the best part of a decade, and even tried to forecast it. There were experiments with magnets and compasses, and explanations about the origin of rainbows. Teachers and writers in surgery and human anatomy began to rely more on evidence from dissections and autopsies and less on the writings of Galen.

Around 1500, these trends began to pick up pace. The reasons for this are much debated. Some people point to the growth of dissent within the Church; others to the widening of horizons through the exploits of voyagers to new lands; others still to the stimulus of early industry and the challenges it posed. A fourth influence was probably the new ease of recording and disseminating ideas through the printed word.

Whatever the reasons, we can look to the period around 1500 as the launching pad for what we now call science. What came after was very different from what went before. So that is when this book begins.

One last point: nearly all the discoveries within this book were made in Europe or in regions to which European influence spread, such as North America. Modern science is rooted in the history and culture of Europe; other parts of the world have begun to make their contribution only recently. Well before 1500, centuries of research and study in India and in Islamic countries had petered out. Some centuries earlier, China, which had been so active in technology and exploration, had gone into eclipse when autocratic central governments shut down the tradition of free enquiry essential for science to flourish.

1500–1550

The World Stage

EUROPE IN 1500 LITTLE RESEMBLED the peaceful and increasingly integrated Europe of the twenty-first century, where science is universally established and supported. Politically, it was far more fragmented and uneasy.

Modern Germany and Italy did not exist; in their places were collections of independent and often fractious principalities and dukedoms. This may have helped the establishment of science, as the many courts provided patronage for early 'natural philosophers'.

Spain and France had central governments controlling all or most of the territory they hold today, though there were rival factions in France, and territory in the east and north was in other hands. England had a troubled administration over Ireland, but Scotland was an independent kingdom. Democracy as we know it hardly existed. Government was autocratic. Power had to be secured and held by force of arms, though a notion of the 'divine right of kings' was beginning to grow.

The greatest influence, and much of the territory, was held by the Habsburg dynasty. From centres in Austria and Spain, it controlled much of fragmented Italy and the 'Low Countries', later to become Belgium and Holland, as well as the German-speaking regions of central Europe, gathered under the banner of the Holy Roman Empire.

From around 1517 Europe was divided again, this time between two great religious factions. The onset of the Reformation, sparked by the German Martin Luther and the Frenchman John Calvin, saw the various states line up either as supporters of the 'Protestants' (mostly lying north of the Rhine) or as upholders of the traditional 'Catholic' faith (mostly to the south). Religious passion and traditional rivalries made a volatile mix. Two centuries of devastating religious conflict, both between states and within them, followed.

The Reformation was a protest against what was seen as corruption in the Roman Church, though the sentiments had been simmering for decades, perhaps centuries. The Church condemned Luther and fought back with the Council of Trent in 1545 to propose its own reforms, and with the founding of the Society of Jesus in 1540 (interestingly, in future centuries, the 'Jesuits' would include prominent scientists among their number).

Yet the motivation to challenge long-established ideas, many derived from the Greek savants and sanctified by the Church, would prove an irresistible force; it was spurred on by the decision of the Church to establish an index of 'forbidden books', which would include the writings of Copernicus and Galileo.

Perhaps this is why rich and powerful Catholic Spain made little contribution to science for centuries and was far outdone by smaller powers in the Low Countries and Scandinavia, as well as England and Scotland, where the Reformation took root. The tide of the Renaissance and the patronage of philosophers by its many princes and dukes, most notably in Florence, sustained new ideas in Catholic Italy for a time, though the punishment handed out to their promoters, most notably to Galileo (**1633**), caused that tide to ebb again. Nothing much happened in Catholic France until rulers like Louis XIV began to patronise science as a source of national greatness a century later.

External factors also contributed to instability in Europe. One was the push for empire, led by Spain in the New World of the 'Americas' and by Portugal in Africa and the 'Far East'. This led inevitably to tensions over trade, with other states—particularly England and Holland—willing to seize their share. The conflict between Spain and England that culminated in the 'Spanish Armada' came partly from religious conviction and partly from commercial considerations.

The other threat lay to the south-east. The Ottoman Empire had been expanding into Europe since the fall of Constantinople in the previous century. Scholars fleeing from that catastrophe had come into Italy and helped to sustain and revitalise the Renaissance and to stimulate science. But the Turks had now spread through the Balkans and Hungary and besieged Vienna in 1529.

It was in this complex and unstable environment that modern science was born, beginning in Italy and spreading north and west. From simple beginnings its growth has been spectacular.

ARTS AND IDEAS

This was the high water mark of the philosophy of humanism, typified by the Dutchman Erasmus in his book, *Education of a Christian Prince*. The Italian politician Niccolo Machiavelli set down his more cynical 'rules for rulers' in *The Prince* in 1513, while in 1516 his English counterpart Thomas More outlined his vision of a perfect state in *Utopia* (the name means 'nowhere'). Martin Luther instigated the Reformation the next year. The Bible was translated from Latin into everyday speech (into English by William Tyndale, into German by Luther). The English *Book of Common Prayer* was published in 1540.

Among the contrasts in literature of the time were poetic romances of the Italian Locovico Aristo and the English morality play *Everyman*. The music commonly heard in church and court included compositions by the English Robert Fayrfax and the Flemish Josquin de Pres.

With the flood of the Renaissance still to ebb, Italians were the masters in the visual arts. Michelangelo sculpted his *David* in 1504 and completed decorating the ceiling of the Sistine Chapel in 1512. Da Vinci painted the *Mona Lisa* in 1505. Titian (*Venus of Urbino* 1538), Raphael, Tintoretto and Correggio were among the many others also busy, as was the sculptor Cellini. German painter Hans Holbein was famous for his paintings of royalty and nobility, including Henry VIII of England and his court. His countryman Albrecht Durer was equally famous for his woodcuts of scenes of everyday life.

EXPLORATION

The early sixteenth century was a period of intense activity as European explorers continued to push both east and west across the planet in search of new territory and opportunity. Portuguese navigators explored the coast to Brazil, Spaniards discovered Florida and Mexico's Yucatan Peninsula, penetrated into Peru and, to the north, found the Grand Canyon.

One adventurer sailed the length of the Amazon River from the Andes to the Atlantic. Vasco Balboa crossed the Isthmus of Panama to see for the first time the Pacific Ocean, which he named. In 1507 German mapmaker Martin Waldseemuller proposed that the whole region should be called *America* after Spanish explorer Amerigo Vespucci.

The Portuguese, meanwhile, followed the increasingly familiar route around the bottom of Africa and across the Indian Ocean. They reached past India to New Guinea and the 'spice islands' of the Malaccas, setting up trading posts in Sri Lanka and China and making contact with Japan.

The world was first circumnavigated in 1521 by ships initially commanded by the Spaniard Ferdinand Magellan. They sailed around the tip of South America and across the Pacific, though Magellan himself did not survive the discovery of the Philippines.

In his division of the world in 1494, the Pope had given exclusive rights to Portugal in Africa and Asia, and to Spain in the western hemisphere. The dividing line cut the nose off South America to form Portuguese-controlled Brazil and placed the still-to-be-discovered Philippines in the Spanish realm.

First Steps in Science

Though science as we know it had barely begun, there was significant progress in a wide range of areas in the 50 years to 1550: in medicine with the work in anatomy of Vesalius and his many contemporaries, and Fracastoro opening up new thoughts on the transmission of disease; in astronomy with the challenge made by Copernicus to the long-established understanding of the cosmos; and in mathematics. Paracelsus introduced the first stirrings of modernity in chemistry, Agricola in geology and Gesner in biology, with his encyclopaedias of plants and animals. The often inspired insights of Leonardo da Vinci impacted on many fields.

Those active tended to be mostly medical men or mathematicians, and many enjoyed patronage. Italy was the prime location, either as a homeland or as a place for work or study.

Leonardus Camillis: Starting with Stones

Speculum Lapidum (the 'Mirror of Stones'), first published in 1502, was an early compilation of knowledge about minerals, especially precious and semi-precious gems. It was comprehensive, covering essentially all of the 250 minerals known at that time, describing their physical properties such as colour and cleavage.

1502

At this time ideas about chemical composition were very ill formed, barely existent, so Camillus said very little about what these stones contain, but much about their power to cure various conditions and diseases. The colour of the stone was thought to be relevant. Conventional wisdom, handed down from ancient times, claimed that red stones such as rubies and garnets were good for controlling bleeding and infection, while sapphires cured boils.

Like many 'natural philosophers' of the day, the book's author, Leonardus Camillis, was a doctor. He served as court physician to the warlord Cesare Borgia, to whom he dedicated the work. The book reported, among many other things, that amber (solidified tree resin) was most useful for fixing loose teeth, as an antidote to poison and as a remedy for all disorders of the throat. This indicates how far science had to go, since there would have been little evidence for such claims. To test them against experience would have been thought disrespectful to the great minds of the past. Science could not progress until this reverence for old ideas was replaced by observation and experiment.

Leonardo da Vinci: First in Science, Too

1517

In this year, Leonardo da Vinci, one of the most extraordinary talents in recorded human history, died. Born into the full flood of the late Italian Renaissance, his immense achievements in painting, sculpture, literature, music, military strategy, invention and engineering have been much celebrated.

Leonardo also made powerful contributions to many areas of early science, especially in his later years. But he hardly ever finished any investigation he began and committed very few of his discoveries to paper. Many of his most profound ideas were mere speculations with little evidence to back them at the time. Yet he had a grasp of many important principles that were to be the foundations of whole areas of science.

> *There is no higher or lower knowledge,*
> *but one only, flowing out of experimentation.*
>
> LEONARDO DA VINCI

Geology was one. In his service as an engineer, he was familiar with the layered structure of the rocks in northern Italy. He deduced that rocks had been laid down underwater one layer at a time, with the oldest rocks at the bottom. This was perhaps the first statement on the 'law of superposition', rediscovered more than a century later by Nicholas Steno (**1667**).

Leonardo also noted that distinctive 'fossils' occurred in the various rock layers, enabling them to be traced and identified over large distances. Three hundred years later William Smith (**1799**) would use the same notion to found modern stratigraphy and geological mapping. Leonardo believed that fossils such as shells were the remains of once-living creatures and could be used to retell the story of the Earth.

His writings contain tantalising references to the use of mirrors and lenses to give enlarged views of the planets. Whether or not he ever used such a 'telescope', he knew that the Moon shines not by its own light, but by reflecting light from the Sun. Just after a new Moon, light reflected by the Earth dimly illuminates the unlit part of the Moon, so creating 'the old Moon in the young Moon's arms'. He also mused on the radical possibility that the Earth orbited the Sun, rather than the reverse, and that the Earth was just a planet like the five others known.

To a painter and sculptor like Leonardo, a proper understanding of the way the human body was constructed was key knowledge. So he was a pioneer in anatomy, including dissection of the body, and made highly detailed drawings. He appears to have understood the way the valves of the heart worked and speculated that blood circulated through the body rather than ebbing and flowing back and forth as was commonly believed at the time. A century later William Harvey (**1616**) presented the mature understanding of that idea.

Leonardo was a pioneer in physics as well, making early observations that helped in time to bring down the still widely held views of motion descended from Aristotle. He suggested that objects could keep moving steadily without being continually pushed and that the application of a force was needed only to increase or decrease an object's speed or change its direction. He postulated that: 'Every object has a weight in the direction of its motion'. Here he anticipated Galileo (**1604**) and then Isaac Newton (**1686**).

Paracelsus: A Life of Conflict

The sixteenth-century physician who styled himself Paracelsus was one of the first to openly challenge the received wisdom in science and medicine that had reached his time via the Arabs from the Greeks. Though never properly trained, he began a medical and surgical practice in Basle in Switzerland around 1520 but was usually in such conflict with his colleagues and the authorities that he spent much of his life wandering Europe and died at the age of 48.

1520

Paracelsus was violently opposed to the generally accepted ideas of his time. He publicly burnt (or at least urged the burning of) the iconic books of Galen and the Arab physician Avicenna. Conventional wisdom attributed human illness to an imbalance between the four bodily 'humours'. Paracelsus looked instead to external causes and advocated specific cures for specific diseases.

Paracelsus had his own system of 'elements': 'sulphur' represented fire, 'mercury' represented movability, and 'salt' represented solidity, but these did not always refer to the common substances with those names. Nonetheless, many of the drugs he prescribed contained sulphur, mercury and salt, as well as arsenic, antimony and iron, in place of the herbs preferred in contemporary medicines. In this, he foreshadowed a new study—chemical medicine or 'iatro-chemistry'. He also promoted the medical benefits of opium. It was he who first referred to 'spirits of wine' as 'alcohol', and he made the first recorded reference to a metal called zinc.

Paracelsus was a paradoxical personality. He remained convinced of the truth of much of alchemy, while at the same time advocating some experimental methods

> *Medicine is not only a science; it is also an art. It does not consist of compounding pills and plasters; it deals with the very processes of life, which must be understood before they may be guided.*
>
> PARACELSUS

and new explanations. He also proposed that alchemy might be employed more usefully seeking treatments for human diseases than trying to change lead into gold.

In his human dealings he was not helped by an overbearing, conceited manner. He chose the name Paracelsus to indicate at least parity with the famous Roman physician Celsus. His real surname (he was baptised Theophrastus Bombastus) gives a further hint to his flawed character. Nevertheless, his place in the early history of science is secure.

Niccolo Tartaglia:
The Mathematical Triumphs of the Stammerer

1537 Mostly self-taught in mathematics, Italian engineer and mathematician Niccolo Tartaglia won fame through the solutions he proposed to difficult mathematical equations. At the time, intense rivalry flowed between mathematicians over their methods, which they often tried to keep secret. Ill feeling towards some of his competitors, notably Gerolamo Cardano, affected him for much of his life.

Tartaglia used his mathematical insights to analyse the motion of projectiles such as cannon balls and bullets (like many researchers in those uncertain times, he had an interest in fortifications and other military matters). He found that, once fired, a ball or bullet followed a path called a parabola, which could be described mathematically. He published this insight in his 1537 book *A New Science*, and it enabled gunners to aim more accurately (using 'firing tables').

This later fitted neatly with the laws that Galileo (**1604**) devised to explain how objects behave under the effect of gravity.

It was also intriguing to discover that the parabola is the shape exposed when a solid cone is sliced through along a line parallel to its sloping sides. This helped reinforce the growing realisation that mathematics is deeply entwined with the workings of the world. As Galileo was to remark memorably: 'Mathematics is the language in which the book of nature is written'.

In later life Tartaglia always wore a beard to hide the terrible scars on his jaw caused by injuries he had suffered as a youth from the swords of French invaders of his village. The injuries also affected his speech, hence his name, which means 'the stammerer'.

Konrad Gesner: So Much in One Lifetime

1541 Swiss-born Konrad Gesner lived only 49 years (he died of the plague), but found time to practise as a physician and surgeon, to pioneer mountain-climbing in the nearby Alps, to compile a comprehensive list in Latin, Greek and Hebrew of all the important literature in all languages (a 'universal library', as he called it), to complete a comparative study of 130 languages and to found zoology and botany as scientific studies. He knew something about almost everything and a great deal about many subjects. Some likened him to the Roman Pliny, who 1500 years earlier had compiled encyclopaedias containing everything known about the natural world.

His two great works were encyclopaedias of all known plants and animals (at least those known in the Old World). His *Historia animalium* ran to 4500 pages in five volumes. It included mythical and imaginary creatures, particularly sea monsters, in order to be comprehensive. In addition to descriptions, the book was illustrated with woodcuts commissioned by Gesner from leading artists of his time; this was a first for a book of this kind. The illustrations were not always accurate, but certainly made the book much more useful, and set the pattern for the future.

In his commentaries he drew on the works of all the relevant ancient authors and his contemporaries. He included proverbs, Bible references, pagan mythology, even Egyptian hieroglyphs. He did not include many direct observations of his own; that crucial element of science did not always get much prominence in his day. His *Historia plantarum*, published in 1541, was equally comprehensive and the first modern work in botany. He anticipated by nearly 300 years the work of Carl Linnaeus in setting up a system of classification (**1753**).

Gesner's other influential writings included *The New Jewel of Health*, which described how many useful medications could be distilled from plants. In his book *On Fossil Objects* (the full title is *A Book on Fossil Objects, Chiefly Stones and Gems, their Shapes and Appearances*) he made a powerful case, supported by illustrations, that many fossils were the remains of long-dead organisms. Like Leonardo da Vinci, he failed to convince most people, since he could not explain how the tissues of plants and animals could be turned into stone.

Andreas Vesalius: Exploring the Human Body

One of the two epoch-making scientific books published in 1543 (the other being by Copernicus) was *The Fabric of the Human Body* by the Flemish-born doctor Andreas Vesalius. In this book—indeed in all his teachings—Vesalius challenged the authority of the Greek writer Galen, whose writings had been the standard in medicine and anatomy for more than 1000 years.

1543

While Galen appears to have never dissected a human cadaver and instead based his ideas on the bodies of pigs and apes, Vesalius relied only on his own observations of the internal structures of the human body revealed by dissection. He found so many errors in Galen's descriptions that the authority of the old Greek ought to have been totally undermined. But Vesalius had made powerful enemies—even in his own university of Padua—among the medical establishment, to whom Galen was still supreme. Many in the Catholic Church suspected him of sympathy with the Lutherans and Calvinists.

Such conflicts filled much of his life even though he had been appointed physician and apothecary to the Holy Roman Emperor Charles V and his son Phillip ll of Spain. Called before the Inquisition, Vesalius was required to set out on a pilgrimage to the Holy Land at the age of 50, from which he never returned.

Vesalius did more than correct the many errors found in Galen (and passed on by teachers who did little more than cite his authority). He founded the modern tradition of anatomy based on precise observation and provided the first sound descriptions of the structure of the brain and the details of the human digestive system. Like Agricola (**1546**) in geology, Vesalius was in large measure the prototype of the modern scientific investigator, making the theory fit the facts and not the other way round.

While Vesalius and other anatomists of his day (**1552**) provided an accurate and comprehensive description of how the human body looked, there was little understanding as to how it worked. Real progress there would have to await inputs from physics and chemistry.

Nicholaus Copernicus: A Revolution in the Heavens

1543

In 1543, the year that Andreas Vesalius disputed the millennium-long authority of the Greek physician Galen, the Polish astronomer Nicholaus Copernicus, in his book *The Revolutions of the Heavenly Spheres*, challenged the even more venerable views of Ptolemy and Aristotle.

At issue was the very structure of the known universe: the Sun, the Moon and the five planets, together with the Earth, placed in the midst of a vast sphere of 'fixed stars'. The authority of Aristotle and Ptolemy, bolstered by that of the Church, granted the Earth, as the 'abode of Man', the central place. There it lay unmoving, with all the other celestial objects and the starry heavens wheeling around it.

In his new vision of the celestial order Copernicus boldly moved the Sun into the centre. Henceforth the Earth was merely one of now six planets in orbit around the Sun. He claimed that all the complex patterns of movement of the heavens could be explained by the daily rotation of the Earth on its axis and the yearly movement of the Earth around the Sun.

The idea was not new; some ancient Greek thinkers had placed the Sun at the centre of the world, so requiring the Earth to move. It was already known that the

> *To know that we know what we know, and to know that we do not know what we do not know, that is true knowledge.*
>
> NICHOLAUS COPERNICUS

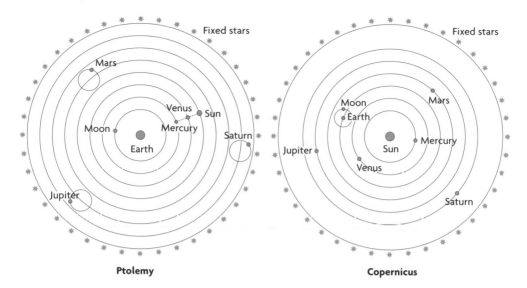

Ptolemy **Copernicus**

In Ptolemy's model of the solar system everything rotated around a motionless Earth. Copernicus devised the Sun–centred system to replace it. Ptolemy's scheme was in fact far more complex than is shown here. To explain why Mercury and Venus always stay close to the Sun in the sky, Ptolemy required the three bodies to be connected in some unexplained way and so to move around the Earth as a group. Copernicus had a much simpler explanation: the two planets are closer to the Sun than the Earth is.

To explain 'retrograding', the apparent backward movement of Mars, Jupiter and Saturn against the more distant stars for a few months each year, Ptolemy suggested that the planets make yearly trips around small circles ('epicycles'), the centres of which go around the Earth. For Copernicus this phenomenon was an illusion, the inevitable result of the simultaneous movement of the Earth and the other planets around the Sun.

Sun was much larger than the Earth; that made it all the more likely that the Sun was at the centre. The Earth-centred view, however, had more powerful supporters. Furthermore, the new system was no better than Ptolemy's as a way of predicting where the planets would appear in the sky, and it had its own complexities, because Copernicus stuck with the old idea that the orbits of the planets had to be perfect circles (**1609**).

Copernicus was himself a devout man, but by challenging the Earth-centred universe, which most people saw as supported by Scripture, he was sure to stir up opposition among ecclesiastical authorities. To head off trouble, a colleague wrote a preface (without the author's approval or even knowledge) stating that the 'model' was only an aid to calculation and did not necessarily match reality.

Copernicus had been in no hurry to publish, refining and revising his views for 30 years. The book may not have come out when it did had the young German mathematician Georg Rheticus, his house guest for two years, not urged him into print. (In **1684** Edmond Halley would perform the same service for a reluctant Isaac Newton.)

After early studies in Italy, Copernicus served many years as a church official in Silesia (now in Poland) and was active in public life, including as a diplomat and even an economist. A copy of his great work was placed in his hands only a day before he died following a stroke.

Girolamo Fracastoro: The Seeds of Disease

1546

Sixteenth-century Italy overflowed with 'Renaissance men'—men who were skilled and knowledgeable in just about every known art and science.

Leonardo da Vinci was merely the greatest among many. Girolamo Fracastoro was by employment a doctor, but he wrote poetry, was active in music and produced books on geography, astronomy and theology. He was the confidant of popes and other great men of his time.

Born in Verona, he was educated in Padua, where he shared classes with Nicholaus Copernicus and became a close friend. When the builders of a new fortress in Verona found fossils that looked like shellfish in the rocks being excavated, he put forward the radical view (shared by da Vinci) that these were the remains of animals that had once lived in the area.

Fracastoro's legacy includes the first statement of a 'germ theory' of disease, 100 years before any such organisms were seen (**1673**) and 300 years before Louis Pasteur and Robert Koch proved any such thing (**1876**). Fracastoro noted how quickly contagion could spread through human and animal populations (he had witnessed an epidemic of what was probably foot and mouth disease in local cattle). In 1546 he suggested that such outbreaks were carried by 'seeds of disease', passed on by direct contact, contaminated clothing and utensils or even through the air. At that time, common opinion maintained that epidemics were the result of God's wrath at human wickedness.

Fracastoro had no substantial evidence for his proposal. It was speculation and was generally rejected. Still, his influence was sufficient to induce the Pope to move the Council of Trent in 1545 to Bologna to avoid an outbreak of disease. The disease may have been typhus, which he first described soon after.

In 1530 Fracastoro wrote a poem about a 'French disease', sexually transmitted, that we now call syphilis, the name coming from the title of the poem. Four centuries later, the medical journal *The Lancet* would describe Fracastoro as the man who did most 'to spread knowledge of the origin, clinical details and available treatments of the disease throughout a troubled Europe'.

Agricola: The First Geologist

If anyone can be said to have founded the methodical study of rocks, later to be called 'geology', it was Georg Bauer from Saxony (Germany), better known as Agricola. He was a modern scientist in the sense that he relied on his own observations and experiments or on information from very reliable sources. His most famous book, published in 1556 after his death, was *De Re Metallica*, a textbook in mining and metallurgy used for the next two centuries. Among other things, it describes the production of nitric and sulphuric acids.

Two of his books, both published in 1546, were seminal in the study of geological structures and 'fossils', which at the time meant anything dug out of the ground. Agricola described wind and water as major forces shaping the appearance of the Earth's surface, and suggested that earthquakes and volcanoes were signs of the Earth's internal heat.

Agricola reported that among the fossils found in rocks which had been split open were shells which resembled those of living things. However, he rarely stated that fossils represented organisms once living, since that was a matter of strong debate at the time and would not be settled for another 100 years.

Like many natural philosophers of his time, Agricola started off in medicine. He was later drawn to study rocks and minerals, largely because he was practising in the town of Joachimstal, site of the famous silver mines of the time. He also filled public posts in local government and in the court of the Duke of Saxony.

Clocks, Guns and Knitting Machines

SIXTEENTH-CENTURY TECHNOLOGY

Sixteenth-century technology lay in the hands of craftworkers and engineers. It would be centuries before science would have much impact. Clocks were the most advanced machines of the day, decorating the towers of churches and public buildings. In 1515 a German clockmaker devised the first pocket watch.

Small arms for the many battles were advanced in 1515 by the invention of the 'wheel lock', which scraped flint on steel to ignite the gunpowder. For generations after, a gun could be broken down into three parts: the firing mechanism, the wooden rest against the shoulder and the metal tube at the front end; hence 'lock, stock and barrel'. The accuracy of small arms increased after 1520 with the introduction of rifling—metal grooves in the barrel to spin the pellets and stop them from tumbling.

If there were struggles over land claims, boundaries could be more accurately set through the use of the new technologies of surveying: triangulation and the theodolite. Astronomy, allied with mathematics, showed its usefulness in devising a new calendar, which Pope Gregory XIII introduced in 1582 (hence the *Gregorian calendar*). This adjusted the occurrence of leap years to more accurately reflect the length of the year. Catholic states adopted it at once; Protestant regions delayed, often for centuries.

Most men of science, as well as artists, had to take an interest in military affairs, not least to serve the interests of their patrons, who were often at war. They had to double as engineers to build fortifications, and early mathematicians often studied the trajectories of shells and shot to improve accuracy. Metals were needed for swords, spears and gun barrels, as well as for ploughs and other peaceful implements, so handbooks for mining and smelting, such as Agricola's *De Re Metallica* (**1546**) were popular.

Finding new lands for exploration and conquest was still a lively endeavour, hence the first circumnavigations of the globe by the ships of Spaniard Ferdinand Magellan, completed in 1521, and the *Golden Hind* under Englishman Francis Drake, completed in 1580. Maps and compasses were key elements of technology, advances being made in the former by cartographers in the Low Countries and studies of the latter (**1581**) provoking an interest in what magnetism was (**1600**).

The first glimpse of the age of machines and industry came with the invention by English clergyman William Dee of a 'knitting machine' to make stockings. Unable to secure a patent in England he set up work in France. The invention of the first flushing toilet (or 'water closet') by the English courtier Sir John Harrington, banned from Elizabeth's court for telling risqué stories, was a step forward for quality of life.

1551–1600

The World Stage

EUROPE CONTINUED TO BE TORN by sectarian strife between Catholic and Protestant parties and states, rivalry over trade and access to foreign lands, and struggles to secure or retain borders and fragments of territory. When the Habsburg Emperor Charles V retired to a monastery in 1556, his son, Phillip II of Spain, inherited the south of Italy, the Netherlands and the conquered overseas territories in the New World. His brother Frederick gained the Austrian and German heartland and the title of Holy Roman Emperor.

The Low Countries (later Holland and Belgium) were a cockpit of ambitions, with the locals' own drive for independence from Spain meshing with interference by France and England. English and Spanish rivalry, exacerbated by religious differences, the execution of the Catholic Mary Queen of Scots and raids by English 'privateers' on Spanish treasure fleets, provoked the 'Spanish Armada' in 1588.

Religious and political tensions continued within states. Spain took control of Portugal and persecuted its Muslims (Moors). The accession of Elizabeth to the throne in England in 1558 ultimately reduced overt hostility between Catholics and Protestants. Factions struggled for control in France. When Henry IV, leader of the Protestant Huguenots and the first of the Bourbons, won power, he granted his followers religious freedom under the Edict of Nantes in 1598, ending for a time the persecution typified by the 1572 Massacre of St Bartholomew's Day.

Russia, under its first czar, began to push both east and west, beginning the colonisation of Siberia, fighting with its neighbour Poland and invading what are now the Baltic states. That conflict also drew in Sweden. The Ottoman Turks continued to chafe against Europe in the south-east, battling on land against Austria, and by sea with Spain and Venice. Defeat in the naval battle of Lepanto in 1571 set back their ambitions, though they later took Cyprus.

ARTS AND IDEAS

Painting continued to flourish in Italy with, for example, works from Veronese (*The Marriage in Cana* 1562) and Caravaggio. New centres of activity were now developing in the Low Countries, with Pieter Brueghel (*The Tower of Babel* 1563), and in Spain, with El Greco.

Great new talents flowered in music, such as Italians Giovanni Palestrina, Giovanni Gabrieli and Claudio Monteverdi; the Flemish composer Roland de Lassus; and Englishmen Thomas Tallis, William Byrd and John Dowland. Most of them excelled in secular as well as in sacred music. Of course, any educated person could sing madrigals and play an instrument.

Literature was mostly made up of poetry, plays and books of history and geography; there were few works of prose fiction. The Frenchman Nostradamus produced his books of prophecies; plays and tales came from the pen of the 'mastersinger of Nuremberg', Hans Sachs. Late in the century, the early plays of Englishman William Shakespeare (*Romeo and Juliet* 1595) were performed in London's new Globe Theatre, as were the fruits of the short life of Christopher Marlowe (*Doctor Faustus* 1588), who was murdered in a tavern brawl. Poet Edmund Spenser wrote the *Faerie Queen* in 1596.

EXPLORATION

New players, such as France, Holland and England, entered the race to explore and exploit new territory and the opportunities for trade. They set up trading companies such as the East India Company, with private shareholders but often royal patronage as well. French expeditions tried, ultimately unsuccessfully, to establish colonies in the New World. Englishman Francis Drake completed the second circumnavigation of the world in the *Golden Hind*, sailing along the west coast of what is now North America and claiming California in 1578.

By 1600 mapmakers could draw the outlines of five continents (Europe, Asia, Africa, North American and South America) but without the detail, especially in the interiors and towards the poles. Many suspected that a sixth continent existed in the far south and it was sketched in on many maps, but remained elusive. By rounding South America, Drake showed that open ocean lay between South America and a 'Great South Land', if there was one. The vast, unexplored expanses of Siberia began to come within the knowledge and control of Russian forces, which were advancing slowly eastwards.

Expanding Horizons

From 1550 the centre of activity in science began to move north and west from Italy. Passing over France for the time being, it settled in countries where the Reformation had made a major impact: the Low Countries, Denmark and England. Major progress was made in astronomy, with increasing challenges to old ideas on the structure of the universe. New concepts were emerging in physics; concepts handed down from Aristotle were under increasing doubt.

Italy continued to contribute to the understanding of human anatomy, but physics was finding a home elsewhere. For example, Simon Stevin was Dutch and Robert Norman was English. The exception was, of course, Galileo. His emergence in the later decades of the century began one of the most productive periods of discovery by a single individual.

Significant discoveries were often made by practical men, such as makers of maps and compasses, and engineers. Patronage helped some: Tycho Brahe was astronomer to the Danish king; William Gilbert was a royal physician.

Gabrielle Fallopius and Bartholomew Eustachio: More Body Parts

Spurred on by the labours of Andreas Vesalius (**1543**) to accurately describe human anatomy, other physicians continued to probe the architecture of our bodies. Many of these came from Italy, where anatomy, particularly at the University of Padua, led the world. Gabrielle Fallopius and Bartholomew Eustachio were two such, and with Vesalius they essentially founded the modern science of human anatomy.

1552

Fallopius, a pupil and successor of Vesalius, was one of the stars at Padua, being Professor of both Anatomy and Surgery and superintendent of the botanical garden. He died young at 39 but not before he had discovered the very fine tubes or 'oviducts' which connect the ovaries to the uterus in the human female. These tubes now carry his name. He also worked out the operation of the tiny bones in the ear that connect the eardrum to the inner ear, where the sense of hearing lies.

Eustachio, younger than Fallopius, flourished in Rome. He may have been the equal of Vesalius as an anatomist but many of his findings remained little known for 100 years. In 1552 he announced two important discoveries: one in the ear, namely the tube which connects the inner ear to the back of the throat, and one in the heart, a valve in the wall of the right auricle. Both of these are called 'Eustachian' after him. He also gave the first comprehensive descriptions of the adrenal gland, the thoracic duct and the structure of teeth.

Gerard Mercator and Abraham Ortelius: Mapping the World

1569

During the sixteenth century a major upsurge in exploration saw European ships sailing east and west. The new discoveries needed to be set down on maps. The skills of cartographers, particularly those from the Low Countries (Holland and Belgium), like Gerard Mercator and Abraham Ortelius, grew to meet the need.

The challenge was how to represent the curved surface of the Earth on flat charts in ways that minimised the inevitable distortions. In 1569 Mercator produced a map of the known world using his famous 'projection', which placed the lines of latitude and longitude at right angles. This distorts the size and shape of landmasses near the poles but accurately represents those near the equator and so has been the most widely used of all projections. It allows compass courses to be drawn on the chart as straight lines, a great aid in navigation at sea.

The accuracy and value of the maps depended on the data they used, which in turn came from the captains and navigators on ships. Techniques for finding latitude and longitude at sea were still crude. Latitude could be estimated by measuring the height of the Sun above the horizon at noon or of stars at night using a primitive instrument like the cross staff; the much more precise sextant was still 200 years in the future (**1731**). Longitude was largely dead reckoning, based on the estimated speed of the ship. Not until the eighteenth century and the development of the chronometer (**1765**) were acceptably reliable longitudes available.

Tycho Brahe: A New Star

1572

Tycho Brahe was official astronomer to the King of Denmark. At the Royal Observatory at Uraniburg he used the best instruments available before the invention of the telescope to measure the positions and motions of the planets and stars. For a time, his chief assistant was the young German Johannes Kepler, who would in time achieve his own fame (**1604**, **1609**).

In 1572 Brahe saw a 'new star' in the constellation of Cassiopeia. It had not been there previously. Prevailing wisdom held that the heavens were eternal and unchanging, so the 'nova' had to lie within the Earth's atmosphere, where nature was subject to change and decay. Brahe's precise measurements showed that the new star did not change its position relative to the stars around it during the year it was visible. So it had to be out among the stars, not close at hand. The heavens were not unchanging after all, and were probably subject to the same laws as natural events on Earth. It was a momentous discovery.

Brahe was a colourful character. He is said to have worn a gold and silver prosthesis on his nose to disguise the damage done by the sword of a rival as they duelled over primacy in mathematics. He reportedly died from a urinary tract infection contracted at a banquet. He had drunk heavily, but was inhibited by etiquette from leaving the table to urinate before his host. ➤➤**1577**

Tycho Brahe: Not Quite With Copernicus

Following the revelation of his 'new star' in **1572**, the Danish astronomer Tycho Brahe continued to challenge the accepted model of the universe, as Copernicus had done (**1543**).

1577

In 1577 he followed the path of a comet with his precise instruments, showing that it (and presumably all comets) were distant objects, lying far beyond the Moon, and not just phenomena in our own atmosphere. Its track seemed to cut across the orbits of the planets, which could therefore not be hung from vast crystalline spheres as many still believed. Nor were the orbits of the comets perfect circles, but greatly elongated. Bit by bit, the 1500-year-old universe of Aristotle and Ptolemy was being dismembered.

But Brahe did not go all the way with Copernicus. If the Earth really did orbit the Sun, all the stars should have shown some degree of 'parallax', moving back and forth by different amounts as the viewpoint from Earth changed. Brahe could see no such movement. This meant that either the Earth did not move at all or the stars were too far away for the parallax to be observed. He could not imagine such vastness and so settled for the first option. In his model of the universe, which was very popular at the time, the five known planets orbited the Sun, but the Sun (and the Moon) orbited a motionless Earth.

Others went beyond Copernicus. Thomas Digges from England doubted that all the stars were fixed to a sphere. He thought they were at different distances, stretching away infinitely. He also suggested, even more radically, that the stars were other suns, reduced to points of light by their great distance. Perhaps many of them had planets like our Sun does. That would take the Earth even further from the centre of the universe.

Italian monk Giordano Bruno took another step, postulating not only a plurality of worlds but the possibility of life on those worlds. Bruno was burned at the stake for heresy in 1600; his support of Copernicus was one of the charges against him.

Brahe delivered to his successors a vast accumulation of precise measurements about the movements of planets and stars. Without this treasure trove, his one-time assistant, Johannes Kepler, may not have been able to unravel the laws of planetary motion (**1609**).

Galileo Galilei: The Law of the Pendulum

1581

The Italian Galileo Galilei, like Napoleon, is better known to posterity by his first name only. Galileo's father Vincenzio (who was also his first biographer) was a musician. He had by experiment found the mathematical relationship between the tension in a stretched string and the pitch of the musical note it produced. So the young Galileo had an early role model in experimentation and in the use of mathematics to express the laws of nature.

He made his first major discovery in 1581, when only 17. According to the popular story, Galileo was standing in the cathedral at Pisa, watching the great lamp in front of the sanctuary swing from side to side and timing the swings against the beat of his pulse. He was familiar with the pulse, having begun to study medicine. Each swing of the pendulum took an equal number of heart-beats. Later his friend, the doctor Santorio, would turn the discovery on its head. He used a small pendulum called a 'pusilometer' to beat out equal periods of time against which to measure a patient's pulse.

Galileo had found what is formally called 'the isochronism of the pendulum': an object hanging from a chain or rope takes equal periods of time to swing from side to side, regardless of whether the swings are small or large. Only the length of the chain determines the 'period' of the swing. Though only really true for relatively small swings, this 'isochronism' was to become the basis of the pendulum clock, making possible for the first time the accurate measurement of small time intervals. Galileo proposed the idea just before he died; the Dutchman Christiaan Huygens would build the first such device, in **1656**.

A pendulum is about as simple as any bit of scientific apparatus can be, but it would generate some very important science in coming centuries. This included early evidence about the shape of the Earth (**1736**) and the first real proof that the Earth turns (**1851**). ➤➤**1589**

> *In questions of science, the authority of a thousand is not worth the humble reasoning of a single individual.*
>
> GALILEO GALILEI

Robert Norman: The Attractions of Magnets

1581

The ancient Greeks had known about 'lodestones', strange lumps of rock with the power to attract iron. The stones seemed able to transfer that attractive power to pieces of iron, which were named 'magnets' after the city of Magnesia (now in Turkey), where lodestones had first been found.

One thousand years later the Chinese knew that a magnet floating on straw in a bowl of water always pointed roughly north and south. From this came the magnetic compass, which reached Europe in the thirteenth century. Peter Peregrinus had improved the compass by having the needle swing around a vertical pivot, and it became indispensable during the more daring voyages of the late fifteenth century.

Sailors noted two things about the compass. Firstly, it did not always point due north but often east or west of north, depending on where you were. This difference became known as 'magnetic declination' or the 'variation of the compass'. It would be first methodically charted by the astronomer Edmond Halley (**1702**).

Secondly, the needle did not stay horizontal; the northern end would be pulled downwards. A piece would have to be snipped off the northern end of the needle to make it balance and swing smoothly. This phenomenon was 'magnetic dip', first explored by Georg Harman of Nuremberg in 1544 and in more detail by Robert Norman of England, who published his findings in *The Newe Attractive* in 1581.

Norman was a maker of compasses and once ruined a compass needle by snipping off too much in trying to make it balance. To find out more, he balanced the needle around a horizontal axis so it could swing freely up and down. He found that the northern end pointed steeply down into the Earth. Dip was usually greater the further north you went. These findings, based on perhaps the first ever methodical study of such a phenomenon, were very important to his countryman William Gilbert when he wrote his major study about the magnetism of the Earth (**1600**).

Simon Stevin: Forces and Pressures

Hardly anything is known of the life of Simon Stevin (or Stevinus), an engineer who flourished in Flanders (modern Belgium) in the late sixteenth century. He may have been a merchant's clerk to begin with, but

1586

he later travelled widely and rose to be a man of some influence. The sluices he designed to allow areas of Holland to be flooded were part of the national defence against invasion.

His fame in science lies partly in anticipating Galileo (**1589**) by arguing that objects fall towards the Earth's surface at speeds that do not depend on their weight. He probably dropped various objects to show that this was true, so countering the long powerful ideas of Aristotle (this mirrors Galileo's 'Leaning Tower' experiment, which he may never have done).

He was also in the vanguard of the new sciences that dealt with forces and movements in liquids and gases. He proclaimed the principle, later restated by the Frenchman Blaise Pascal (**1653**), that the pressure exerted by a liquid on the walls of its container depended only on the depth of the liquid and not at all on the size and shape of the vessel. He based this insight on experiment, using weights to pull up a flat plate resting on the bottom of a vessel filled to different depths, and published it in 1586.

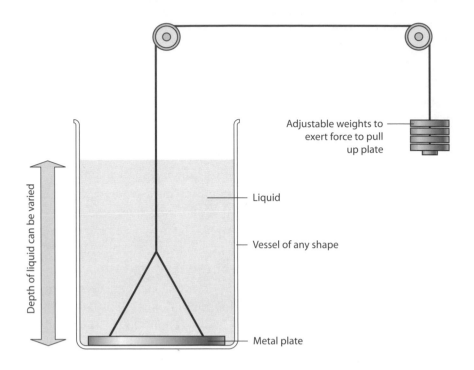

Depth of liquid can be varied

Adjustable weights to exert force to pull up plate

Liquid

Vessel of any shape

Metal plate

With this apparatus, Simon Stevin proved that the pressure on the bottom of the vessel (and, more generally, the pressure at any point in the liquid) depends only on the depth of the liquid (and its density) and not on any other factor.

Stevin had influence when it came to studying forces in general, devising the 'triangle of forces' and 'parallelogram of forces' to make easier mathematical study of networks of forces in motion and at rest. These methods are still widely used in school laboratories.

His achievements were diverse, and not all scientific. For example, he popularised the system of 'double entry' book-keeping, which had been invented in Italy, and used it to keep the household accounts of the Prince of Orange. It was perhaps he who first suggested that the tides in the ocean were due to some attraction by the Moon. He promoted the use of decimal notation in place of unwieldy fractions. And he strongly urged scientists to use the vernacular to report their findings, in place of Latin, which was commonly seen as the 'right language' for science.

Galileo Challenges Aristotle on Falling Objects

In 1589 Galileo Galilei, then only 25 years old, took up a lecturer's post at the University of Pisa. His two years there were productive. For one thing, he experimented with a crude thermometer called a 'thermoscope', which used the expansion of air to measure temperature.

More importantly, he began a search to uncover the laws controlling the movement of objects, such as when they are dropped. At the time, the authority in this matter was of course Aristotle. He had maintained that motion was either 'natural' or 'violent'. 'Natural motion' expressed the tendency of the different types of matter to seek their 'proper place' in nature. 'Earth' and 'water', being endowed with 'gravity', tended to fall; levity-rich 'air' and 'fire' wanted to rise. Since all objects were varying mixtures of these elements, each would move up or down at its own speed depending on its weight.

Galileo doubted this. He reputedly dropped two objects of very different weights (a cannon ball and a musket ball) from the Leaning Tower of Pisa to show that they hit the ground together. This story was probably invented by his biographers, his father Vincenzio or his student Viviani. But the Dutchman Simon Stevin (**1586**) had done something very similar a few years earlier, so there is little doubt that Galileo believed it to be true, provided the effects of air resistance were neglected. Nearly 400 years later, an astronaut on the airless surface of the Moon dropped a hammer and a feather. They took the same time to fall.

So what did control the rate at which objects fell? Could that be linked to the other sort of motion Aristotle spoke of, which he called 'violent' motion? The answer to that question would come only from years of experimenting (**1604**).

There is a fascinating link between all this and the discoveries Galileo had already made about pendulums (**1581**). The time a pendulum takes to swing does not depend on how heavy it is, just as heavy and light objects, once released, reach the same speed. The reason is the same. Both the falling ball and the pendulum move in response to the pull of gravity. The pull of gravity on an object depends on how heavy it is (its 'gravitational mass'); how fast it moves when pulled depends on a resistance to motion known as inertia, which also varies with its mass. The findings with pendulums and falling objects show that gravitational mass equals 'inertial mass'. It is not immediately clear why this should be so, but it is. This 'Principle of Equivalence' has now been tested to a very high degree of accuracy. ➤➤**1604**

William Gilbert: Magnets and Electrics

English doctor William Gilbert came to prominence late in life. Not until 1600, three years before his death, was he appointed a personal physician to Queen Elizabeth I and made President of the Royal College of

1600

Physicians. That year he also made his mark on science with a most influential book. *On the Magnet* was the first methodical and comprehensive study of 'magnetism' and the compass, which had been used for many years to guide ships across the open ocean.

Much was already known from observation. The compass could not always be trusted. The needles usually showed 'declination', that is, they pointed a little east or west of true north. As Robert Norman had found in **1581**, they also showed 'dip', pointing down at an angle into the Earth. Both declination and dip varied from place to place over hundreds or thousands of kilometres.

To Gilbert, as for Norman, these observations suggested that the Earth itself was a giant magnet (he went so far as to suggest that magnetism was the 'soul' of the Earth). He made a sphere of lodestone, which he called a 'terrella' or 'little Earth', and demonstrated that a small compass moved across its surface showed the same variations in declination and dip as a compass in the real world.

If the model was right, the Earth itself had two 'poles', like the ends of a bar magnet, and these attracted the ends of the compass needle. These 'magnetic poles' were not aligned with the geographic poles, which explained why the compass did not always point due north.

Gilbert had studied magnetism for 20 years before publishing his book. He knew that heating a magnet destroyed its influence, but (contrary to popular opinion) garlic did not.

Gilbert also busied himself with what he termed 'electrics', that is, pieces of amber, which when rubbed with cloth were able to attract dirt, hairs and other light objects. The ancient Greeks had known about this: Gilbert created the name from the Greek word for amber. He found other substances that behaved in the same way— glass, for example. To Gilbert, electrics and magnets were different phenomena; the vital link between them would not show up for more than 200 years (**1820**).

1601–1650

The World Stage

THE SEVENTEENTH CENTURY WAS THE century of religious intolerance, enmeshed with rivalries between states over territory and trade. For 30 years from 1618, Europe was tormented by war as nearly all the continental states allied themselves with either the Catholic or the Protestant cause. Much of Europe was laid waste by war, famine and disease; in places 30 per cent of the population perished. Later struggles for supremacy pitted France against Spain, with battlefields in the Low Countries. On the eastern fringe of Europe, Poland and Russia scrapped over the territory between them.

England endured a civil war in the 1640s. The struggle was as much a battle between the king and the parliament as between faiths. The political settlement and cultural renaissance that followed brought a flowering of science without precedent. An undeclared civil war also gripped France, between the Catholic forces of the king and the Protestant Huguenots.

By 1650, the Thirty Years War was over, leaving central and northern Europe exhausted. Protestants were in power in England, the Catholic king having been executed. The Dutch had their freedom; the Romanovs ruled in Russia.

The first enduring European colonies were established on the North American seaboard by Britain, France, Holland and Spain. These seeds would grow into Canada and the United States of America. The United States would become the superpower in Western science three centuries on.

ARTS AND IDEAS

In music the Renaissance began to give way to the baroque. Italian Claudio Monteverdi wrote his famous *Vespers* in 1610; he was also pioneering a new art form called opera (*The Coronation of Poppea* 1642). In England Orlando Gibbons and William Byrd were active.

Flemish and Dutch painters, now using the new technology of oil paint, were very influential. They included Peter Paul Rubens (*The Raising of the Cross* 1608), Franz Halls (*The Laughing Cavalier* 1624) and the young Rembrandt van Rijn (*The Night Watch* 1642). Antony van Dyke become Court Painter to England's King Charles 1. In Italy, Caravaggio was pioneering a dramatic new style. France was beginning its rise to prominence in painting (Nicholas Poussin *The Rape of the Sabine Women* 1637).

Theatre remained lively, with the English playwrights William Shakespeare (*Hamlet* 1603, *The Tempest* 1611) and Ben Jonson (*Volpone* 1605) and the Frenchman Pierre Corneille (*Le Cid* 1637). The famous *Passion Play* at Oberammagau in Germany was first staged in 1634. Spaniard Manuel Cervantes' novel *Don Quixote* first appeared in 1605. English poet and cleric John Donne published *Of the Progress of the Soul* in 1612.

In the realm of ideas, Dutch lawyer Hugo Grotius wrote *On the Laws of War and Peace* in 1625. Soon after, Thomas Hobbes in England was musing on his defence of autocratic government, to be published in *Leviathan*.

EXPLORATION

The nations of Europe battled offshore as well as onshore, seeking influence, trade and souls in the 'new worlds' to the west and in the Far East, and later in Africa. Piecemeal exploration slowly filled the blanks in the maps. Explorers returned home with exotic plants: potatoes, tomatoes, coffee, maize, tea, tobacco, chocolate and rubber, as well as copious gold and silver. Syphilis, smallpox and influenza went the other way. The trade in slaves was another gift of civilisation.

Spaniard Luis de Torres first sighted land south of New Guinea in 1606. Other finds in this region were made mostly by accident, by ships heading east from below Africa along the great belt of winds and missing the turn north to the Spice Islands. In 1642 Dutchman Abel Tasman named New Zealand after his homeland and found Tasmania. Three years later, he explored the north-west coast of what now appeared to be a sixth continent, linking the findings of other Dutch and Portuguese navigators.

In the far north, explorers sought the 'North-west Passage' to the resources of the east, avoiding Spanish- and Portuguese-controlled seas. There was no easy path though the ice-choked seas. English navigator Henry Hudson entered Hudson Bay (Canada) and perished soon after. His countryman Henry Baffin reached within 13 degrees of the North Pole between Newfoundland and Greenland; this was further north than anyone travelled for the next 200 years.

THE AGE OF GALILEO

Through the early seventeenth century the traditional view of the universe and our place in it, supported both by ancient wisdom and the authority of the Church, continued to unravel. Changes and imperfections became more evident in regions of the cosmos previously thought to be perfect and eternal.

The great figure throughout this period was the Italian Galileo, who was immensely influential in both astronomy and physics. His continual challenge to Aristotle, and so to the Church, brought him undone; he died under house arrest imposed by the Inquisition. His students, notably Torricelli, made significant discoveries, but after them Italian science fell largely silent for more than a century.

The German Johannes Kepler corrected Copernicus about the orbits of the planets and discovered the laws by which the planets move. The transition from alchemy to chemistry advanced with the discoveries of Dutchman Jan van Helmont. Anatomy moved towards physiology when William Harvey uncovered the truth about the circulation of the blood. Many researchers focused on light: How do we see? Why does light bend? How fast does light travel? Others explored the arrangement of lenses that brought distant objects close and made tiny objects large.

Not all great figures of the time were experimenters. The philosophers Francis Bacon of England and René Descartes of France wrote on how science should be done, and visualised the underlying stuff of the universe. Anticipating both learned societies and scientific journals, Marin Mersenne of France acted as the scientific 'mailbox' of Europe.

French names began to figure prominently in the story; not only Descartes and Mersenne but also Pierre Gassendi, first to revive the ancient Greek belief in atoms, on which much of the science to come would build. Science in England was a little subdued (other than Harvey), ahead of the great flowering in the next half-century.

The Inconstant Heavens

As the seventeenth century unrolled, astronomers became more comfortable with the notion that the heavens were not eternal and unchanging. The 'new star' ('supernova' as we would call it today) seen by

1604

Tycho Brahe in **1572** and therefore called 'Tycho's Star', was matched by 'Kepler's Star', which appeared in the constellation Serpens in 1604. It was closely watched by Tycho's one-time assistant, the German Johannes Kepler. It, too, flared up quickly to a brilliance visible in broad daylight, before fading away over the following months. It, too, showed no parallax and so lay far beyond the Moon in the realm of the 'fixed' stars. Though the two new stars burst into view only 30 years apart, none so bright would be seen in the next 400 years.

Change in the heavens of a different kind was confirmed by David Faber (or Fabricius). This north German astronomer had previously seen a star in the constellation Cetus the Whale appear from nowhere and then fade again from view. When he saw it again in 1609, he knew that something new was happening among the stars, a rhythmical change rather than a once-only one. Later observers found it coming into view roughly once a year. It was dubbed Mira, meaning 'wonderful'; its behaviour would be unexplained for centuries, even after others like it were found.

Around 1667 Geminario Montanari of Italy found another type of variable star. He noted a star in the constellation of Perseus—called Algol in Arabic—fluctuating noticeably in brightness every few days. The changes were nowhere near as great as with Mira, as Algol remained visible. The star's name in several languages means 'demon' or 'ghoul', suggesting that its inexplicable behaviour had long been known. European astronomers who had seen it may have failed to say so for fear of contravening the orthodoxy that nothing in the heavens changed.

Galileo Galilei: The Truth About Motion

1604

From 1592 Galileo Galilei was Professor of Mathematics at the University of Padua. There he began to carefully study motion, partly to build the case against Aristotle's theories, partly to find out what was really going on. He based his conclusions on careful experiments and expressed them in numbers, not just words. Quantity mattered as much as quality.

Galileo's first guess at the problem was that the speed reached by a falling object depended on how far it had fallen. Certainly the further it fell, the faster it went. But a little thought showed that this could not be so. For a start, an object left to drop would not get moving at all. Until it had fallen even a minute distance, it could not have any speed and therefore could not fall. And if it could get started, it would fall through any distance in the same time. Clearly this theory was wrong.

The other answer was the speed depended on the time taken to fall. The longer it fell, the faster it would go. If this was so, the speed of any object falling from rest would be twice the average speed up to that moment, and the average could be found by dividing the distance covered by the time it had taken.

Around 1604 Galileo began a crucial series of experiments to settle the matter. He needed evidence of the relationship between times and distances. The challenge was somehow to slow down the motion of falling objects so that those factors could be measured. Falling objects simply went too fast, so he rolled balls down a gently sloping plank up to 6 metres long, arguing that the same rules apply as if the object

> Long experience has taught me this about the status
> of mankind with regard to matters requiring thought: the less people
> know and understand about them, the more positively they attempt
> to argue concerning them, while on the other hand to know and
> understand a multitude of things renders men cautious
> in passing judgement upon anything new.
>
> GALILEO GALILEI

had simply been dropped. To measure the time intervals he used a water clock: water was run into a bowl, with flow started and stopped in time with the motion; the weight of the water in the bowl was proportional to the time interval.

Crude as it was, it was good enough to find the mathematical rules governing motion. For an object picking up speed at a constant rate, the increase in speed was proportional to the time taken; the distance covered was proportional to the time squared. If a rolling ball covered 1 metre in the first second, it would cover 4 metres in two seconds and 9 metres in three seconds.

There was something more in this. Aristotle had claimed that, in contrast to 'natural motion' (**1589**), 'violent' motion came from some action—a push or a pull. If you stopped pushing or pulling an object it would cease moving. Not so, said Galileo (and others before him). He found his balls would continue to roll along a horizontal plank until they were deliberately halted. If there was no friction, he saw no reason why they should not keep rolling for ever.

From this, Galileo derived his Law of Inertia. All objects resist a change in their movement (speed and/or direction), a property dubbed 'inertia'; any change requires some impetus or force. Galileo published this understanding in 1638 near the end of his life, blind and under house arrest (**1633**), though it was more clearly stated by René Descartes in **1644** and Christiaan Huygens in 1673. Fifty years later (**1686**), Isaac Newton would turn the Law of Inertia into his First Law of Motion and acknowledge his debt to Galileo. ➤➤**1610**

Johannes Kepler: How Do We See?

To explain the phenomenon of sight the ancients had proposed that a beam of light goes out from the eye seeking the object to be seen, much as we reach out with our hand to sense something by touch. By the time German mathematician Johannes Kepler came to set down his ideas about light in 1604, the old idea had given way to a rather better one: we see because light from objects in the world gets into our eyes.

1604

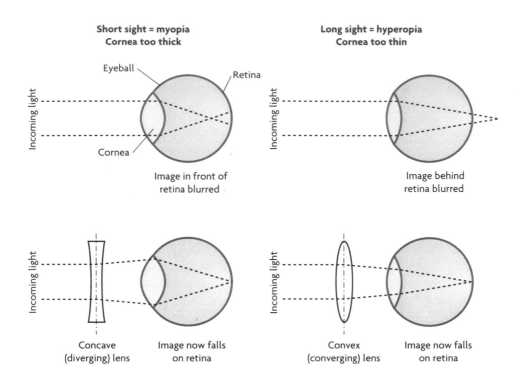

The diagram shows how Johannes Kepler explained common defects in vision and how they can be corrected by using the appropriate form of lens in spectacles.

But how did the eye work? Kepler had studied lenses, and knew that a convex lens, one which is thicker in the middle than at the edges, can cast an upside-down image on a blank surface. He thought the tissue at the front of the eye, the cornea, did the same, so the image on the retina at the back of the eye must also be upside-down. Somehow our minds made sense of that image, since we see the world right way up. Many thought the idea nonsense, but proof came years later when René Descartes dissected the eye of a dead bull and scraped away the retina so he could see the upside-down image cast by the cornea.

Kepler correctly attributed 'short sight' and 'long sight' to something wrong with the lens of the eye. It was too thick or too thin in the middle, so that the image reached focus in front of or behind the retina rather than on it. This explained something long known—that lenses of various thicknesses could be used as 'spectacles' to correct defects of vision.

Noting, as everyone did, that light grows apparently fainter the further it travels from its source, Kepler proposed that this weakening followed an 'inverse square' law: the light 2 metres from a lamp is not half as bright as at 1 metre, but only a quarter as bright. In time, other phenomena, such as magnetism, electrical attraction and gravity, would be found to follow the same law. ➤➤**1609**

New Eyes on Nature: Telescopes and Microscopes

No one knows who first invented the telescope. Various writers over the centuries (including da Vinci) had referred to the use of lenses and mirrors to make distant objects appear larger and closer, but it seems no one successfully built such a device.

In 1608 Hans Lippershey, a Dutchman who made lenses for use in spectacles, applied for a patent on a telescope. His bid failed, mostly because other people had had the same idea at much the same time. His embryonic telescope had a pair of lenses mounted in a tube, a convex or converging lens (thicker in the middle than at the edges) at the top of the tube to make an image of a distant object, and a concave or diverging lens (thinner in the middle) to enlarge the image and examine it.

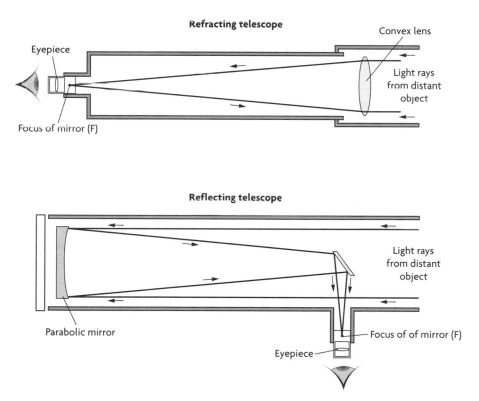

*Telescopes receive light rays from distant objects. The rays are brought together (focussed) by a convex lens (in refracting telescopes) or a parabolic mirror (in reflecting telescopes) to form a bright image at point F. The image is enlarged by the eyepiece, with the light passing into the observer's eye, a camera or some other light-measuring instrument. Refracting telescopes were first seriously used by Galileo (**1610**), and reflecting telescopes by Newton (**1672**), though not quite in the form shown here.*

It has been rumoured that the combination of lenses was found by accident, perhaps by Lippershey's children playing in his workshop. But it certainly worked. Within a few months, news of this invention reached Galileo in Italy, and he turned a handmade 'optic tube' onto the heavens in **1610**. This opened the path to a century of stunning astronomical discoveries.

Around the same time, or earlier, someone noted that the same combination of lenses could make very small objects nearby appear large. A Dutchman called Hans Janssen may have put one together around 1590. But Galileo certainly did in 1609, calling his device the 'occhiolino', meaning in Italian 'an eye for the very small'. The word 'microscope' (by analogy with 'telescope') did not arrive until decades later.

Dutchman Cornelius Drebbell, 'court inventor' to England's James I (and reputedly builder of the first submarine), made major improvements to the microscope in the 1620s. Following ideas put forward by Johannes Kepler, Drebbell used a pair of convex lenses, rather than one convex and one concave, in his version of the 'compound microscope'. It was both more powerful and easier to use and did for the 'microcosm' of small objects what the telescope did for the 'macrocosm' of the universe. Into view would come sights previously unimagined (**1665**).

Johannes Kepler: How the Planets Move

1609

The vision of the universe expounded in **1543** by Nicholaus Copernicus was revolutionary in several senses, but it clung to one old idea. The orbits of his planets were still perfect circles, as they had been for Ptolemy. But the model could not accurately predict the positions of the planets in the sky, as measured with increasing accuracy by Tycho Brahe (**1572**) among others. Brahe's one-time assistant Johannes Kepler, now court mathematician to the Hapsburg Emperor Rudolph II, saw the way out.

Analysing the mounds of data, he recognised the subtle change needed. The orbits had to be very slightly elongated—ellipses rather than circles. Kepler set down the first two of his Laws of Planetary Motion in 1609 (the year Galileo first looked at the stars through his telescope) in his book, *New Astronomy*, and the third law in 1619.

According to Kepler, planets in elliptical orbits will travel fastest when closest to the Sun (at perihelion) and slower when further away (that changing speed had been

> *The diversity of the phenomena of nature is so great, and the treasures hidden in the heavens so rich, precisely in order that the human mind shall never be lacking in fresh nourishment.*
>
> JOHANNES KEPLER

one of the puzzles). The size of the orbit is linked by a mathematical formula to the time taken to go around once. Jupiter takes 12 years to the Earth's one, and so must be just over five times further away; since 12 squared (144) is a bit more than 5 cubed (125). Time would show that Kepler's laws also applied to the movements of the various moons (then being seen for the first time), and even to the apparently unpredictable comets.

Kepler had no idea why the planets behaved so. He knew only from observation that they did. Over the next 70 years astronomers began to see that this behaviour had to involve some pull by the Sun on the planet and some other influence resisting that pull. Many people glimpsed the answer, but it would take the genius of Isaac Newton to set it all down (**1687**).

Kepler called his 1619 book *The Harmony of the World*. Throughout his life, he was fascinated by the idea that the patterns in the universe, for example the distances between the planets, reflected some deeper symmetry, such as the relationships between various sorts of regular solid objects, each of which had all its sides the same shape. Try as he might, he could not make this idea fit the facts. As a good scientist should, he ultimately abandoned it, though with reluctance.

Galileo: The Messenger from the Stars

Italian physicist Galileo Galilei did not invent the telescope. He may not even have been first to look at the night sky with one, but he was certainly the first to tell the world what he saw. Hearing of the new 'optic tube' invented in **1608** in Holland, he built his own. It made celestial objects appear 30 times closer and 1000 times larger. What he reported in his book *The Starry Messenger* in 1610 was frankly astounding; it profoundly challenged the commonly held and officially sanctioned views of the cosmos and our place in it.

Galileo saw that the Moon, which under the old image of the universe should have been smooth and perfect, was covered with mountain ranges and craters, much like the Earth. Far beyond the Moon, the telescope pulled into view many more stars than the unaided eye could discern. The 'seven sisters' in the Pleiades became 30; the Milky Way's band of light resolved into countless stars. Clearly, the universe was vast.

The real shocks lay beyond the Moon but in front of the stars. The planet Venus, star-like to the unaided eye, had a visible disc and showed a phase, as did the Moon. Four tiny points of light showed that the more distant Jupiter had a retinue of moons, objects that clearly did not orbit the Earth. To Galileo, these discoveries supported the new image of the universe put forward by Nicholaus Copernicus in **1543**. By so deciding, he brought on an inevitable conflict with the Church (**1633**).

Soon astronomers all over Europe were peering through telescopes. As these got better, new wonders came into view. The planets showed discs, with faint markings that moved as they rotated. Clearly these were worlds like our own planet. The Moon's rugged face was mapped and its markings erroneously given the names of oceans and seas. The Sun revealed its spotty face (**1611**); Jupiter showed its Great Red Spot; Saturn gained moons and its stunning system of rings (seen but not understood by Galileo).

Among the growing numbers of stars, astronomers glimpsed many a faint misty patch of light dubbed a nebula (Latin for 'cloud'). A nebula in the constellation Andromeda, the Maiden in Chains, known to Arab astronomers, was the first to be examined by telescope. It would in time precipitate another profound revolution in our understanding of the cosmos (**1924**). ➤➤**1633**

Blemishes on the Face of the Sun

1611 This was the first year of the 'spotty Sun'. The orthodox view of the universe at the time required that the Sun, lying as it does beyond the Moon, be unblemished and eternal, made of very different stuff from the decay-prone Earth that we inhabit.

So it was a shock to find the bright face of the Sun disfigured with small black spots; these were easily seen when the Sun's light, collected with the newly invented telescope, was thrown onto a wall. Who first saw them is a matter of some dispute (in Europe, that is; Chinese astronomers were most likely already familiar with them). The claimants include: Thomas Harriot in England; father and son David and Johann Fabricius in Germany; the Jesuit Christolph Scheiner and Galileo Galilei in Italy.

Certainly, sunspots were soon common knowledge. Within a year, they were the subject of letters and tracts. Of particular interest was the steady movement of the spots across the Sun's disc, evidence that the Sun turned on its axis about once a month. Watchers also began to count the number of spots visible at a time. Within a decade or so of the first discoveries, the numbers of sunspots began to dwindle almost to none, though not many people were keeping watch. Between about 1640 and 1720 sunspot sightings were scarce.

Sunspot numbers revived early in the eighteenth century, and a more regular pattern emerged. Numbers seemed to peak every 11 years or so, a rhythm that can be glimpsed in the earliest data. By the nineteenth century the 'sunspot cycle' was well recognised, but an explanation of it, and indeed of sunspots themselves, took a little longer.

William Harvey:
Our Blood Goes Around and Around

By all accounts William Harvey was a very good doctor by the standards of the time. He was personal physician to two English kings, James I and Charles I, and a leading light of the Royal College of Physicians. He married well, to the daughter of the physician to Elizabeth I, and so had the means to pursue his research.

1616

It was that research, exploring how blood moved around the human body, that got him into trouble. The accepted wisdom of the day came from the Greek physician Galen, 1500 years before. The two sorts of blood, dark red blood flowing in veins and bright red blood in the arteries, originated in different organs; venous blood in the liver and arterial blood in the heart. Carried to various parts of the body, they were consumed.

Harvey had very different ideas, derived possibly from the teachings of Arab doctors, as well as perhaps from the sixteenth-century Spaniard Michael Servetus, who had been burnt as a heretic. For Harvey, blood circulated around the body in great loops, driven by the heart, to which it kept returning. One loop took bright red blood to muscles and organs, returning it darkened. The other loop passed the dark red blood through the lungs, which restored its bright colour.

Harvey first expounded these heretical views in 1616 and published them in 1628 in *On the Motion of the Heart and Blood* (*De Motu Cordis* in Latin). They were never widely accepted during his life, the medical establishment clinging to Galen. Harvey had to fight off strident critics. His defence was always the same—his model fitted the available evidence; the old view did not.

Harvey did have a problem; his system was not quite complete. The arteries plainly divided and redivided into tiny vessels, and small veins joined together to become large blood vessels once more. But he could not explain how blood crossed the gap from the smallest arteries to the tiniest veins, and vice versa, though he was sure it happened.

Vindication came years later. Through the microscope Marcello Malpighi (**1661**) and others saw blood moving through a network of minute capillaries in the muscles and lungs to complete the circulation. Always a careful observer, Harvey had gone beyond observation to an explanation we still accept today.

Willebrord Snell: Why Does Light Bend?

Light does not always travel in straight lines. A beam of light reflects off shiny surfaces and bends (refracts) on passing from one transparent medium to another, such as from water to air, or air to glass. Such refraction causes pools of water to look shallower than they are, and a stick in water to appear bent. It also explains the behaviour of lenses, which can focus a beam of light to a point or make it spread out.

The principle that tells by how much a ray of light will bend is usually credited to the Dutch mathematician Willebrord Snell in 1621 and is commonly called Snell's Law. But others knew about it before he did, including the brilliant Englishman Thomas Harriot as early as 1601, and Arab scientists even earlier. Frenchman René Descartes beat Harriot and Snell into print with it in 1637.

Snell's Law links the angles the light ray makes with the boundary between two media (or, more precisely, the sines of those angles) before and after it crosses over. The ratio, the same for any 'angle of incidence', is called the refractive index and is a property of the medium.

Just why does the light bend? The answer involves the speed of light, though this had not been measured in Snell's day. Light would presumably travel faster in air than in glass, and so bend one way or the other when changing medium. Consider

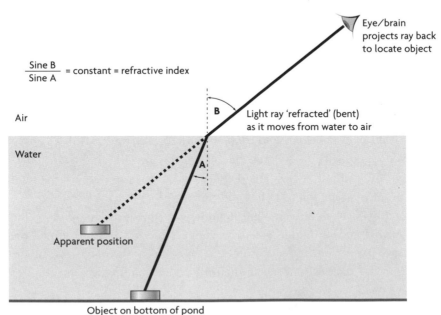

The bending of the light ray from the object on the bottom of the pond as the ray leaves the water fools the eye and the brain, which apply the principle that light always travels in straight lines.

an oil drum rolling diagonally across a hill; differing slopes will make it speed up or slow down, and so change direction.

French mathematician Pierre Fermat (author of the famous 'Last Theorem') had a more profound explanation. A ray of light, he said, will follow a path which takes the least time, whatever medium it passes through. Figuring the different speeds of light and the distances covered, this delivers Snell's Law. Fermat's Principle also explains the law of reflection; making 'the angle of incidence equal to the angle of reflection' gives the shortest path and therefore the least travel time.

Marin Mersenne: The Great Networker

1623

Not all the leading players in seventeenth-century science made their major contributions through research. René Descartes, Francis Bacon and John Locke did hardly any experiments between them, but their ponderings and writings about the origins of knowledge, the organisation of nature and the role of science were all very influential.

Marin Mersenne was a French clergyman and monk from a poor background, but he established himself as the 'scientific mailbox of Europe'. For more than 25 years from 1623, he carried on a prodigious correspondence with a carefully selected group of savants—leading researchers and thinkers from all over the continent—and as far away as North Africa and Constantinople.

Many of these came to meetings at his monastery in Paris. It was at one of these that the young Blaise Pascal met Descartes; Pierre Gassendi, Pierre Fermat and Jan van Helmont were also in his circle. He corresponded with Galileo and Torricelli, among many others. At his death he reputedly had correspondence in his cell from 78 such people.

Mersenne did do some quite important research. His particular interest was musical acoustics and sound. By observation he linked the number of vibrations in a string (its frequency) with the pitch of the note the string produced. With his friend Pierre Gassendi (**1624**), he made the first measurement of the speed of sound and showed that it was independent of frequency. He studied mathematics as well, particularly prime numbers (numbers divisible only by themselves and 1). Numbers of the form $2n - 1$, where n is a prime, are still called Mersenne primes.

But his greatest contribution was in keeping the researchers of his day in touch with each other, bringing them together, adding his own thoughts to the mix of ideas. In time this role would be taken over by the scientific societies (**1662**); Mersenne founded one of the first of those. He was, as we would say today, 'a great networker'.

Pierre Gassendi: The Atomic Revival

1624

The curriculum vitae of Frenchman Pierre Gassendi reads like those of many of his contemporaries. Brilliant in his youth, with a flair for mathematics, he entered the Church and gained high office. But he had to straddle the growing divide between the dogmas of the Church and the findings of embryonic science. His achievements were diverse, from the first observation of a transit of the planet Mercury across the face of the Sun in 1631 to ideas about the origins of living things that sound very much like 'natural selection' as expounded more than two centuries later by Charles Darwin (**1859**).

Gassendi was part of the growing groundswell of opposition to the ideas and methods of Aristotle, writing his first book on the matter in 1624. Aristotle's world view, by then 2000 years old, had become welded to the theology of the Church through the writings of Thomas Aquinas. For Aristotle, knowledge was deduced from basic ideas that everyone accepted as being true. Gassendi, like most early scientists, favoured 'induction', looking for regularities in observations of the natural world from which general principles or 'laws' could be inferred.

Gassendi also revived the ancient Greek idea of 'atoms', countless tiny indestructible particles that moved through space. Atoms were the building blocks of matter and explained its behaviour. Aristotle had rejected 'atomism', believing that matter was continuous and 'empty space' was impossible. The Greek writer Epicurus was an early advocate of atomism, but he was also an atheist. However, Gassendi's version of the theory was acceptable to a man of faith.

But there was more. Aristotle thought that God, having made the world, had to intervene to keep it running smoothly. But for atomists such as Gassendi, the atoms, once created, would simply go on their way eternally, like a clock set going by its maker. This was the germ of the 'mechanical' view of nature as regular and predictable, free from the capricious acts of gods or demons or other agents. It was to triumph in the 'world system' of Isaac Newton (**1687**) and remain unchallenged for 300 years, until the rise of quantum physics in the twentieth century.

Francis Bacon: 'Knowledge Is Power'

1626

Englishman Francis Bacon was not really a scientist. He did very few experiments, though a legend persists that he caught a fatal chill one freezing day in 1626 by stepping from his carriage to see if a chicken could

> *It is well to observe the force and virtue and consequence of discoveries,*
> *and these are to be seen nowhere more conspicuously than in printing,*
> *gunpowder and the magnet. For these three have changed the whole*
> *face and state of things throughout the world, in so much that no*
> *empire, no sect, no star, seems to have exerted greater power and*
> *influence on human affairs than these mechanical discoveries.*
>
> FRANCIS BACON

be preserved by being stuffed with snow. Bacon was instead a lawyer, career public servant and diplomat. He was well connected at court, a favourite of Good Queen Bess when a child, an associate of Walter Raleigh and the Earl of Essex, and holder of many high offices, including Lord Chancellor under James I. His career ended in disgrace, with accusations of bribery.

Bacon has his firm place in the history of science through his writings. From around 1605 he expounded with vigour (he might have been called a 'propagandist') the new philosophy that reliable knowledge about the workings of the world comes from observation, experiment and hypothesis, rather than being deduced from the preserved words of ancient sages. Such knowledge has a purpose. 'Knowledge is power', he proclaimed, by which he meant power over nature. But 'nature, to be commanded, must be obeyed'. In other words, we must understand how nature operates if we are to shape it to meet our needs.

In his utopian fantasy *The New Atlantis*, published just before his death, Bacon dreamt of 'The House of Solomon', a premonition of a modern research laboratory where the problems of the world would be addressed and solved by experiment and invention. Here the wise men of science would gather to share ideas and experimental results, much as would happen 40 years later with the founding in **1662** of the Royal Society, which owed much to Bacon's vision.

In 1620 Bacon noted how the east coast of South America could fit snugly into the west coast of Africa. Such things, he said, 'do not happen without a reason'. Three hundred years later, Alfred Wegener would use that and other evidence to propose the theory of continental drift (**1912**).

Galileo and the Inquisition

The conflict between the Italian physicist Galileo Galilei and the Roman Catholic Church is one of the great sagas of early science. At issue was the new view of the universe expounded by Nicholaus Copernicus in **1543**, 90 years before Galileo's troubles climaxed in 1633.

1633

> *I do not feel obliged to believe that the same God who has endowed*
> *us with sense, reason and intellect has intended us to forgo their use.*
>
> GALILEO GALILEI

In that year, the Inquisition condemned Galileo for agreeing with Copernicus that the Earth 'moved', rotating once a day on its axis, and travelling in an annual orbit around a stationary Sun. The Church cited the combined authority of Holy Scripture, Aristotle, Ptolemy and common sense to proclaim a motionless Earth at the centre of all things. Galileo, who had been censured once before and forbidden to teach such heresy, endured house arrest for the last decade of his life. His punishment scared contemporaries such as Descartes and Torricelli away from Copernican ideas and even from astronomy altogether.

Legend insists that Galileo muttered 'but it does move', even as he knelt to recant. If he did, it was almost an act of faith; the case for Copernicus was not strong. No one could prove the Earth turned. Galileo, decades before, had ascribed the ocean tides to the Earth rattling as it turned, like water slopping in a shaken bowl, but it was not convincing. The phases of Venus, found in **1610**, did not prove Copernicus right; Venus would show a phase with the Earth at the centre. And Galileo supported circular orbits for the planets even though these made the Copernican model no better at predicting where the planets would be than the system it sought to replace. The universe of Copernicus had an appealing simplicity, but was that enough?

Galileo has, of course, been vindicated by history, but centuries passed before the the Church withdrew its condemnation. In **1687** Isaac Newton set out a theory of gravity that explained Copernicus and made the complexities of Ptolemy physically impossible. It is fitting that in this Newton drew much from Galileo, one of the 'giants' on whose shoulders he stood. ➤➤**1638**

Galileo, Light and the Parabola

1638

Italian Galileo Galilei was first to attempt to measure the speed of light. Opinion was divided as to how fast light travelled. Certainly it was very speedy; the uncovering of a lamp filled a room with light immediately, or so it seemed. Johannes Kepler and later René Descartes thought the speed infinite, so that light moved instantaneously from place to place. Others, like the thirteenth-century English monk Roger Bacon, thought the speed was finite, and so measurable, at least in principle. But no one really knew.

As he reported in his *Two New Sciences* in 1638, Galileo set up two observers, each with a lamp, on nearby hills. One uncovered his lamp; the other, seeing the light, did the same. The first observer tried to time the interval between uncovering

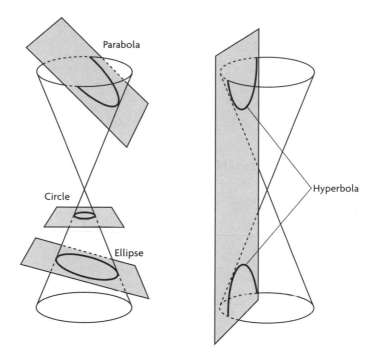

Conic sections are a special group of curved lines revealed when a flat surface cuts through a cone at different angles. These curves are found in nature. The orbits of the planets around the Sun are ellipses (a circle is a special case of an ellipse). A thrown object moving under gravity traces out a parabola.

his lamp and seeing the other light. The results proved nothing, measuring in reality just the reaction times of the two observers. More than 200 years would pass before the speed of light could be measured within the confines of the Earth's atmosphere (**1849**). The first reliable measurements, made by Ole Roemer in **1676**, would rely on the vastness of space.

Galileo's 1638 book is full of interesting things, indeed many profound ones. For example, he was able to prove what others before him had suspected (**1537**); a projectile, such as a cannon ball, moving freely, follows precisely a curved path

> *Philosophy is written in the great book which is before our eyes, meaning the universe. That book is written in the mathematical language, without whose help it is humanly impossible to comprehend a single word of it, and without which one wanders in vain through a dark labyrinth.*
> GALILEO GALILEI

known as a parabola. His laws of motion (**1604**) required that it be so. This curve is well known: it is the shape formed when a cone is sliced in a particular way. Other ways of cutting the cone reveal other 'conic sections', well known to the ancient Greeks—circles, ellipses and hyperbolas. All these curves turn up elsewhere in the natural world, particularly in the paths of various objects, like the planets and comets as they move through space under the influence of the pull of the Sun (**1609**).

Evangelista Torricelli: The Weight of the Air

1643 Evangelista Torricelli was one of a generation of Italian scientists inspired by the work and writings of Galileo Galilei. Born in 1608, trained by the Jesuits and skilled in mathematics, he became Galileo's secretary in 1641. Galileo, by then old and blind, was under house arrest by order of the Inquisition (**1633**) and was to die the next year. Typhoid fever killed Torricelli in 1649.

Torricelli's major contribution to science was the invention of what was later called the barometer (**1659**). He had sought an answer to a puzzle: as any engineer knew, water could not be lifted up by a suction pump to a height of more than 10 metres. Why was that? Galileo thought there was a limit to the 'power of the vacuum'; Torricelli had other ideas.

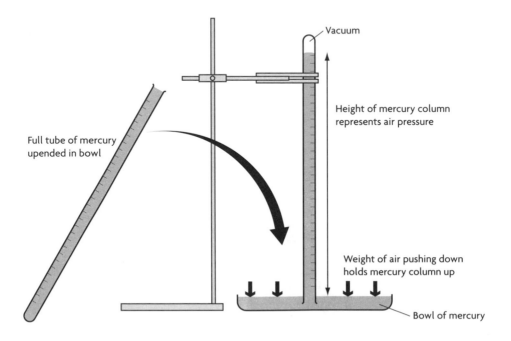

Vacuum

Height of mercury column
represents air pressure

Full tube of mercury
upended in bowl

Weight of air pushing down
holds mercury column up

Bowl of mercury

In Torricelli's simple but profound experiment, an upended tube filled with mercury not only demonstrates that air has weight and exerts pressure, but also enables small variations in air pressure, such as those associated with weather events, to be measured.

Experimenting in 1643, he replaced water with mercury, 14 times heavier, so his apparatus could be more compact. Assisted by another Galileo protégé, Viviani, he tipped a long tube, closed at one end and filled with mercury, into a bowl of mercury and held it upright. The level of mercury in the tube fell, leaving a space above. Torricelli reasoned that the space could not contain any air and so was a 'vacuum', something still believed by many not to exist anywhere.

Not all the mercury ran out of the tube; Torricelli reasoned that the rest was kept in place by a balancing force, the weight of the atmosphere pushing on the surface of the mercury in the bowl. So air had weight, another shock for the followers of Aristotle, who had maintained that air possessed 'levity' and was intrinsically light.

The weight of air was normally balanced by about 760 millimetres of mercury in the tube. Ten metres of water weighed the same, which set the limit on what a suction pump (which created an imperfect vacuum to receive the water) could do. Torricelli observed slight changes in the mercury level from time to time. So the pressure of the air was not quite constant. Others would find out why (**1676 Halley**).

Jan von Helmont: The Old with the New

Jan von Helmont was perhaps the last alchemist and the first chemist, though the latter term did not exist in his day. Certainly he combined the old and the new; his forward-looking ideas on techniques and applications shared mind-space with antique mystical notions and the belief that he could turn mercury into gold.

Von Helmont came from Flemish aristocracy, and took to medicine from a desire to relieve human suffering. He treated his patients for no charge, and having already given up his share of the family fortune, was able to pursue years of private research only by marrying a wealthy woman. His laboratory, equipped with furnace, crucible and retort, was the source of many of the medicines he prescribed, as it had been for his kindred spirit Paracelsus a century earlier (**1520**).

There he also assembled evidence against the old notion of 'air' as an element. Many forms of 'air' existed—he maintained he had a list of 15 such—and he coined a new general term for them—'gases'—from a Flemish word for 'chaos'. The term was still full of mystical meaning. One gas he called 'gas silvestre' because it came from burning wood (and so was mostly carbon dioxide). But he also identified (but gave his own names to) hydrogen sulphide or 'rotten egg gas', which gives the aroma to flatulence, hydrogen chloride ('spirits of salt') and nitrogen oxide.

Von Helmont had troubles with the establishment and the Church. Some of his propositions were condemned by the Spanish Inquisition as 'arrogant and heretical' and he spent time in ecclesiastical custody. The medical faculty at Louvain objected to his following of the still unacceptable views of Paracelsus. These were trying times for new ideas.

In 1644 he reported his most famous experiment—a tree grown in a pot with nothing added to it but water. After five years, the tree weighed more than 50 kilograms, while the weight of the soil was almost unchanged. Van Helmont argued that water alone made the tree grow, that the water had been converted into wood. A better answer would have to wait nearly a century (**1727**).

René Descartes and the 'Mechanical Philosophy'

Frenchman René Descartes is most famous for saying 'I think, therefore I am'. He carried out very few experiments, and relied more on systems of ideas to generate knowledge. So Descartes was not a scientist in the modern sense. But his union in 1637 of algebra and geometry, known as 'analytical geometry' or 'coordinate geometry', would prove a powerful tool for science, especially when allied with the soon-to-be-invented calculus. The sets of numbers used to locate a point in two- or three-dimensional space are still called Cartesian coordinates.

Descartes was at the forefront of the new philosophy that likened nature to a machine, rather than seeing it as an organism or a playground of arbitrary influences. He was 40 years ahead of Isaac Newton in attempting the task of explaining the behaviour of the greater universe on the basis of mechanics, expressed in mathematics.

Newton's version, published in **1686–87** in his *Principia*, was a triumph, and survives largely intact to the present day. Descartes published his attempt in 1644, and it was not so long-lived. In his universe, all-but-undetectable 'subtle matter' filled the spaces between ordinary objects. Great whirlpools or 'vortices' in this fluid carried the planets and comets in their orbits around the Sun, and the various moons around their planets.

The image was popular. People could grasp the idea of a vortex; they had seen whirlpools in water carry floating debris around and around. In contrast, the notion that some force of 'gravity' might act over great distances through empty space, as Isaac Newton was later to propose, seemed to be nonsense.

But the scheme did not answer some important questions, such as why the planets do not all orbit the Sun in the same plane and why the Moon's orbit is parallel neither to the orbit of the Earth nor to its equator. So after about a century, Descartes' vortices lost their lure. But his ideas about 'subtle matter' (aka the 'aether') persisted into the nineteenth century (**1887**). And he had put the 'nature as machine' model into a winning position.

> *One cannot conceive anything so strange and so implausible that it has not already been said by one philosopher or another.*
>
> RENÉ DESCARTES

Matches, Hand Grenades and Fountain Pens

SEVENTEENTH-CENTURY TECHNOLOGY

I n the seventeenth century the 'technology of knowledge' improved rapidly—telescopes and microscopes, barometers and early thermometers, mathematical aids such as logarithms, slide rules and primitive calculating machines, accurate clocks and vacuum pumps, precise measuring instruments such as vernier callipers and micrometers. But while these advanced science, they had very little impact on everyday life. Most aspects of society, and industry such as it was, were little affected. Traditional methods continued to be used to grow food, construct buildings, spin fibres and weave cloth, move about and communicate, and even to keep healthy and ward off illness.

The new 'learned societies', such as the Royal Society of London, had charters to improve 'practical arts and sciences', and many researchers of the time were very inventive, but their ideas did not reach far beyond their meeting rooms and laboratories. Any piecemeal improvements came mostly from the insights and experience of craftworkers. For example, ironworkers found better ways to make wire and chain, but the first accurate metal-turning lathes were introduced only at the very end of the century.

There were some signs of what was to come. Both coke and flammable gas were produced by heating crushed coal in a closed container, though no use of either was made at the time. Experiments with the pressure of gases such as air and steam led Frenchman Dennis Papin to make a 'steam digester' (we would call it a 'pressure cooker' today), and Englishman Thomas Savery to build a primitive steam pump to raise water out of mines. The first of these worked by raising pressure, the second by reducing it. It took time to realise that such increased and reduced pressures could do useful mechanical work, as was proven by the steam engine in the next century.

Engineering works consisted mostly of buildings, such as those that had to be reconstructed after the disastrous fire in London in 1666, or of drainage projects, such as the network of canals that converted the English fens to dry land. For obvious reasons, the Dutch were the experts there. Other machinery supplied water to the fountains in gardens of great houses, such as the palace at Versailles near Paris.

There were, of course, new ideas: the first use of wallpaper around 1645, fountain pens (1657) to replace quills, springs to soften the ride in coaches on the largely unmade roads of the time (1663), hand grenades as weapons (1667), new mechanisms for clocks and watches (1670), flexible fire hoses (1670s), flint glass, which was much harder than traditional soda glass (1675), pocket matches using the newly discovered phosphorus (1680), techniques for making plate glass (1688) and the new fabric called calico from India (1690).

But the pace of change in technology was slow compared both with later centuries and with the growth of science at the time. The Industrial Revolution was yet to forge a mutually beneficial bond between the practical and the scientific.

1651–1700

The World Stage

THE SUN KING LOUIS XIV, who seized virtual absolute power in France in 1661, pursued his quest for national glory by almost continual warfare against his neighbours for power and territory, but also by promoting the arts and sciences. In England the monarchy was restored and public life revived, but by the end of the century the power of the Crown was greatly constrained, unlike in France.

Spain, France, Holland, Belgium and Britain, together with Prussia and the other German states, were regularly at war, with the other states seeking to curb French ambitions. At the other end of Europe, Austria, Russia and Venice continued to push back the Ottoman Turks, who failed in another siege of Vienna in 1683. The glory days of science and technology in Muslim lands were now long past.

ARTS AND IDEAS

France's Louis XIV established the fashion for opulent new palaces such as Versailles, which took 100 years to complete. Its gardens were laid out in 1662 and the Hall of Mirrors designed in 1678. In London, Christopher Wren began rebuilding dozens of churches, including St Paul's Cathedral, destroyed by the fire of 1666. A new style of architecture, the baroque, began to spread from Italy across Europe.

The new musical form known as opera, which Italian Claudio Monteverdi had pioneered, continued under the Frenchman Jean Lully. Leading the way into the baroque in music were the Englishman Henry Purcell (*Dido and Aeneas* 1688) and the Italian Arcangelo Corelli. In Cremona, the master Antonio Stradivari was crafting his unsurpassed violins and other stringed instruments.

The satirical plays of Molière were first seen on French stages (*The Misanthrope* 1667). With the London theatres reopening upon the return of King Charles II to the throne, the 'restoration playwrights' were flourishing. John Milton published his epic *Paradise Lost* in 1667, Samuel Pepys began his famous diary in 1669 and John Bunyan's *A Pilgrim's Progress* first appeared in 1678.

THE AGE OF NEWTON

During these 50 years the pace of discovery quickened markedly. For the first time, we knew the speed of light and the size of the solar system. We discovered oxygen (though it was not yet called by that name), saw microbes in water and red cells in blood, appreciated the power of air pressure, understood the nature of colour and the meaning of the order of rock layers, gave chemistry a name, defined a chemical element, realised that plants can breathe, built clocks that kept good time, found the secret of Saturn's rings and marvelled at the properties of phosphorus. Battle was joined over the nature of light (particles or waves?), the purpose of spermatozoa in semen (parasites of the sex organs or essential for conception?) and the significance of fossils (mere rocky growths or memorials to past life?).

Over all towered Isaac Newton, born in 1642, the year Galileo died. His summation of the laws of the Earth and the heavens in the *Principia* was the outstanding achievement of science in the seventeenth, and perhaps any, century. Newton was surrounded by men of only slightly less talent: the immensely versatile Robert Hooke; Christopher Wren and Edmond Halley, active in astronomy, physics and other fields; Neminiah Grew and John Ray in biology; John Mayow and Robert Boyle in chemistry. English science rose powerfully in this half-century; all these men were leading lights in the Royal Society, founded in 1662.

Learned societies and national observatories were founded, but internationalism was strong. The Paris Observatory, established under an Italian, attracted leading figures from Holland and Denmark. The use of telescopes and microscopes advanced: in the former, Newton, Halley, Wren, Jean Cassini of France and Christiaan Huygens from Holland were all prominent; in the latter, it was Hooke, Dutchmen Jan Swammerdam and Anthony von Leeuwenhoek, and Italian Marcello Malphigi. Englishman John Locke built a platform of philosophy to support the growing edifice of science.

Blaise Pascal: The Pensive Scientist

The life of this French mathematician was short and overshadowed by ill health. He died aged only 39, when a malignant growth in his stomach spread to his brain. As a result of life experiences, Pascal was intensely religious and often gave up his scientific work to concentrate on spiritual matters. In his *Pensées*, written late in life, Pascal argued that everyone should believe in God; it is a wager on which we can lose nothing.

1653

Beyond religion, he spent most of his time on mathematics, establishing the study of probability and preparing the way for the invention by others of calculus. He also invented a calculating machine called the Pascaline, apparently to help his father, who was a collector of taxes.

In physics his name lives on in Pascal's Principle (or Law), published in 1653. This is a crucial understanding in the field of 'hydrostatics' or the science of fluids at rest. Apply pressure to any part of a fluid and it is transmitted everywhere through the liquid undiminished. This is the basis of the 'hydraulic press'. Pressure equals force divided by area; so a small force applied to a small area becomes greatly magnified when it is transmitted onto a much larger surface.

His interest in pressure lasted only a few years, but his discoveries were important. Picking up from Torricelli (**1643**), he convinced himself that a vacuum must exist above the mercury column in a barometer. Since air pressure decreased with altitude—a fact he proved in 1648 by having a barometer carried up a local hill—at some point above our heads air pressure would fall to zero. Therefore a vacuum (a 'nothingness') must exist beyond the atmosphere in the region between the Sun and the planets we now call 'space'. This was controversial, since many scientists, such as his countryman Descartes, believed, as Aristotle had, that 'nature abhorred a vacuum'.

The unit of measurement of pressure is now the pascal.

PASCAL'S PRINCIPLE

Pressure applied at any point (including at the surface) in a fluid (gas or liquid) at rest is transmitted equally and in all directions to all other parts of the fluid without loss.

PASCAL

The pascal is the unit of pressure. A force of 1 newton applied to an area of 1 square metre will generate 1 pascal of pressure. The pressure of the atmosphere at the surface of the Earth is roughly 100 000 pascals (1000 hectopascals).

Christiaan Huygens:
The Clocks and the Heavens

1656

Dutch-born mathematician and physicist Christiaan Huygens spent much of his working life in France under the patronage of Louis XIV. His diplomat father's wide contacts among mathematicians and natural philosophers helped set the younger Huygens on his course early. His life was a quiet one, and biographers had not much to report other than his significant achievement in science.

As a man of private means, and with a royal pension, Huygens was free to pursue a wide range of interesting topics. These included astronomy. He advanced the art of telescope construction and made important discoveries, such as the Great Nebula in Orion, surface features on Mars, the major moon of Saturn, and the true nature of Saturn's rings. His discovery of Saturn's largest moon, Titan, would be commemorated more than 300 years later in the name of the probe that made the first ever landing on its surface in 2005. Later in life he pioneered the use of very long telescopes to overcome some of the defects of the lenses of the time; one built towards the end of his life was 80 metres long.

As astronomy grew more precise and sophisticated, accurate, reliable clocks became essential. Nearly a century before, Galileo (**1581**) had found that a swinging weight keeps up a steady beat even as its swing diminishes. Around 1656 Huygens used this fact to make the first clock regulated by a pendulum.

Over the next 30 years he continued to refine his clocks and to work out their mathematics. He made clocks with compound pendulums and moved from pendulums to balance springs in search of a clock which could keep good time on a ship at sea. Such a 'marine clock' could solve the vexing problem of longitude by preserving the time at the port of departure and allowing it to be compared with local time as determined by the Sun. A final solution to that challenge would be found in another century and another country (**1765**). ➤➤**1690**

Otto von Guericke: Horsepower vs Air Pressure

1657

Otto von Guericke, as his name indicates, was of the German nobility, both by birth and by attainment. He was mayor of his hometown of Magdeburg for 30 years towards the end of his life. He also had studied law and served as a diplomat.

But it is his pioneering work in physics that preserves his name today. At university he went to lectures in mathematics and engineering, and from around 1650 he became interested in air pressure, perhaps through reading the writings of Evangelista Torricelli (**1643**) and Blaise Pascal (**1653**). They had shown that air had weight and could exert force. Von Guericke invented a piston-like pump that could for the first time extract much of the air from a closed vessel. He soon found that any shape other than a sphere would collapse under the unbalanced pressure of the air.

For his famous 'Magdeburg hemispheres' experiment in 1657 he made two closely fitting copper hemispheres, each about 30 centimetres in diameter. He pumped out most of the air, so creating a partial vacuum inside. It took 16 horses in two teams to pull the two halves apart, a powerful demonstration of the pressure of the air and of the reality of the vacuum. (The same imbalance of air pressure holds a suction cup to a wall or plastic cling wrap to the edge of a bowl.) Von Guericke later claimed his vacuum pump to be his greatest invention.

Von Guericke was the first to note that the ringing of a bell inside a vessel grew fainter as the air was extracted, but the attractive effect of a 'lodestone' (magnet) and an 'electric' (a piece of amber that had been rubbed with cloth) was not affected. Clearly, sound needed air to be carried but electricity and magnetism did not. He also noted that animals died if left too long in an evacuated chamber and that a flame went out without air.

Of major practical significance was his observation that a piston in a cylinder could be pushed strongly one way by air pressure if a partial vacuum was created on the other side of the piston. This was the first step on the long road towards the steam engine and the Industrial Revolution. ➡1663

Robert Boyle: Experimenter with Air

Irish-born to a wealthy family, educated at Eton, free to devote much time and energy to research, Robert Boyle was one of the great minds of his time. For some years he lived in Oxford, associating there with the scientists of the 'Invisible College', a forerunner of the Royal Society. He went to lectures but exercised the nobility's privilege of leaving without taking a degree. For the last 20 years of his life he lived with his sister in London, dying unmarried at the age of 64. Devoutly religious, he worked as hard on theology as on science, paid for translations of the Bible and left money to fund lectures on the place of Christian belief in the new age of science. For Boyle and many others of his time, there was no conflict between the two.

1659

In 1659, while at Oxford, Boyle read of Otto von Guericke's **1657** work on air pumps and partial vacuums, and determined to repeat the work with an air pump that was more efficient and easier to use. This he did with the help of the young and very able Robert Hooke (**1665**), who became his protégé. The air pump was connected to a glass jar shaped like a bell, into which he put various objects and observed their behaviour.

Boyle noted that as the air was pumped out, the sound of a ticking watch inside the bell died away (though the watch could still be seen), a burning candle or lump of charcoal would go out and small animals such as birds or mice struggled for breath and soon died. So he found, as von Guericke had, that sound needed air to travel; he added the observation that light did not. Neither respiration in animals nor combustion could go ahead without air.

It was Boyle who gave the familiar name 'barometer' to the mercury-filled 'Torricelli tube', which in **1643** had demonstrated that air has weight; the word means a way of measuring weight. More famously, in 1662 he demonstrated what is commonly called Boyle's Law. He made a simple piece of apparatus; a piece of glass tubing bent into a J, with the stubby end sealed off. When he poured mercury into the long open arm of the J, a small bubble of air was trapped in the toe. The more mercury he poured in, the smaller the bubble became. The difference between the mercury levels in the two arms of the J represented the pressure squeezing the gas

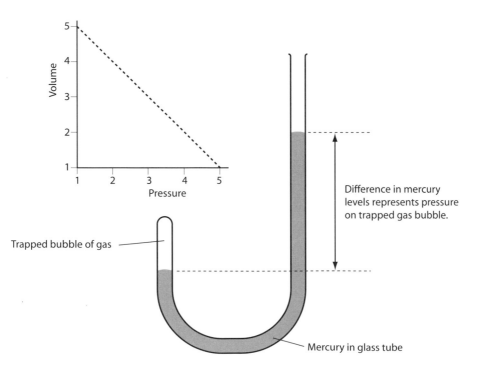

With this simple device, Robert Boyle proved that as the pressure on the bubble of gas increases, its volume decreases in the same ratio. This is true only if the temperature remains unchanged.

bubble. So as the pressure went up, the volume went down—Boyle's Law is an 'inverse' relationship.

Though Boyle did not say so (he may have just assumed it), the relationship held only while the temperature was constant. The French physicist Edme Mariotte rediscovered the connection 20 years later (as a result it is Mariotte's Law in Europe), and stated explicitly that the temperature had to be steady. The link between temperature and volume (and pressure) took another century to be tidied up (**1783**).

Behind all this lay another big question. Why could gases be easily squeezed into smaller spaces, when solids and liquids could not (or very little)? Why were gases 'spongy' or 'springy'? Boyle drew on the increasingly popular notion, an ancient one revived by Pierre Gassendi of France (**1623**), that all matter is made up of tiny particles (Boyle called them 'corpuscles'). In a gas, there is space between the corpuscles, which Boyle visualised as filled with tiny springs like, say, coils of wool. So the atoms can be pushed closer together and the gas compressed. In solids and liquids the particles are already in close contact, and further squeezing is hard or impossible. When the pressure was released, the atoms would spring apart again.

Here, then, was support for the 'atomic' view of nature, showing that it was more than an interesting idea. It could actually explain why matter behaved as it did.
➡**1661**

John Ray: God Revealed through Nature

English clergyman John Ray was arguably the leading naturalist of his time. For 40 years from around 1660 he published detailed systematic studies of plants, mammals, fish and insects, collecting widely in England and also in Europe. The son of a blacksmith, Ray may have gained his love of nature from his herbalist mother.

1660

His outstanding work generated a powerful reputation and he was among the early Fellows of the Royal Society. But his chief aim in life was to glorify his God through the study of His creation. He saw everywhere in nature signs of divine planning and purpose, and set down his beliefs in 1691 in *The Wisdom of God Manifested in the Works of Nature*.

Ray sought a system of classification of plants and animals that would also reflect the Divine Order. To transcend barriers of language he proposed using Latin. He defined a species much as we do today, and separated flowering plants into two major classes ('monocots' and 'dicots'), depending on whether one leaf or two first emerged from the germinating seed. His books contain descriptions of 18 000 plants. His work inspired generations of collectors and classifiers, including the great Carl Linnaeus (**1730**).

In Ray's time, naturalists debated the nature and origin of 'fossils'—which were being dug up in increasing numbers—especially those that resembled present-day plants or animals. Many saw them as 'seeds' of organisms that never come to life, or as remains of creatures destroyed in the Great Flood. Like the Dane Nicholas Steno (**1667**), Ray believed that such fossils had once been living organisms, but that not all of them had perished in the Flood. Like da Vinci (**1517**), he questioned why the Flood would have washed fossils onto the land, rather than away from it, and why they would have been deposited in beds rather than at random.

Late in life, Ray even wondered whether some fossils represented creatures that had become extinct over time. This challenged his vision of a perfect Creation, but opened the way for new ideas about the age and history of the Earth.

Marcello Malpighi: Master of the Microscope

1661

Nothing further advanced biology in the seventeenth century than the methodical use of the microscope—invented at the start of the century—to explore objects, both living and non-living, too small for the unaided eye. Among the leading practitioners in the new study, along with Englishman Robert Hooke (**1665**) and Dutchmen Jan Swammerdam (**1670**) and Anthony van Leeuwenhoek (**1673**), was the Italian Marcello Malpighi. He concentrated on studying living things and made some major discoveries.

Malpighi was a doctor who rose to become personal physician to the Pope and, in 1669, an honorary member of the Royal Society of London. In 1661 his microscope revealed blood passing through tiny blood vessels or capillaries connecting arteries and veins and vice versa. This gave final vindication to William Harvey's **1616** theory on the circulation of the blood. In 1666 he first saw the red cells in blood, and attributed the colour of blood to them.

Almost every human tissue yielded to his probing. He discovered the taste buds, the minute tubes within the kidneys, the cell-forming layers of the skin, the structure of the optic nerve. More fundamentally, he recognised that tissues have structures that relate to their function. Even the largest organs are made up of glands that secrete fluids. This made him the first histologist; medical research was profoundly changed.

Of similar importance was his work on the embryo of the chicken, identifying the minute structures that were later to become part of the heart or the nervous system. His work on the economically important silkworm, including the discovery that silkworms have no lungs and breathe through holes in their skin, aided the growth of the industry.

His later life was tumultuous. His radical new ideas, judged heretical by many, aroused controversy, envy and incomprehension. In 1684, his house was sacked and burnt, his microscopes smashed, his papers scattered and destroyed. The papal appointment, and other honours, restored his fortunes and endorsed his place as one of the great researchers of his time.

Robert Boyle: Chemistry Gets a Name

Irish aristocrat and prolific experimenter Robert Boyle called his seminal 1661 book *The Sceptical Chymist*. In so doing, he invented a new name for his subject ('chemistry'), as well as for his pursuit, by dropping the first syllable of the increasingly discredited 'alchemy'. Boyle based his ideas on experiment and observation, in the manner of the new age, rather than logic and conjecture. He carefully weighed substances involved in reactions, so showing, for example, that substances got heavier when they burnt. He discovered that plant extracts such as 'syrup of violets' showed different colours in acids than in alkalies (the first use of indicators).

1661

Like a growing number of researchers, Boyle believed that matter was made up of 'corpuscles', aka 'atoms', though he doubted these 'smallest bits of nature' could ever be seen. In his vision, different substances contained different clusters of corpuscles; chemical reactions were the result of these clusters being rearranged. The properties of substances reflected the characteristics of the atoms; for example, acids had a sharp taste and burnt the skin because their corpuscles had sharp points.

Boyle proclaimed himself 'sceptical' in refusing to blindly accept old ideas, such as the doctrine of the four 'elements'. If someone declared a substance to be an 'element', they must prove that it cannot be broken down into anything simpler. For the first time, Boyle defined elements practically, not philosophically.

Boyle thought that some materials would be 'mixtures', able to be easily separated, just as particles of dirt can be filtered from muddy water, leaving pure water. Others might be 'compounds', two or more elements intimately associated but

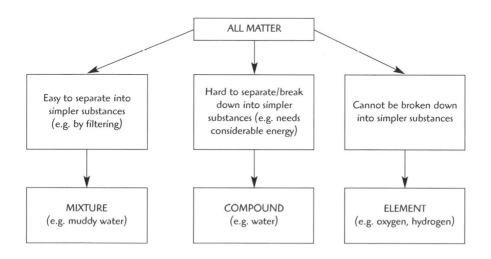

The table shows the chemical classification of matter as devised by Robert Boyle. At the time, Boyle did not know that hydrogen and oxygen were among the elements.

still able to be separated, though with much more difficulty. Both elements and compounds could be regarded as 'pure substances', in contrast to mixtures; mixtures might contain both elements and compounds.

This first separation of substances into elements, mixtures and compounds was a defining moment in the growth of chemistry. The idea took time to be accepted, but by the end of the century a number of substances fitting his definition would be known: the well-known metals gold, silver, copper, tin, iron, lead and mercury, and the more recently discovered arsenic and antimony, together with the 'non-metals' carbon, sulphur and phosphorus.

The last of these was newly discovered in Boyle's time. In 1656 the German alchemist Henning Brand was seeking a way to make gold from silver, as alchemists often did. He repeatedly distilled, of all things, human urine, until he was left with just a remarkable white substance that glowed in the dark and burnt brilliantly. He named it phosphorus, meaning 'bearer of light'. Perhaps hoping to profit from its marvellous powers, Brand kept it a secret. It was Boyle who found phosphorous a second time, in 1680. Being a chemist rather than an alchemist, Boyle told the world.

Interestingly, Boyle himself could not say for certain which substances were elements, or how many there might be. He wrongly thought, for example, that metals were not elements and could be transmuted one into another (lead into gold, for example). Even in the sceptical Boyle, some echoes of alchemy lingered.

The Rise of 'Learned Societies'

1662
As modern science developed, its early practitioners (not to be called 'scientists' for another 200 years) felt the need to get together, to exchange ideas and hear of each other's discoveries. The earliest such association that survives to the present day was the Royal Society of London, given its Charter by the restored King Charles II in 1662.

It was not the first such organisation; that honour belongs to The Lincean Academy, created in Italy in 1603. Galileo had been a member. His student Viviani had set up the Academia di Cimento in Florence in 1657. In England, men of scientific persuasion had been meeting for some years, often in secret, in what were termed 'invisible colleges', during the English Civil War. The concept of such a society dated back several decades to Francis Bacon's (**1626**) ideas about how scientific research should be organised for the betterment of society.

The founding of 'The Royal' made science respectable and brought it (and its practitioners) into the mainstream of social and political life. Over the next century similar groups began to meet in other leading capitals: in Paris (1666), Naples (1695), Berlin (1700), St Petersburg (1724) and Philadelphia (1769). The academies in France, Germany and Russia were small, with select groups of scientists salaried by the king and dependent on royal patronage.

> The business and design of the Royal Society is to improve
> the knowledge of natural things, and all useful Arts, Manufactures,
> Mechanick practices, Engynes and Inventions by experiment
> (not meddling with Divinity, Metaphysicks, Morals, Politicks,
> Grammar, Rhetorick or Logick), to advance the glory of God,
> the honour of the King, the benefit of his Kingdom and the
> general good of mankind.
>
> ROBERT HOOKE

On the other hand, anyone with the right credentials could be elected a Fellow of the Royal Society—and hundreds were, often when they were still in their twenties, and including researchers from overseas. In the growing bourgeois manner, they paid their own way.

'Papers' describing the latest findings and ideas were read at these gatherings and later published in the regular journal of the society, called its *Philosophical Transactions*—the first ever publication just for science. This greatly speeded up the spread of new knowledge, which previously had relied on the appearance of books, often years after the discoveries had been made, or on informal exchanges by letter.

From the start, the Royal Society featured all the great names of the time and many lesser ones. Robert Hooke, Christopher Wren, Robert Boyle, John Flamsteed, John Ray, Neminiah Grew and Edmond Halley were prominent, as were non-scientists like the diarist Samuel Pepys. Isaac Newton was President of the Royal Society from 1703 until his death in 1727.

Otto von Guericke: The First 'Electric Machine'

German nobleman and experimenter Otto von Guericke was already well known for his work on air pressure (**1657**) by the time he got involved with 'electrics'. He knew the prickly sensation caused by the rising of hairs on

1663

his arm when it was brought close to an amber rod that had been rubbed with a cloth. At the time this effect was thought to be like a small breeze, so it was a surprise to find, as he had done, that the effect could be felt through a partial vacuum.

In investigating magnetism in **1600**, William Gilbert had likened the Earth to a magnet, and had made a small globe out of lodestone to experiment with. Von Guericke thought that perhaps the Earth was an 'electric' too, and made a sphere out of sulphur to study the effect. He chose sulphur because it was one of the three elements of the alchemists, along with salt and mercury; ideas from alchemy were still influential.

Rubbing his hands on the globe (which in later models was spun by a crank) he saw all the effects we now associate with 'static electricity': tiny sparks, a crackling noise, the attraction of light objects. In particular he noted that a light object such as a piece of paper was first attracted to the electric and then, after it had touched it, repelled.

Von Guericke had made an 'electric machine', the first of many that were to fascinate researchers for the next century. But he had no idea of what he had found. His successors were still arguing about the nature of 'electricity' 100 years later.

Francesco Grimaldi: Light Goes Around Corners

1665 Italian Jesuit Francesco Grimaldi was one of many in his order interested in science. At a time when other forces in the Catholic Church were actively impeding new ideas in science and persecuting the early researchers who promoted those ideas on the grounds that they conflicted with official teachings (**1633**), a number of Jesuits were advancing knowledge. Grimaldi was a leader among these, and indeed a pioneer in the study of light. He anticipated Isaac Newton, who turned to the study of light after reading Grimaldi's book, published in 1665.

His best known experiment was very simple. He let light fall through holes cut in two screens—the second hole slightly larger than the first—onto a third screen. If light travels in straight lines, as was generally believed, the edges of the resulting pool of light should have been sharp, and determined precisely by the sizes of the two holes. The reality was different. Light seemed to spread beyond the allowed area into a region that should have been completely dark. It was as if light had somehow 'gone around the corner'. And the edges were not sharp. They seemed to dissolve into a series of light and dark rings, tinged with rainbow-like colours.

The pattern of light and dark reminded Grimaldi of the circle of ripples created when a stone is thrown into a pond. This set him wondering if light itself was not some sort of wave motion. He may have been the first to set this idea down. It was certainly taken up vigorously by other researchers, like Dutchman Christiaan Huygens (**1690**) and Englishman Robert Hooke (**1665),** and would ultimately become the accepted view, despite the opposition of Newton, who had different ideas (**1672**).

In another experiment, Grimaldi shone a light through two holes cut in the one screen to fall onto a surface beyond the screen. He was puzzled to see that where both beams fell on the same spot, the surface in places was darker than with only one source of light. Again Grimaldi was ahead of his time. He was in fact providing more evidence for the wave-like nature of light, evidence that would be confirmed and explained when Englishman Thomas Young repeated this experiment 150 years later (**1802**).

Robert Hooke: The World of the Very Small

By the time the diminutive English genius Robert Hooke published his sensational book *Micrographia* in 1665, the microscope had been in spasmodic use for half a century; Galileo had made one in 1609. It had,

however, come nowhere near rivalling its near cousin the telescope in revealing new wonders in creation. Indeed, few people seriously imagined that it would. No one suspected there might exist things too small to be seen with unaided sight, other than perhaps atoms, which no one ever expected to see. By his observations, Hooke penetrated more deeply into nature than was thought possible.

Hooke had already gained a reputation as assistant to Robert Boyle in Oxford, working in the years before **1659** on gases and air pumps. In 1662 he became, with Boyle's endorsement, Curator of Experiments at the newly formed Royal Society, though he was not yet a Fellow. *Micrographia* was to make him famous. It sold very briskly; the diarist Samuel Pepys called it 'the most ingenious book I have ever read', and could not go to bed until he had finished it.

Hooke filled his book with vivid, accessible prose and his own stunning engravings of the newly revealed micro-world—a flea enlarged to the size of a bird, a fine needle point looking like a rough carrot, a fragment of ash resembling a lump of coal, the box-like structures in a piece of cork, for which he coined the enduring term 'cells'. Without doubt, *Micrographia* ranks among the greatest creations of the new scientific age of observation and experiment.

Hooke certainly knew the power of his striking images, and went further to ponder the nature of the light that had created them. He explained the bands of colour he saw in thin films of air or mica by proposing that light travelled as some form of wave, like ripples on a lake. The idea was not new (**1665 Grimaldi**), but Hooke's backing gave it new force. One consequence was growing conflict between Hooke and his better known contemporary Isaac Newton, who had a very different view on light (**1672**). ➤**1678**

Isaac Newton: Why Doesn't the Moon Fall?

The enduring popular image of Isaac Newton involves an apple. Legend says it hit him on the head while he sat under a tree, so setting in train thoughts about gravity that ultimately encompassed the whole universe.

If the event actually occurred—and there is some doubt—it was around 1666. Newton had been evacuated from Cambridge University along with fellow students because of an outbreak of the plague, then raging through England. He was at home at Woolsthorpe Manor in Lincolnshire; his family were middle-class farmers.

It was there that Newton began to ponder why an object like an apple would fall straight down and not, for example, move sideways, and also why the Moon did not fall. Did the apple fall because the Earth pulled it, and did not the same attraction extend as far as the Moon? In time Newton would say 'yes' to both questions, asserting that the same laws of gravitation controlled objects on Earth and in the heavens. But it would be 20 years before he set that down in his *Principia* (**1686**).

The young Newton was also working on a new and powerful form of mathematics that he called the 'method of fluxions', but which we call calculus. With its aid, Newton could calculate the behaviour of moving objects and other systems undergoing change. The German mathematician Gottfried Leibniz (**1716**) had similar and complementary ideas; the two were to wrangle for decades over who was first. Leibniz later asserted that Newton had done more for mathematics than everyone else who had preceded him. Under the circumstances, that was a powerful endorsement.

In 1669 Newton was elected Lucasian Professor of Mathematics at Cambridge, though he was only 27 years old. A dazzling career was beginning to unfold. He became a Fellow of Trinity College, although he was a closet Unitarian and rejected the doctrine of the Trinity. He was indeed a complex and perplexing individual, believing the study of alchemy and theology to be at least as valuable as mathematics and physics. Nonetheless, he would dominate science in Europe for the next 50 years and remain immensely influential to the present day. ➤➤**1672**

Christopher Wren: Much More than an Architect

1666   Christopher Wren is best known today as the creator of the 'new' St Paul's Cathedral and much else in London devastated by the Great Fire of 1666, at which time he was 34. But architecture was only one of his talents. A man of means and sound education, Wren dabbled in many areas of science then taking shape, though he published little. In astronomy he made maps of the Moon and tried to explain the odd shape of the planet Saturn which, through Galileo's primitive telescope, had looked like a 'triple planet'. The Dutchman Christiaan Huygens (**1656**) trumped him by proposing (correctly) that Saturn was surrounded by a ring.

Wren did some anatomy while studying at Oxford, and looked at insects through a microscope (of his own making) even before Robert Hooke did (**1665**). Hooke was both a friend of Wren and his assistant in the post-fire years; together they designed the famous monument to the fire. Wren's time at Oxford during the rule of Oliver

Cromwell brought him in touch with the secretive circle of scientists that was the seed of the Royal Society, given its charter in **1662**. Wren was a founder and leader of the society, and served as its president for three years.

Like Hooke and Newton, Wren was very practical and inventive. He designed (and sometimes made) an astonishing array of machines, devices and instruments: water pumps, ways to harness the power of gunpowder, equipment for surveying, offensive and defensive military engines. He made the first rain gauge (meteorology was another interest), as well as various devices to attach to telescopes, so that the positions and separations of stars and planets could be accurately determined. In his hands, the telescope became an instrument for measuring, not just for looking.

Nicholas Steno: The Order in the Rocks

Danish-born Nicholas Steno (or Neils Stensen) abandoned science after his conversion to the Catholic faith at the age of 29 and spent the rest of his life in the Church. He wrote only one small book but his legacy was **1667** substantial. With him, the study of rocks and fossils began to develop some order and method.

Steno started in medicine, so he knew his anatomy. Living in Florence in the 1660s (hence the Latin version of his name), he noted the great similarity between the teeth of a shark he had dissected and certain stony objects recently dug up. These 'bodies within bodies' were controversial. Some observers said they had fallen from the sky, others that they grew naturally within rocks.

Steno thought they were what they look like; teeth or bones or other parts of plants and animals long dead and altered over time to resemble rock. Few people accepted so radical an idea at the time, though da Vinci (**1517**) had suggested something similar a century earlier. It would be another century before the notion of fossils as the remains of bygone life became respectable.

Studying the rock layers in which such fossils were encased, Steno claimed to see a pattern that reflected the story of the rocks. Most rock layers or 'strata', he said, had been deposited from water, presumably from the Great Flood in which all the devout believed. Such layers would therefore be horizontal to start with; if they were not, they had been lifted or folded by subterranean forces.

If one stratum sat upon another, the upper layer was the younger. This 'Law of Superposition'—as obvious as it seems today—was fundamental to a fledgling science. Here was a first notion of 'geological history', relative if not absolute. Steno had no idea how long rock layers might take to form and would not have contemplated a greater time than the Bible allowed.

Steno's 1667 book (the first part of a larger work that never appeared) was widely read and translated into English. With it, 'geology' took its first firm steps and Steno gained the title 'Father of Stratigraphy'.

Jean Cassini and John Flamsteed: Their Nations' Astronomers

1669

By the 1670s, 60 years after the first use of the telescope by Galileo and others in **1610**, enlightened opinion acknowledged the practical value of astronomy, particularly to help ships locate their position at sea. To get the most useful knowledge about the movements of the Sun, Moon, planets and stars, leading countries began to set up national observatories financed by the government or the Crown.

The French were first. Once established in 1671, the Paris Observatory, with its strong royal patronage, attracted leading astronomers, including the Dane Ole Roemer, who was to measure the speed of light in **1676**, and the Dutchman Christiaan Huygens (**1656**). The first director, Giovanni (Jean) Cassini was Italian, the first in a four-generation dynasty of influential astronomers.

Cassini is commemorated in the major gap that he found in the rings of Saturn, known ever since as the Cassini Division, and in the name of the spacecraft that reached Saturn from Earth in 2004. It was Cassini who found that the rings rotated around the planet and could not be solid. By tracking surface markings, he measured the length of the day on Mars, Jupiter (and incorrectly) Venus. He first reported the polar caps on Mars and the Great Red Spot on Jupiter, a cyclone 50 000 kilometres wide that has been visible ever since. Perhaps most importantly, he organised the experiment in **1672** that told us how big the solar system is.

The English had led the French by setting up the Royal Society in **1662**, but they trailed in the matter of a national observatory. The Royal Observatory at Greenwich was not established until 1675, in buildings designed by Christopher Wren, the creator of the new St Paul's Cathedral. Heading the observatory was John Flamsteed, the first Astronomer Royal; he held the post for over 40 years, also serving as parson in a nearby parish.

Flamsteed was a tireless and meticulous observer and a builder of precise instruments. He saw the planet Uranus 100 years before its official discovery but mistook it for a star, the known solar system then being thought complete. He later feuded with Isaac Newton, then President of the Royal Society, who had published some of Flamsteed's observations without his permission. That led to a falling out with Edmond Halley, who was to succeed Flamsteed at Greenwich. Halley was a close associate of Newton (**1684**). There was no getting away from personalities, even in science. ➤➤**1672 CASSINI**

Jan Swammerdam: Exploring the Natural World

1670

Dutch naturalist and microscopist Jan Swammerdam was originally destined for the Church but preferred medicine. His preoccupation with research caused him to neglect his medical practice, and so to fall out with his father, who cut off financial support. Swammerdam suffered great hardship; his health never really recovered.

His major discoveries in human anatomy include valves in the lymph vessels, follicles in the ovaries and the mechanism for the erection of the penis. He invented ways to study the circulation of blood by injecting dyes. His work on the relationship between nerves and muscles in frogs left no place for 'vital spirits' in the make-up of living things.

Around 1670 he began to study insects, especially their metamorphosis from juvenile to adult. He developed a method for classifying insects on the basis of their modes of development that is still used today. His studies of mayflies, bees and silkworms were classics; his *Natural History of Insects* was the first important book on ontomology, describing hundreds of insects, as well as worms and spiders.

His work, and that of Italian Francesco Redi, had another consequence. Careful studies undermined the still-common notion of 'spontaneous generation'; insects do not suddenly appear from non-living matter, but hatch from eggs laid by a female of their species. That finding, and the debunking of 'vital spirits', helped bring biology within the fold of the 'mechanical model', already well established in physics and astronomy. Nature operated by intelligible, material laws. Nothing was beyond study.

For his microscope studies, Swammerdam, like his compatriot Anthony van Leeuwenhoek (**1673**), preferred single-lens instruments, which he made himself. These were powerful but difficult to use, requiring immense patience and the best natural light. With no cameras to record what he saw, he made hundreds of drawings, many later engraved for printing.

Swammerdam's scientific career was interrupted in his late 30s by an enthusiasm for religious contemplation, and his later writings are full of spiritual allusions. But he did return to science to observe red blood cells, which he thought contained fat Weakened by malaria, he died before completing his 'great work', the magnificently illustrated *Book of Nature*, not published for another 50 years.

Jean Cassini: How Big is the Solar System?

By the mid-seventeenth century, astronomers had a good sense of how the universe (as they knew it) was organised, even if the Church did not agree. By common agreement, the Sun was in the centre, with the six known planets (Mercury, Venus, Earth, Mars, Jupiter and Saturn) going round it. Several of the planets had their own orbiting moons.

1672

But how big was all this? How much space did it occupy? The laws of planetary motion, propounded first in **1609** by Johannes Kepler, tell us how big the orbits of the other planets are compared with the Earth's. For example, Saturn is about nine times further from the Sun than the Earth. But nothing was absolute, though the ancient Greeks had made some estimates.

That changed in 1672. Giovanni (Jean) Cassini, Italian-born Director of the newly founded Paris Observatory, sent his colleague Jean Richer to South America (French Guiana) while he stayed home. They both observed Mars on the same day. Since they were thousands of kilometres apart, Mars appeared to each of them in a slightly different position relative to the background of the much more distant 'fixed stars' (no one had more than a guess as to how far away they were!).

From this parallax, they could figure out the distance from the Earth to Mars and so calculate the distance between the Earth and the Sun. This 'astronomical unit' was calculated as about 140 million kilometres. So the Earth's orbit around the Sun was about 100 times the size of the Earth itself. That put Saturn, the outermost planet, a bit more than a billion kilometres from the Sun; it made the solar system some two billion kilometres across, with nearly all of that empty space.

Over the coming century this figure (which is a bit smaller than the value accepted today) would be refined, including through observations of the Transits of Venus, for example in **1761**. ➤➤**1740**

Isaac Newton: Light of Many Colours

1672

Seeking an explanation for the spray of colours that fly from a cut diamond or fill out a rainbow, one early school of thought argued that all colours were a mixture of the two primaries, white and black. Others thought that colour appeared when light was 'degraded' by passing through glass or water.

The twenty-something-year-old Isaac Newton, a new star at Cambridge, soon applied his already prodigious talents to the matter. By his own account, he bought a prism (a triangular slab of glass) at a country fair to study 'the celebrated phenomenon of colours'. As expected, the prism broke a ray of sunlight into a spectrum of colours, from red to violet. But a second prism, set the other way round, did not 'degrade' the colours further, instead blending them back into white light.

Newton saw here proof that 'white light' is a mixture of colours, separated when light was bent or 'refracted' by a lens or prism. Assuming (wrongly, as it turned out) that the light passing through the lenses in telescopes or microscopes would always be smeared with confusing colour, he built the first 'Newtonian' telescope, using a mirror rather than a lens to collect and concentrate the light. Some years before, James Gregory had proposed, but not built, a similar instrument.

Gregory also found another way to break up white light. Colours appeared when light passed through very fine slits or openings (in his case, through the fronds of a feather). Explaining such 'diffraction' was a challenge for Newton's preferred theory of light, namely, that light was a stream of tiny particles or 'corpuscles'. A rival theory, that light was a wave, had been elaborated by Robert Hooke in **1665**, and would be followed up by Christiaan Huygens in **1690**. It explained diffraction and other phenomena as well or better than Newton's theory, so battle was joined.

Hooke and Newton never got on. When Newton first announced his discovery to the Royal Society in 1672, Hooke criticised it so severely (as being either plagiarised from him or simply wrong) that a mightily offended Newton refused to publish anything more on the subject while Hooke lived. Hooke died in 1703; Newton's *Opticks* appeared soon after (**1704**). ➤1686

Anthony van Leeuwenhoek: The Living Microworld

Without any scientific training, Anthony van Leeuwenhoek, a cloth merchant from Delft in Holland, became one of the most famous researchers of his day. For more than 50 years from 1673 he reported his revolutionary findings in a voluminous correspondence with the Royal Society and the French Academy.

1673

Van Leeuwenhoek did not 'invent the microscope', as is often claimed. But he certainly greatly refined the simple magnifying glass, known for centuries and used by him to count the threads in cloth. It is likely he was inspired by reading Robert Hooke's *Micrographia* during his only visit to London. The book had been published in **1665**, and was filled with images of objects normally too small to be clearly seen.

Hooke had used a 'compound microscope', a combination of two lenses invented around 1600. These were hard to make and use and could not magnify more than 30 or 40 times. Van Leeuwenhoek's tiny magnifying lenses, which he ground himself from glass and mounted on frames, were so well made and he was so skilful an observer that objects appeared enlarged as much as 270 times. A whole new world of previously unknown phenomena was revealed.

Van Leeuwenhoek was first to report the existence of tiny living things swimming 'nimbly' in water. He called them 'animalcules'; they now go by names such as protozoa, rotifers and nematodes. He estimated that a hundred of them laid end to

> *Whenever I found out anything remarkable, I have thought it my duty to put down my discovery on paper, so that all ingenious people might be informed thereof.*
>
> ANTHONY VAN LEEUWENHOEK

end would hardly stretch across a grain of sand, and thought they were carried from place to place in the air, rather than being 'spontaneously generated'.

He found the organisms we now call bacteria living in the plaque between his teeth (and much more abundantly in the plaque of two old men who never cleaned their teeth). He observed the shapes and structures of fossils and crystals, and of algae and other microscopic plants, and noted the red cells in blood, though he had no idea what they did. Not everything he did needed the microscope; for example, he mapped the life cycle of an ant, from egg to larva to pupa to adult.

In this own estimation, his greatest achievement was the discovery of spermatozoa in semen. The name means 'animals in semen', and he delayed publishing for fear of causing disgust or scandal among those who thought they were parasites. Over 40 years van Leeuwenhoek became expert in the sperm of all sorts of creatures, and was probably first to suggest that new life has its origin in the act of fertilisation, when an egg is penetrated by a sperm.

John Mayow: The Spirit in the Air

1674 The life of John Mayow was short—barely 35 years—but this son of Cornish gentry still left a deep impression on early science. Mayow studied law but moved to medicine, practising in Bath. Robert Hooke proposed him for membership of the Royal Society in 1678; the following year Mayow married, but he died shortly after.

Around 1674 Mayow built on the work of Robert Boyle, who had shown in **1659** that nothing would burn in a vessel that had been emptied of air. Mayow put a burning candle and a mouse into a closed jar. The candle soon went out; the mouse died soon after. But without the candle, the mouse lived twice as long. Clearly, he argued, something in the air that makes a candle burn is also necessary for life. So what happens in living things, generating heat and powering motion, is a bit like burning.

But there was more. If a burning candle floating on water was covered with a jar, water rose some way into the jar as the candle burnt out. Mayow saw in this evidence that only a small part of air helped burning. The greater bulk of the air was useless for that purpose. Experiments with expiring mice showed the same thing. He also found that this same component of air was absorbed when a metal is heated and develops a 'calx' or crust.

Mayow called this 'active and subtle' part of common air 'nitro-aerial spirit'; he believed it was found in the chemical nitre, without which gunpowder will not burn. Another title was 'igneo-aerial spirit', meaning roughly 'fire air'. Whatever the name, Mayow had discovered oxygen, a full century before its better known finders Priestley (**1774**), Scheele and Lavoisier (they called it, among other things, 'fire air'). It was a remarkable achievement. He also got the combustion–respiration link right in essence, so founding what we later called physiology. What else might he have discovered had he lived longer?

Edmond Halley: The Complete Astronomer

For most people, the name Edmond Halley is associated with the best-known comet, but like so many in his time, he was active across a broad swathe of science in its early days.

1676

Son of a wealthy maker of soap and salt, Halley went to Oxford, but left in 1676 without a degree, as the wealthy often did. Soon after, Halley took his telescope to the South Atlantic island of St Helena. His cataloguing there of 3000 southern hemisphere stars earned him a Master's degree by Royal Decree and the Fellowship of the Royal Society, both by the age of 25.

Halley's voyages also awakened an interest in the origin of the great ocean-spanning winds. In 1686 he published the first maps showing the easterly trade winds, the higher latitude westerlies and the six-monthly cycle of the tropical monsoon. He correctly argued that this grand global circulation of air was driven primarily by the heat of the Sun, the monsoons being the consequence of the apparent movement of the Sun in the sky north and south with the seasons. He linked the trade winds to the turning of the Earth, arguing that these blew towards a point directly under the Sun, but that this moved westward across the ground as the Earth rotated to the east. This was a good first go, later improved on by George Hadley (**1735**).

Careful use of the barometer clarified the link between air pressure and weather that had been noted by Toricelli (**1643**): for example, rapidly falling pressure warned that a storm was coming; a rising barometer foretold fine weather. In **1702** Halley published maps of the pattern of the Earth's magnetic field, similar to those he had made of the winds, and later still he detected the large-scale motion of the stars (**1718**).

Halley also argued that the saltiness of the sea was the result of the accumulation of salts slowly eroded from rocks on land. He was indeed versatile.

Halley's visit to Isaac Newton at Cambridge in **1684** was the catalyst for the publication of Newton's *Principia*, that great summation of the new understanding of the workings of the physical world. There was a reward; the *Principia* provided Halley with the tools for his great insight in **1705** into the behaviour of the comet on which his popular fame rests. ➤➤**1684**

Ole Roemer: Jupiter and the Speed of Light

In **1610** the Italian Galileo Galilei announced he had found four moons orbiting the planet Jupiter, each taking a certain predictable time to circle the planet. The Jupiter system was a sort of celestial clock; by consulting the right tables, an observer anywhere could use the positions of the moons to tell the time anywhere on Earth that Jupiter was visible.

Finding the longitude of a ship at sea was then a serious problem. Many a ship was lost or wrecked as a result. The King of Spain had offered a prize for a solution. Galileo argued that the captain could use Jupiter's moons to calculate midday in his home port, and compare that with local noon, when the Sun was highest in the sky. Each hour's difference in time meant 15 degrees difference in longitude, east or west.

The method was unreliable and tricky to use at sea, so it never really caught on. But it did encourage careful observations of Jupiter's moons. Astronomers timed the precise moment a moon was 'eclipsed' by passing into Jupiter's shadow. The results were puzzling. At times the eclipses came earlier than expected, at other times later, in a regular pattern.

In 1676 Danish astronomer Ole Roemer was working in Paris, where among other things he tutored the Dauphin (heir to the throne of France) and helped design the fountains at Versailles. Roemer grasped the answer. The difference in eclipse timings represented the varying gap between Jupiter and the Earth as each orbited the Sun. There was more or less distance for the light bringing the information to travel.

The sizes of the planets' orbits had just been worked out (**1672**) and Roemer could calculate that light, once thought to travel instantaneously, actually moved through space at a finite, if startling, speed. It was a major achievement, though his answer was a bit less than today's figure of around 300 000 kilometres per second.

Robert Hooke: England's Leonardo

The name of Englishman Robert Hooke is most commonly associated with his law, first published in 1678. Hooke's Law links the force needed to stretch a wire or a spring with the amount of stretch. Double the force and you double the extension. Remove the stress and the object will return to its original shape, that is unless the force has been too great. Exceed the 'elastic limit' and the wire is permanently deformed.

But there was so much more to Hooke than his law. He was one of the great original thinkers of his day, and so innovative and creative that many saw him as an English da Vinci. Hooke could write, draw, invent, investigate and imagine. In most of these he was da Vinci's equal, in some his master. Da Vinci dreamed, Hooke built.

Hooke's innovative investigations illuminated most of the developing areas of science, though he never dwelt long on any one topic, often leaving others to take the credit. He was Robert Boyle's collaborator at Oxford in the study of gases (**1659**) and a pioneer in the use of the microscope (**1665**). He often anticipated other great minds. Isaac Newton, for example, drew much from Hooke in his work on gravity, though he did not give Hooke full credit. Some see in Newton's famous comment, 'If I have seen further than other men, it is because I have been standing on the shoulders of giants' as a slighting reference to Hooke's small stature.

Hooke and Newton were alike in many ways. Both had lower middle-class backgrounds (farming for Newton, the Church for Hooke), were sickly in their youth but highly skilled with their hands, suspicious and secretive by nature and even more by experience.

Hooke was overshadowed by Newton's brilliance, both at the time and since, but he had an illustrious career. Boyle's recommendation secured him the post of Curator of Experiments at the newly founded Royal Society. He held this post, ideally suited to his talents, for 40 years, until his death in 1703. Like Newton, he never married.

Neminiah Grew: How Do Plants Work?

The Anatomy of Plants, published by English doctor Neminiah Grew in 1682, helped to make the new science of botany more than just a matter of classification, important as that was. He took it beyond a preoccupation with the usefulness of plants; most previous books had been 'herbals', concerned with medical applications.

1682

Grew was concerned to see how plants functioned and reproduced. He was among the first to suggest that plants, like animals, have two sexes, or at least two sorts of sexual organs, and that each contributes something (through the male pollen and the female ovule) to the next generation.

As a doctor, Grew was interested in intestines, publishing his studies on many animals in his *Comparative Anatomy of Stomachs and Guts*. As with plants, he wanted to know how the form of these organs affected their function, allowing them to do what they did. It was a relatively new idea.

Grew had another notable achievement. It was he who first explained the beneficial affects of the springs at Epsom Downs north of London. He found these 'healing waters' to be rich in a mineral soon dubbed Epsom Salts and known today as magnesium sulphate. Grew made good use of its medicinal properties in his practice.

Like most English researchers of his time, Grew was active in the Royal Society, serving for several years as its secretary. For a time he held the post of Curator of the Anatomy of Plants and made use of microscopes provided by Robert Hooke, the Curator of Experiments. To pay for the publication of his books, he organised a public subscription, the first time such a thing had been done.

Edmond Halley: 'Write It Down, Isaac'

1684

The *Principia* of Isaac Newton is arguably the most powerful and profound product yet of human intelligence.

But the *Principia* may never have been written but for the intervention of Edmond Halley, later best known for 'his' comet but in fact one of the most versatile scientists of the time. In August 1684 Halley visited Professor Newton in Cambridge to discuss a nagging problem: what is the nature of the law that defines the strength of the gravitational attraction between, say, the Earth and the Sun, or the Sun and the Moon? Most people believed that the force of 'gravity' depended on the distance between the two objects, and many that the relationship followed an 'inverse square' rule, like the one Johannes Kepler had found for light (**1604**). Halve the separation and the attraction increased four-fold; double it and the attraction fell to a quarter of its original value.

There was another issue. In **1609** Kepler had shown that the planets move in elliptical orbits around the Sun. Did these two things go together? Did an inverse square law decree elliptical orbits and vice versa? Halley and others suspected so, but did not have the mathematical skills to prove it. Newton did, and in fact had, nearly 20 years before. He promised to show Halley the calculations.

Halley wanted more. He urged Newton to write down all that he knew about the problem to show the Royal Society and then for publication. Reluctant at first, Newton set to work with increasing intensity, extending the basic ideas to new fields. Halley kept him at it, encouraging him, checking the proofs and ultimately paying for the printing. When the grand synthesis appeared in **1686–87**, after 18 months of prodigious effort, it was Newton's work but also Halley's achievement. **➤➤1705**

Isaac Newton: The Power of the *Principia*

1686

Isaac Newton's *Principia*, the first two books of which were published in 1686, makes tough reading. Even when translated from the original Latin, the prose is dense and long-winded, the geometrical arguments complex and the diagrams daunting. The usual title is part of a longer one in Latin, translated as *The Mathematical Principles of Natural Philosophy*. Newton strives to apply the burgeoning power of mathematics to what we now call physics. However obscure, it is a triumph, a treasure house of wondrous insights.

NEWTON'S LAWS OF MOTION

First Law: Every object continues to stay at rest or to move at a steady speed in a straight line (that is, at a constant velocity) unless some force acts on it.

Second Law: When a force acts on an object, the rate of change of the object's motion (that is, its acceleration) is proportional to the force applied and takes place in the direction the force acts.

Third Law: Every action is opposed by an equal reaction; the mutual actions of two bodies are always equal and act in opposite directions.

Newton does not fail to acknowledge his predecessors, those 'giants' on whose shoulders he claimed to stand. The first two of his three laws of motion, central to the first section of the book, had been developed by Galileo, beginning in **1604**. They enshrine the concept of inertia, the property that keeps an object moving at a constant speed in a constant direction unless acted on by a force. The change in speed and/or direction of motion (together these make up its 'velocity') is proportional to the applied force and in the same direction.

To those two laws, Newton added a third which says that however two bodies interact, be it by collision or attraction, what happens to one will happen to the other, but in an opposite sense. Thereby something crucial is 'conserved'. Newton called it 'motion'; we would say 'momentum', which is mass multiplied by velocity. The total momentum of the bodies before and after the interaction must be the same. This was not a new idea; Christopher Wren and Christiaan Huygens, among others, had figured it must be so. Newton made it part of a greater understanding.

$$F = ma$$
(Newton's second law)

$$v = u + at$$

$$s = ut + \frac{1}{2}at^2$$

$$v^2 - u^2 = 2as$$

F = force (newtons)
m = mass (kilograms)
a = acceleration (metres per second per second)
u = initial velocity (metres per second)
v = final velocity (metres per second)
s = distance covered (metres)
t = time elapsed (seconds)

These equations, inspired firstly by the work of Galileo, connect velocities, distances covered and time elapsed for an object being uniformly accelerated, that is, subject to a constant force such as gravity.

The laws of motion are a single page in the *Principia*'s 400, but the rest flows inevitably from them, thanks to Newton's mathematical genius. He explains how objects move under the influence of gravity, how they can be made to move in circles and how objects attract one another. The whole second part of the book explores moving objects fighting the resistance of air or water.

It is detailed, comprehensive and compelling. Nothing like it had been seen before and there has been little like it since. **➡➡1687**

Isaac Newton: *The System of the World*

1687

The third book in Newton's epochal *Principia*, called *The System of the World*, first appeared in 1687. It was an audacious title. Newton planned to show not only how the larger cosmos moved but also why. His explanation involved the universal dynamic he called 'gravity', now seen as a force of attraction between all objects everywhere. It united the fall of an apple to the ground with the progress of the distant planets.

As Newton laid it out, the pull of gravity depended firstly on the amount of matter in objects. The more massive a pair of objects, the more fiercely they tugged at one another. A large and a small object attracted each other equally, but the larger object resisted more and moved less. The Sun's vast mass allowed it to hold its place near the centre of our solar system and required the Earth and the other planets to go around it. So Nicolaus Copernicus was vindicated (**1543**).

The other determinant was distance. As many before him had suspected— including Robert Hooke who had suggested as much in a letter to Newton—gravity followed an 'inverse square' law. Doubling the separation of two objects weakened their mutual attraction by three-quarters. In a striking result, Newton showed that only mass and separation mattered, not size. Symmetrical objects, however large, behaved as if all their mass was concentrated at a single point at the heart of each.

Thus empowered, Newton could now prove that objects moving freely under gravity traced explicable paths: ellipses if the orbits were closed, parabolas if the ends did not meet. The planets, moons and comets held their courses because gravity exactly balanced their tendency to fly onwards in a straight line, and so pulled them into predictable orbits.

NEWTON'S LAW OF GRAVITY

The gravitational attraction between any two bodies is directly proportional to the two masses multiplied together, and inversely proportional to the distance between their centres.

$$F = G \frac{m_1 \times m_2}{d^2}$$

Newton's Law of Gravity links the force of attraction F (in newtons) between two objects to the masses m_1 and m_2 (in kilograms) of the two objects and the distance d (in metres) between their centres. The 'gravitational constant' G is a very small number (6×10^{-11}), indicating that gravity is a very weak force.

Universal gravity explained in detail much else: the ocean tides raised by the pulls of the Sun and Moon; the shape of the Earth; the Moon's complex motions; the 'precession of the equinoxes'—which changes the pattern of the night sky over thousands of years. All these and more were elements of *The System of the World*. On one point only did Newton admit ignorance. He could not explain gravity itself. That simply was. **➤➤1701**

John Locke: An Empiricist's Universe

The English philosopher John Locke has some small claim to be a scientist. His training was in medicine and he did work for a time as a physician, though he did no research. From early in his life he kept weather records: temperature, pressure, humidity, as much as the crude technology of the day permitted. Some of these were published in the *Philosophical Transactions* of the Royal Society, of which Locke was a Fellow. Locke's place in the history of science stems from his 1689 *Essay on Human Understanding*. This sought to sum up how we learn about the world. His answer fitted very well with the growing spirit of scientific inquiry, especially in England, as it had developed from Francis Bacon onwards. Knowledge comes only from experience, and from meditating on experience. This philosophy of 'empiricism' was at odds with the 'rationalist' views of French philosophers such as Descartes. They thought some ideas about the world were 'a priori', innate in our minds.

Locke's universe, like Newton's, was a mechanical system of material bodies, these being made up in turn of 'corpuscles' or 'atoms'. This was the way the world was being seen more and more across the full spectrum of science: machine-like, intelligible, predictable, even if complex. A real challenge to this vision would be a long time coming (**1927**).

Christiaan Huygens: Light Is a Wave

Christiaan Huygens was already famous for his work with clocks and telescopes (**1656**) when he came to England from Holland in 1689 to meet Isaac Newton. He much admired Newton's masterwork, the *Principia*,

1690

though he thought Newton's notion that the force of gravity could pass though empty space was 'absurd'. As a follower of Descartes (**1644**), Huygens wanted some more believable mechanical explanation. Still, Huygens' work on the forces on an object moving in a circle had helped Newton sort out his own ideas.

Though the regard was mutual, Huygens and Newton were to have another disagreement, this time over the nature of light. Newton had already (in **1672**) expounded his 'corpuscular theory': a beam of light was a chain of minute particles. He would do so more fully in his *Opticks* in **1704**. Huygens took his cue from Descartes, most likely from the Jesuit Francesco Grimaldi (**1665**) and from Newton's rival Robert Hooke. In his **1665** book *Micrographia*, Hooke had pushed an 'undulatory theory'. Light was a sequence of ripples, like waves on the surface of a pond.

Back in Holland, Huygens wrote up the ideas he had been incubating for a decade. His 1690 *Treatise on Light* showed that a 'wave' approach could explain most things about light: the colours seen in an oily film on water, reflection from mirrors, refraction by lenses and prisms, the newly discovered 'diffraction' or bending of light around small objects. Waves on water did all that, or something analogous, so the theory was plausible. His methods are enshrined in Huygens' Principle, still taught today.

The wave theory did not explain everything, but nor did Newton's. In fact, Newton was already dealing with some of the shortcomings by requiring his particles to make waves in an almost mystical medium called the 'plenum': this was cousin to Frenchman René Descartes' 'subtle matter' (**1644**) or the 'aether' that nineteenth-century physicists would hunt for in vain (**1887**).

Each theory had its strengths and failings. But Newton was Newton, already the most influential scientist of his time, thanks to the *Principia*. His views on most things, including light, generally prevailed. But neither side had a clear-cut victory. People later saw that light and other radiation could be usefully seen as either waves or particles, depending on the circumstances. That debate was to persist into the twentieth century (**1905**).

HUYGENS' PRINCIPLE

Every point on the line of an advancing wave (the 'wave front'), which may be a straight line or the arc of a circle, is a source of secondary waves whose interference determines the behaviour of the wave front thereafter.

1701–1750

The World Stage

IN EUROPE THE EIGHTEENTH CENTURY dawned much as the seventeenth century had ended, with a series of wars involving most states. These, however, had lost most of their religious fervour and were dynastic and territorial, seeking to decide who would sit on which throne. Struggles came in quick succession: the War of the Spanish Succession in 1702, of the Polish Succession in 1733, of the Austrian Succession in 1740.

The major beneficiary was the German state of Prussia; it gained the territory and resources that would make it the core of the future Germany, where so much important science would be done.

ARTS AND IDEAS

Music was in the hands of the great baroque composers, including the German Johann Sebastian Bach (*Brandenburg Concertos* 1721, *Christmas Oratorio* 1734), the English/German George Frederick Handel (*Messiah* 1741, *Royal Fireworks Music* 1749), Italian Antonio Vivaldi (*The Four Seasons* 1725) and the prodigiously prolific German Georg Telemann. John Gay's *The Beggar's Opera* premiered in 1721. The first piano was built in 1709 by Italian Bartolemeo Cristofori, with hammers replacing the plucking of the strings. The French horn first appeared in orchestras about the same time.

Memorable writings came from the pens of Englishman Daniel Defoe (*Robinson Crusoe* 1718) and Jonathan Swift (*Gulliver's Travels* 1726). Frenchman Charles de Montesquieu's *Persian Letters* of 1721 was the first major airing of the ideals of the Enlightenment. American Benjamin Franklin issued his first *Poor Richard's*

Almanac in 1732. The first recognisable novels appeared, written by Frenchman Antoine Provost (*Manon Lescaut* 1731) and Englishman Henry Fielding (*Tom Jones* 1749). In 1702 *The Daily Courant*, the world's first newspaper, appeared on the streets of London.

The Russian Imperial Ballet was established in St Petersburg in 1738. Italian dramatist Carlo Goldini premiered his classic comedy *The Servant of Two Masters* in 1744.

In architecture, the baroque was evolving into the rococo with its fantastic ornaments and scroll-like forms; Sevres china from France and Chippendale furniture from England expressed the same spirit in other mediums. Italian sculptor Nicole Salvi began work on Rome's famous Trevi Fountain in 1732. Three years later, Englishman William Hogarth produced his memorable engravings of *A Rake's Progress*.

THE MACHINERY OF NATURE

No one figure dominated the early eighteenth century in science as Isaac Newton had in the previous 50 years and as Antoine Lavoisier would in the half century after 1750. Many minds were busy in many countries; perhaps the ever-versatile English astronomer Edmond Halley was first among equals, with his studies of comets, the motion of the stars and the magnetism of the Earth.

Physicists concentrated on temperature and electricity. Increasingly accurate thermometers made possible the discovery of latent and specific heats in the half-century to come. With ever more efficient machines to generate electricity, means were discovered to conduct and store it, and we had our first thoughts as to what it was.

Chemistry gained two new tools, one highly valuable, the other a drag on progress. The invention of ways to collect gases for study made possible the immense advances after 1750; the doctrine of phlogiston prevented three generations of chemists from seeing things as they really are. The question 'Why do chemicals react?' was posed but not answered. In biology Carolus Linnaeus led the way to methodically cataloguing the living world. Not far behind was experimenter Stephen Hales, with his early insights into the physics and chemistry of plants and animals.

There were controversies about both the shape and the age of the Earth. The issue of its shape was eventually settled to everyone's satisfaction, but the age argument was just beginning. Beyond the Earth, the farsighted began to envisage the existence of other star systems besides our own. Physics and chemistry began their march into biology. Researchers found it increasingly valuable to regard plants and animals at different times as machines or as sites for chemical reactions.

Perhaps the most important new understanding was the 'need to measure'. Science would advance best by taking precise and careful accounts of phenomena, not merely describing but counting, and that needed appropriate measuring instruments. The challenge included finding where we were on the surface of our planet. The issue of latitude was solved; the problem of longitude was not finally answered until after 1750.

Isaac Newton and Daniel Fahrenheit: Better Thermometers

The eighteenth century in science was the century of measurement. Researchers came to understand the importance of measuring as well as observing, and as a result, new measuring instruments were developed to meet their needs.

1701

Studies in heat and temperature show this clearly. Isaac Newton, the great man in science at the time, led the way again. In 1701 he made the first practical thermometer. He greatly improved existing models by marking 'fixed points' on a scale alongside the sealed and evacuated glass tube (he used oil, but air, water, alcohol and mercury had all been used previously). Newton chose the temperature at which water froze as the lower fixed point and the temperature of the human body or blood, which seemed to be roughly constant, as the upper. The length between the two marks was divided equally into 'degrees'.

The name of Prussian-born Daniel Fahrenheit was to be linked with the measurement of temperature almost to the present day. He had used both alcohol and mercury in his thermometers, but from around 1714 settled on the latter because it could measure a much wider range of temperatures. For his 'fixed points' he used the melting of ice (32 degrees on his scale) and the boiling of water (212 degrees), dividing the interval between them into 180 equal steps. It is not clear why he chose such inelegant numbers, but the zero point on his scale was reputedly set by the lowest temperatures he experienced in Amsterdam, where he had settled.

Fahrenheit used his improved thermometers to establish the boiling points of various liquids, believing them to be constant. In recent times his temperature scale has been almost universally replaced by the Celsius or Centigrade scale (**1742**) but it lingers in use in the United States and among older people in other countries.
➤➤**1704 NEWTON**

Guillaume Amontons: The Possibility of an 'Absolute Zero'

Brilliant French inventor Guillaume Amontons went deaf as a teenager, but he did not consider it a handicap. Rather, it allowed him to concentrate on his science without distraction. With a background in physics, celestial

1702

mechanics and mathematics, as well as drawing, surveying and architecture, he made his living by working as a government employee on a range of public works.

Dozens of bright ideas tumbled from his mind. These included a type of optical telegraph, with messages sent by the flashing of a bright light, visible through a telescope by the operator at the next station in the communication chain. He demonstrated it to the King of France but it was never adopted. He also developed a water-powered clock, a barometer that could be used at sea because it did not need a reservoir of mercury, and a device to measure the amount of moisture in the air using a substance that expanded or contracted as it absorbed or released water vapour.

Amontons has a place in the thermometer story: he improved the primitive air-filled thermoscope that Galileo had invented in 1593 to measure how hot or cold things were. The air in the tube expanded and contracted not only with temperature but also with the external air pressure, but Amontons worked out a way to correct for that. At the time, no one had invented a temperature scale, and the device was not accurate enough for most scientific work, but it did allow Amontons in 1702 to make an important discovery: provided the external pressure did not change, water always boiled at the same temperature.

He also noticed that, as accurately as he could measure it, if the volume of a gas was kept the same by adjusting both the temperature and the pressure, the temperature and pressure went up and down together, step by step. This suggested that if the temperature dropped low enough, the pressure would go down to zero. Here was the first intimation that there might be a temperature below which we cannot go, an 'absolute zero'. Amontons was 150 years ahead of his time (**1848**).

The Complex Behaviour of the Everyday Compass

1702 Over the centuries the growing use of the magnetic compass to point to the north had freed sailors from reliance on coastal landmarks and set them roaming east and west across uncharted oceans. But those voyages showed that the compass was not always reliable; the needle usually pointed a little (and sometimes a lot) east or west of true north as indicated by the position of the Sun at noon.

This 'variation of the compass' was first published as a map in 1702 by the English astronomer Edmond Halley, using measurements he had himself taken while sailing the Atlantic in the 15-metre HMS *Paramour* for two years. The adventure had included dodging icebergs in fog in his tiny craft. With such a chart in hand, sea captains should be able to correct their compass readings and so reliably find their way.

There was also already a problem. In 1635 Henry Gellibrand had shown that the magnetic variation was not constant but slowly changed with time. So the maps would have to be redone every few years to correct growing inaccuracies. This scuppered a hope that Halley and others had clung to that magnetic variation, once charted, could be an indication of where the ship was, especially since the problem of longitude was proving so difficult (**1765**).

There was a second problem. How could the magnetism of the Earth change in this way? No bar magnet behaved so. Halley suggested that the Earth contained a number of separate layers, each independently magnetised. As these slowly rotated with respect to each other, the Earth's magnetic field drifted. The explanation was not right in detail but it did correctly place the origin of the field and its variation deep within the Earth.

More complications arose in 1724. Englishman George Graham found that the compass needle sometimes veered off by a small angle for a day or so, then came back. Anders Celsius in Sweden, originator of the popular temperature scale (**1742**), observed such a shift at the same time as Graham in London. The effect was later found to be worldwide, and dubbed a 'magnetic storm'.

Celsius and his students found that such magnetic disturbances occurred during displays of the 'northern lights' (the aurora). This tied in with an idea Halley had put forward in 1716 that these brilliant natural light shows were due to 'magnetic effluvia' interacting with the Earth's field. Again the details were wrong, but the fundamental idea was sound.

Francis Hawkesbee: Making Light from Electricity

In **1663** German Otto von Guericke had built the first 'electric machine' to concentrate the energies we now call 'static electricity'. But it was the English experimenter Francis Hawkesbee 50 years later who really pushed the subject along. His later versions of the machine substituted a glass ball for the sulphur sphere to be rubbed by the hands of the experimenter to build up the charge.

1702

Following up an observation made years before, Hawkesbee produced the first electric light in 1702. The earlier report said that the mercury in a barometer tube would give off flashes of light when shaken. Hawkesbee put a little mercury in a sealable tube and pumped out some of the air. He charged up the tube by connecting it to his machine. He reported that the tube gave off enough light to 'read large print across the room'. The observation stimulated interest in what else 'electricity' might do. ➤➤**1709**

Isaac Newton: A Man of the World

The English scientific colossus Isaac Newton finally published his *Opticks* in 1704, more than 30 years after his pioneering research into the behaviour of light. The reason for the delay lay back in **1672**. By the time

1704

> *To myself I seem to have been only like a boy playing on the seashore, and diverting myself now and then finding a smoother pebble or a prettier shell than ordinary, whilst the great ocean of truth lay all undiscovered before me.*
>
> ISAAC NEWTON

Opticks appeared (written in English rather in Latin) Newton was famous and a man of the world. When nearly 60, he had left academic seclusion at Cambridge for a post at the Royal Mint, where he oversaw the reform of the currency. The work was important and he took it seriously, though many of his contemporaries thought it unworthy of his talents.

Newton had also mostly left active science behind. He still did some mathematics, but spare time was given over to numerology, theology and biblical chronology, and to research in alchemy. As President of the Royal Society, he remained a powerful force in science and its politics, and he held the post for nearly a quarter of a century until his death. His eminence was confirmed by a knighthood in 1705, the first to a man of science, though it may have been mostly for his work at the mint.

He remained argumentative. His long-term antagonist Robert Hooke was dead, but Gottfried Leibniz still challenged Newton's claim to have invented calculus (**1716**), and the Astronomer Royal, John Flamsteed, complained that Newton had published some of his (Flamsteed's) data without authority. These disputes affected those around Newton; allies of Flamsteed, such as Stephen Gray (**1729**), had a hard time getting into the Royal Society.

Newton never married. His clever and charming niece kept house for him in London. When he finally succumbed at the age of 85, he was honoured with a burial in Westminster Abbey. That too was a first, but a fitting honour for the premier scientist of his age. As the poet Alexander Pope put it:

> *Nature and Nature's Laws were hid in night.*
> *God said 'Let Newton be!', and all was light.*

Edmond Halley: Taking the Terror Out of Comets

1705 Since ancient times, the unexpected appearance of a comet in the night sky had been seen as a bad omen. There was reportedly a comet in the sky over Rome the night before Julius Caesar was assassinated. Others reputedly foretold the Battle of Hastings in 1066 and the fall of Constantinople in 1452. The word 'comet' means a 'hairy star' and refers to the fuzzy coma and wispy tail that comets develop when they draw near the Sun. But they could also be dubbed 'bad stars', which is the origin of our word 'disaster'.

In 1705 English astronomer Edmond Halley began to lance the fear of comets. He drew on the work of Isaac Newton, whose great *Principia* owed much to Halley (**1684**). The path of a comet through space was controlled only by gravity, and Newton showed how such a path could be calculated from just three or four observations. When Halley did the sums on the paths of comets seen in 1452, 1531, 1607 and 1682, the results came out almost the same. He boldly proposed that they were in fact sightings of the same object, which was therefore in a predictable path that would see it come into view again in 1759.

Halley was dead by the time his prediction was fulfilled, and the most famous comet now bears his name. More importantly, he showed that comets were simply lifeless bodies in orbit around the Sun, obeying the same laws as the planets. Their reappearances were therefore not supernatural but predictable and so not to be feared. That such an object might one day collide with the Earth and cause immense destruction was something Halley did not, it seems, contemplate, though at the end of the century (**1796**) Frenchman Pierre Laplace certainly did. ➤➤**1718**

Francis Hawkesbee: Capillary Power

A mat of thin fibres, say a piece of cloth, quickly soaks up water. Before blotting paper (another good example) was invented, ink on a document was dried by dusting with fine sand, the ink being drawn up into tiny spaces between the grains. Why does this happen? It is not like sucking a liquid through a straw into the mouth, a feat made possible by air pressure (**1643**). 'Capillary action' can take place in a vacuum, with no air present. Some other agency must be at work.

In 1709 the English researcher Francis Hawkesbee, who was also very busy with 'electrics' (**1702**), put forward an explanation that suited the new ideas of the times. Matter was imagined to be made of countless tiny particles ('atoms' or 'corpuscles'). These pulled on each other through forces that acted only over short distances, unlike the pull of 'gravity', which reached at least across the solar system and perhaps for ever.

Hawkesbee set out his ideas in his book *Physico-Mechanical Experiments on Various Subjects*. The particles in the walls of a tube pull on the particles in the liquid, drawing them up against the pull of gravity. The forces act only around the edge of the liquid where it meets the wall, so the wider the tube, the more liquid would need to be supported. So liquids rise further in fine tubes than in wider ones, and the upper surface of the liquid sags in the middle (making a 'meniscus').

The capillary effect (the word 'capillary' comes from the Latin for 'hair') pulls liquids into any material with fine holes in it and so is a powerful force in nature. It

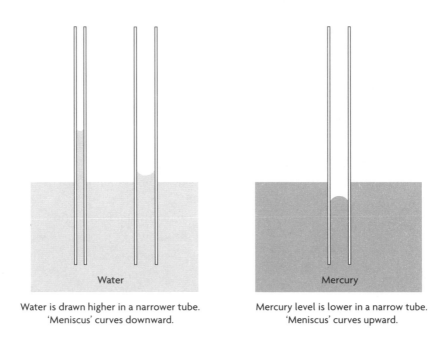

Water	Mercury
Water is drawn higher in a narrower tube. 'Meniscus' curves downward.	Mercury level is lower in a narrow tube. 'Meniscus' curves upward.

In his studies of capillary action, Francis Hawkesbee found that water is drawn up further in a narrow tube than in a wider one. The upper surface of the water in the tube (the 'meniscus') is lowest in the middle. Mercury behaves differently, since the mercury particles are more strongly attracted to each other than to the walls of the tube.

explains 'rising damp', for example. In 'transpiration' (**1733**), water from the soil can be drawn to the tops of trees 50 metres or more above the ground through exceedingly fine pipes (the 'xylem') in the roots and stems.

The forces that drive the effect explain much more. The particles in the liquid pull on each other as well as on the surrounding surface, and the outcome depends on which attraction is stronger. If the pull between the particles (the 'surface tension') dominates, the liquid appears to reject the wall. Water on a greasy surface does this, and so does mercury. That is why the meniscus in mercury curves up in the middle, not down, and mercury in a capillary tube is pulled down, not up.

Leibniz on Newton

1716

The life of the brilliant German Gottfried Leibniz crossed that of the brilliant Englishman Isaac Newton several times. Early in their careers, each had independently invented the powerful new form of mathematics known today as calculus, though they had different approaches and ways of writing it down. A dispute as to who discovered what when persisted until Leibniz died in 1716.

Leibniz also had some problems with Newton's grand vision of the universe in his *Principia* (**1686–87**). To Newton, the world was a great machine, controlled only by forces like gravity. Yet he thought it was unstable, prone to break down or run awry until regularly corrected by the direct intervention of God. Leibniz rejected such an imperfect universe as unworthy of its Creator, and argued that once we understood the universe fully, we would see that it did not need constant tinkering by a divine hand. It took more than a century for the French genius Pierre Laplace (**1796**) to demonstrate that the solar system was indeed stable over time, so provoking his famous response 'I have no need of that hypothesis' when asked by Napoleon why God did not feature in his calculations.

Another challenge for Leibniz was Newton's insistence on 'absolute' space and time, existing whether or not there was any matter in the space or any events to mark the passage of time. Newton claimed that the movement of all objects could be compared with this fixed frame of reference, within which even the Sun was not quite still. There was some theology here; Newton referred to this absolute space and time as the 'Sensorium of God'.

Leibniz saw no reason why this was necessary; others, such as English philosopher and bishop George Berkley, agreed. To them space was simply the measure of the arrangement and separation of objects, and time the order and spacing of events, such as successive ticks of a clock or risings of the Sun. Only 'relative movement' was measurable, that is, movement of one object compared with another. The debate about space and time would continue into the twentieth century, encompassing among other things Albert Einstein and his Special Theory of Relativity (**1904**).

Edmond Halley: The Stars in Motion

In his *Principia* in 1687, Isaac Newton asserted, and indeed proved, that the laws of motion and gravity as we know them here on Earth extend into the heavens. They controlled the Sun, the planets and their moons, the comets—in short the whole 'solar system'. Such was his 'System of the World'.

1718

But what of the so-called fixed stars beyond the solar system? Did they too feel the pull of gravity, and if so, why did they not fall together under its mutual attraction? Newton had an answer. The stars were indeed subject to gravity, but they were so far apart that its effect was negligible. So they held their positions relative to other stars around them.

That answer was not to everyone's liking. Two years before he became the second Astronomer Royal at the age of 62, English 'comet man' Edmond Halley compared the positions of many stars in the sky with those reported 1500 years earlier by the Greek authority Ptolemy. He found that three of them appeared to have moved relative to others nearby. This 'proper motion', he argued, was proof that the stars were not really 'fixed' but rather they were independent objects like the Sun, moving freely in space controlled only by gravity. The modern view of the universe was taking shape.

Georg Stahl: The Legacy of Phlogiston

1723

The idea that everyday matter contains a 'fiery stuff', set free when things burn, is as old as the ancient Greeks. Aristotle's four 'elements' had included 'fire', along with 'air', 'water' and 'earth'. The German John Jacob Becher had revived the notion late in the seventeenth century, calling the fiery element phlogiston, from a Greek word meaning 'to burn'. Becher's compatriot and court physician Georg Stahl expanded and promoted it in his *Fundamentals of Chemistry*, published in 1723.

The idea caught on. For more than 50 years, many early chemists, including great minds such as Joseph Priestley and Henry Cavendish, used the phlogiston theory to explain what they saw happening to matter in their laboratories. For example, when lead was heated strongly, it developed a crusty red or yellow coating or 'calx'. According to the theory, this calx was the 'real lead', the grey metal being a mixture of the calx with phlogiston. Heating had separated them.

The theory must have had some intrinsic appeal, because it did not really fit the facts. In the previous century chemists like Robert Boyle had noted that if metals like tin or lead were strongly heated and developed a calx (presumably as a result of losing their phlogiston), the metal weighed more at the end than at the beginning. The measurements were admittedly crude but they still had to be explained.

How could the loss of its phlogiston make something weigh more? The 'phlogistonists' had a ready answer. Like Aristotle's 'fire', phlogiston was essentially light, not heavy. It had 'negative weight'; when it escaped, things got heavier. It sounds absurd today, but the chemists of the day bought it, and as a result, chemistry was held back for much of the century.

In the long run, such fancy footwork did not save phlogiston. It was not really credible once the facts were better known. Nonetheless, it took the genius of Antoine Lavoisier late in the century to finally put phlogiston in its grave (**1775**).

Herman Boerhaave: Medicine's Most Famous Teacher

1724

When the young Herman Boerhaave came to study medicine at the University of Leyden in Holland, he mostly had to teach himself from books in the library, so depressed was the teaching of medicine there. By the time he retired from the faculty 50 years later, Leyden boasted one of the most famous medical schools in Europe, and Boerhaave himself was the most beloved of teachers. Students, including the sons of royalty and nobility, came from all over the continent to sit at his feet.

Much of his fame and success came from reverting to the methods of the Greek master physician Hippocrates, taking the students to the patient's bedside so they could assess the symptoms for themselves, and then if necessary to the autopsy room. His name is recalled by, among other things, Boerhaave's Syndrome, a rupture of the food pipe due to vomiting while over-eating. It had killed the nation's Grand Admiral.

Despite his eminence in medicine, which led to membership of the Royal Society of London and the French Academy of Science, Boerhaave's greatest passion was chemistry, and his 1724 book *Elements of Chemistry* was his most successful. Translated into German, English and French, it has been called the first modern chemistry textbook. It was, of course, a book of its time. It could not foresee the monumental discoveries that would transform the subject beyond recognition over the next six or seven decades. At almost the same time (**1723**), the German Georg Stahl was promoting the concept of phlogiston, a doctrine that would ultimately succumb (**1775**) to the operation of Boerhaave's own motto: 'The great seal of truth is simplicity'.

Boerhaave's book contains his recipe for extracting pale crystals of a substance called urea from urine. This was the first compound ever identified as being produced inside the human body. Fifty years later, urea was found to give off the pungent gas ammonia when heated. Another 50 years on (**1828**) that discovery led to the making of urea synthetically, so firmly linking the animate and inanimate realms of nature.

Astronomers Count the Stars

Astronomy means in Greek 'measuring the stars', and much time and effort have been spent counting and cataloguing stars. With diligent observers and better equipment, the numbers whose positions were known with precision grew rapidly though the eighteenth century. In 1725 a catalogue of 3000 stars compiled by the first Astronomer Royal, John Flamsteed, was published posthumously. His successor, James Bradley, who among other things measured the speed of light (**1729**), added another 3000. Nicholas Lacaille of France had 10 000 precisely plotted stars in his 1750 catalogue. Bradley topped that with an amazing 60 000 stars in a catalogue published in 1762, 10 times more than are visible with the unaided eye.

Knowledge of the stars grew steadily. In **1718** Edmond Halley (the second Astronomer Royal) had found that some of the so-called fixed stars were moving relative to each other. In 1783, William Herschel, a German-born Englishman and the discoverer of the new planet Uranus in **1781**, found that all the stars were in motion, or so it seemed, creeping away from a point in the constellation Hercules. This was an illusion: it was our Sun that was moving, carrying the Earth with its astronomers through space in that direction. All stars were presumably moving in the same way; our Sun would be nothing special. There was now no such thing as a 'fixed' star.

1725

Another sort of stellar motion came to light. Many stars lie much closer to each other in the line of sight than you expect from chance. The versatile Englishman John Michell (**1783**) suggested that such an alignment was not chance, that the two stars were actually close together. They formed a 'double star' through their mutual attraction. Here again Herschel took the lead. By careful observation over many years, he found that the two stars in many such associations were in fact circling each other. Kepler's laws of planetary motion (**1609**), which explained the movement of the planets and moons in our solar system, covered the motion of these double stars as well, showing that the same force of gravity controlled them.

One such double star was Algol, the 'Demon Star' (**1604**). Its quick changes in brightness, observed for a century at least, were shown to be due to the brighter star of the pair being eclipsed by a dimmer companion that passed in front of it. Another mystery of the heavens was explained.

Finding Food for Plants

1727

What do plants eat? Like animals, they grow and develop, so they must have some source of sustenance. Dutch doctor Jan von Helmont had proclaimed in **1644** that water was all a plant needed. He had given a growing plant nothing but water for five years. However, he took no measure of how much water had been used up.

Following up on this work in 1699, English doctor and researcher John Woodward did keep track. He grew his plants in water rather than soil and added a carefully measured amount of water each day. The results were striking. The amount of water added over three months was very much greater than the weight that the plants gained. Most of the water had, it seemed, passed through the plant and done no good.

Not quite. If the real food for the plants was in soil, as Woodward thought, water was needed to carry nutrients from the roots up into the stem and leaves. But in von Helmont's experiment, the weight of soil had remained almost unchanged while the plant grew massively. So it seems that neither soil nor water was the ultimate source of plant food.

The Italian Marcello Malpighi, discoverer of the vital blood-carrying capillaries (**1661**), had already made a suggestion. European trees drop their leaves in winter, and no part of a plant grows significantly then. The growth spurt starts in spring when the leaves appear. Are leaves therefore the source of food for the rest of the plant? Malpighi reported cutting off the first leaves of a squash seedling. It did not grow thereafter, despite access to water and soil. It was not very convincing evidence, but it did put a new idea into the pot.

By now it was known that leaves are covered with tiny pores, visible through a microscope, and able to open and shut. Neminiah Grew of England (**1682**), among others, argued that these 'stomates' let air and moisture in and out of the leaves. Just about all the water Woodward had given his plants had disappeared that way. Do plants also take in air through the stomates?

In 1727 the English clergyman Stephen Hales published his celebrated *Vegetable Staticks*. In it he reported growing a plant under a glass lid and noting that some of the air around it seemed to disappear. If was as if some part of the air had been 'imbibed into the substance of the plant'. Again the measurement was crude but it did suggest that air was involved in the nutrition and growth of plants. And, as Joseph Priestley was to show in **1771**, as they grow plants also do something to the air.

Stephen Hales: How To Collect Gases

In looking at nature, the venerated Aristotle had bothered only with solids and liquids, basically what he could see. His followers for 2000 years did much the same. Air and other 'vapours' were invisible and easy to overlook. Dutchman Jan van Helmont (**1644**) had found various forms of 'air', for which he coined the term 'gases', but they were still not much taken into account. When a piece of wood burnt and became much lighter ash, no note was taken of any vapours given off. As a piece of iron rusted, it got heavier, but that was because, according to the theories of the day, it had lost 'phlogiston' (which weighed less than nothing). No one thought that perhaps it had taken something in from the air.

To be fair, researchers could not really study 'airs' or 'gases' until there was some way to collect them. That was first described in 1727 by English clergyman Stephen

1727

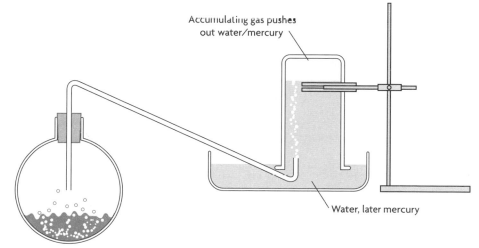

Accumulating gas pushes out water/mercury

Water, later mercury

Chemical reaction releasing gas

Stephen Hales' device was the starting point for modern chemistry. For the first time it was possible to collect the gases released by various chemical reactions for identification and study.

Hales, celebrated for his insights into the physical workings of plants and animals (**1733**). He bubbled a gas into a water-filled jar standing upside down, with the water held in place by air pressure. As the gas collected, it pushed the water out, leaving ultimately a jar full of the gas ready to be upended for study. This device was the 'pneumatic trough' (from the Greek word for air) and it made possible the detailed study of gases dubbed the 'pneumatic revolution', an appropriate title for a major advance in chemistry.

Hales did not do much with the gases he collected; in truth not many with distinctive properties were known. There was air itself, still thought to be a single substance, the 'gas silvestre' van Helmont had found (and which would be studied anew by Joseph Black in **1754**), and a 'flammable air' that Hales and his predecessor Robert Boyle (**1661**) had seen bubbling off pieces of iron put into an acid. The latter, which we now call hydrogen, was first studied seriously by Henry Cavendish (**1766**). This seemed to differ from another explosive 'air' (now known as methane) that was beginning to cause trouble in the new coal mines. But the process of discovery was starting to pick up. Within 50 years a dozen different gases would be identified and modern chemistry could get started. ➡**1733**

The Speed of Light (Again)

1729

James Bradley, the third Astronomer Royal (he succeeded Edmond Halley in 1742), was an astronomer of extraordinary skill and patience. He used the power of the telescopes of his day to uncover minute effects of great importance. One of these let him find the speed of light, measured only once before, by the Dane Ole Roemer (**1676**).

In a series of painstaking observations undertaken with a colleague from 1725, Bradley noted that at different times of the year some stars appeared to be minutely displaced from their usual positions. To have their light fall down his telescope, he had to point it very slightly off-line. Think of sitting in a train with rain falling on a windless day. If the train is still, the rain droplets run vertically down the window pane. If the train begins to move, they are pushed a little off-line, and the faster the train moves the greater the shift in the path the droplets take.

Bradley argued that the minute displacements he saw let him compare the speed of light with the speed of the Earth in its orbit around the Sun. He already knew the latter. The measurements yielded a light speed around 300 000 kilometres per second, more accurate than Roemer's figure. Publishing his finding in 1729, Bradley called this phenomenon 'the aberration of light'. It provided the first firm evidence that the Earth really was in motion around the Sun, as Copernicus had proclaimed (**1543**) and astronomers now believed.

Once he had corrected his observations for the effect of aberration, certain discrepancies remained. After another 20 years he was able to proclaim the reason.

The axis of rotation of the Earth did not hold steady in space but wobbled slightly in a 19-year cycle—called 'nutation'—due to the gravitational pull of the Moon.

Bradley had hoped to find something more: 'stellar parallax'. This yearly back and forth change in the position of stars should be the result of the changing point of view of observers on the Earth as it circles the Sun. But it eluded him (and everyone else for the next 100 years). The parallax of even the nearest star was apparently too small to be detected, and that made the stars very distant indeed.

Stephen Gray: Conducting Electricity

Stephen Gray, an English dyer and well-regarded amateur astronomer, was largely self-taught. He was short of money most of his life. But he had some powerful supporters, such the first Astronomer Royal, John Flamsteed, who secured him a pension for a time and the leisure to do his science.

1729

Experimenting with an 'electric machine' (**1702**), Gray made an important discovery. The 'electric virtue' (we would now say 'charge'), able to attract light objects like hairs and feathers and pieces of paper, could be carried away from the machine by means of wooden rods and wires. In his most spectacular demonstration, Gray carried the 'virtue' more than 250 metres along a thick, slightly damp hemp thread.

Such a thread was therefore a 'conductor' (though Gray did not use the word). Other materials that did not let the charge pass, such as glass or dry silk threads, were 'insulators'. Furthermore, Gray found that only insulators such as amber and glass were 'electrics', able to be charged up by rubbing with cloth. It seemed that conductors such as metals could not be charged.

Before he died destitute, Gray was made a Fellow of the Royal Society and was the first to receive the Society's Copley Medal. By that time, Isaac Newton was dead and could no longer impede allies of his rival, Flamsteed.

The Prehistory of Carolus Linnaeus

In 1730 Carl Linne, then only 23, was appointed assistant to the Professor of Botany at Uppsala University. It was the beginning of a most illustrious career for the Swede, making his name famous as Carl von Linne (once he was

1730

knighted by the King), or as Carolus Linnaeus, since he wrote most of his books in Latin. He is generally credited with devising our modern system of naming plants and animals ('taxonomy' is the technical term). The 'Linnean system' is still used today, with only some modifications and expansions, but it did not emerge from nothing. As Linnaeus himself would acknowledge, many before him had made major contributions.

For example, Andreas Cesalpinus was personal physician to Pope Clement VII, and Professor of Medicine at Pisa (he may have taught Galileo, who was a medical student there at the time). His book *De Plantis*, published in 1583, was the first attempt to list and classify plants since ancient times. He arranged them in a hierarchy from simple to most complex. It was not sophisticated but it was a start.

Forty years on, in 1623, the Swiss botanist Casper Bauhin produced a catalogue of 6000 plants. He gave each of them a double-barrelled name (so inventing a 'binomial system'). The first was later dubbed the 'genus' name and was given to a group of closely related plants; the second was the 'species' name, which belonged to that plant alone.

Next to contribute was the English naturalist John Ray (**1660**). In *Historia Plantarum*, published in four volumes from 1686 onward, he looked at plants from around the world and grouped them through looking at their many different characteristics. His definition of species, that 'each produces only its own kind', remains valid today.

Ray used binomials for animals as well as plants. The names *Felis leo* (lion), *Felis tygris* (tiger) and *Felis cattus* (domestic cat) showed that these animals were closely related, but the species name made each of them distinct. Ray's taxonomic system for animals came out in 1693. It used features like the number of teeth and toes to decide which animals were most closely related. ➤➤**1753**

Where Am I? (1): The Problem of Latitude

1731

In early eighteenth-century Europe, no greater challenge faced science or commerce than helping ships at sea find out where they were. Fortunes hung on the safe arrival of cargoes of raw materials, manufactured goods or slaves, not to mention the lives of the crew. The great maritime nations, such as Britain and France, had set up national observatories to grapple with the problem.

The venerable Greek geographer Ptolemy may have been under challenge from Copernicus and his followers, but his 'global positioning system' was accepted everywhere. Any point on Earth lay at the intersection of two lines; a parallel of latitude running east–west aligned with the equator, and a meridian of longitude, running north–south from pole to pole. Identify the lines you are on, and you are where they cross.

Finding latitude was the lesser challenge. The Earth being generally spherical, travelling north or south changes your view of the sky. It affects how high the Sun, the Moon or a 'fixed star' rises above the horizon to the north or south. For centuries, sea captains had measured the altitude of the Sun at midday, and so figured out their distance north or south of the equator. In 1701 Isaac Newton suggested that with the right equipment, the same idea could be used for any star whose position was precisely known. This was one of the motivations for the great star-cataloguing endeavours that followed (**1725**).

The old methods required a captain (often on a heaving deck) to keep one eye on the Sun or a star and the other on the horizon, and to measure the angle between them. It was tough to do accurately. In 1736 John Hadley of England simplified matters using an instrument with internal mirrors. These brought the horizon and the celestial object into the same line of sight. The first versions could measure angles up to 90 degrees (a 'quadrant') or 45 degrees (an 'octant'); then he settled on 60 degrees (a 'sextant'). John Hadley was the older brother of George Hadley, who later explained the global pattern of the winds (**1735**).

Finding longitude was a much tougher problem, as there was no easy-to-use method. Its ultimate solution by the Yorkshire carpenter John Harrison in **1765** was one of the great inventions of the century.

Stephen Hales: The Need to Measure

Stephen Hales, born in Kent, was a clergyman with a strong interest in science, like many in his time. It was not surprising; the clergy were usually well educated, their pastoral duties often left them with free time to observe and experiment, and the study of the works of nature seemed an appropriate way to glorify God as Creator. Hales took the last point a stage further. He thought that the Creator would have been very careful and precise in His work, that He would have 'observed the most exact proportions of number, weight and measure in the make of all things'. So the best way to gain insight into the operations of nature was 'to number, weigh and measure'.

1733

In 1733 Hales applied his philosophy to studying the flow of blood in humans and other animals, something that had never before been attempted. In his book *Haemastaticks*, he reported experiments to measure the rate of flow of blood in veins and arteries. He measured the blood pressure in a horse by directly connecting a measuring device to an artery. He saw how the blood pressure varied with exertion, and calculated how much blood the heart sent out in each stroke.

His book *Vegetable Staticks* (**1727 Food for Plants**) had looked at the physics of plants, regarding them, in some senses, as machines. This was pioneering work again; until then plants had been the domain of collectors and classifiers and seekers after herbal remedies. Hales measured the rates of growth of stems and leaves and the 'root force' that held the plant in the soil. He also studied the movement of water from the ground through the plant tissues and out through the openings in the leaves, now called 'transpiration'. Hales accepted that 'capillary action', found by his countryman Francis Hawkesbee (**1709**), kept the water moving.

Charles Dufay and Jean Nollet: Playing with Electricity

By the 1730s the novelty of electricity had become a great entertainment around the courts and drawing rooms of Europe. Frenchman Charles Dufay, supervisor of the Royal Gardens at Versailles, with his friend the

1733

priest Jean Nollet, loved to stage the popular 'flying boy' demonstration. A young boy was suspended above the floor by silk threads and charged up with an electric machine (**1702**). He could then attract small light objects into hands and deliver shocks to those who touched him.

But it was more than fun and games. In 1733 Dufay and Nollet pushed the growing study of electricity along by announcing that there were two forms of 'electric fluid'. One form, produced when glass or crystal was rubbed with silk, they called 'vitreous' electricity; the other 'resinous' form was the result of rubbing amber and some other materials (like paper) with flannel.

Dufay and Nollet further found that the two forms of electricity interacted. Two light bodies charged alike with either form would repel each other, while one charged with vitreous electricity would attract another with a resinous charge. So like charges repelled, unlike charges attracted.

The publication in 1753 of a book by the American Benjamin Franklin (**1751**) describing his experiments in electricity caused quite a stir. Nollet reportedly could not believe that such high-quality work could have been done in the 'New World'. Nonetheless he disagreed with the conclusions. Franklin thought there was only one type of electricity, not two as Nollet had argued in his own book published the same year. To try to set the American straight, Nollet wrote a series of long letters to Franklin and later published them as well.

Nollet was one of the first 'popularisers' of science, writing books that anyone could understand. He was later the first Professor of Experimental Physics at the University of Paris (a century before such a position was available at any English university).

George Hadley: Explaining the Winds

1735 As ships from Europe sailed more boldly across the oceans of the world in search of territory, trade and plunder, understanding of the global pattern of the winds grew. English astronomer Edmond Halley had drawn the first maps of the winds of the world and offered the first explanations of what drove them (**1676**).

In 1735 English scientist George Hadley improved on Halley's work. In explaining why the trade winds blew towards the west as they neared the equator, he produced the first serious (if not conclusive) evidence that the Earth rotates on its axis once each day. Galileo had angered the Church by asserting that the Earth did turn, but he had no real proof (**1633**).

Hadley argued that the surface of the Earth moves more quickly at the equator than further north or south, since a point near the equator has further to go to get back to its starting point a day later. So winds blowing towards the equator would get 'left behind' by the turning Earth and appear to arrive more and more from the east, as the

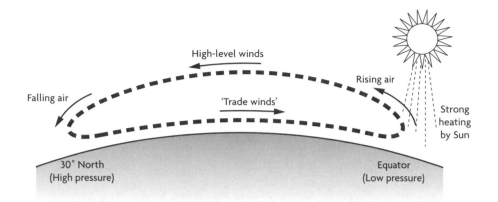

*The workings of the Hadley Cell drive the winds in the tropics of our planet. Air heated by the Sun near the equator rises, cools and flows north (and south) several kilometres above the ground. At 30 degrees latitude, it sinks again and returns to the Equator just above the ground to form the 'trade winds'. If the Earth did not spin, the winds would blow directly north and south. In fact they are bent by the Earth's rotation so that the trade winds come mostly from the east, and the high-level winds become westerlies (**1835**).*

trade winds do. This phenomenon was later called the Coriolis Effect (**1835**).

Hadley's name is remembered in the Hadley Cell, which links the trade winds at the surface to winds several kilometres up blowing away from the equator. To unite the two flows, hot air rises at the equator, flows north and south well above the ground and sinks when cooled at about 30 degrees north or south latitude. Hadley knew these higher level winds must exist to make his system work, but they were not found for another 100 years.

What Shape Is the Earth?

For 2000 years most people had assumed that the Earth was spherical (shaped like a marble). The shape of the Earth's shadow falling on the Moon during an eclipse suggested as much. But was it exactly a sphere? Perhaps it was more like an egg, a bit longer from north to south, or like a pumpkin, flattened at the poles?

Observations using a pendulum gave some clues. The exact time taken by a pendulum of known length to swing once depends on the pull of gravity. In **1672**, as part of an experiment to measure the distance from the Earth to the Sun, Jean Richter in South America found that a standard pendulum beat more slowly there than in Paris, suggesting that the pull of gravity was weaker. Isaac Newton, of course, had an immediate answer. In his *Principia* (**1686**) he claimed that the Earth bulged out along its equator due to its spin; a spot on the surface of the Earth near the equator would therefore be further from the centre and that would weaken gravity.

Even Newton could not convince everyone. An egg-shaped Earth remained popular, especially in France. Efforts there to measure the distance across the ground equivalent to one degree of latitude seem to get a bigger answer to the north of Paris than to the south, as you would expect with an egg.

Settling the argument was a major task. French expeditions to Peru and Lapland took years to organise and complete; the one to Peru involved a decade of travel and privation. The answer came through in 1736. A degree of latitude covered more ground in Peru than in Lapland. The Earth was shaped like a pumpkin; Newton was right. Based on that, Alexis-Claude Clairaut (also of France) worked out in 1743 how to calculate the force of gravity at any point on the Earth's surface.

And there is more. Newton had argued for an equatorial bulge because he, and most people since Copernicus (**1543**) and Galileo (**1633**), believed the Earth rotated on its axis once a day. However, at the time no one had any real proof of that. But now that people were sure the Earth bulged, they could be almost as sure that it was spinning. It was not yet proof but it was more than plausible.

Daniel Bernoulli: Fluids in Motion

1738　Daniel Bernoulli came from a family of Swiss geniuses who over three generations produced eight brilliant mathematicians. He was arguably the brightest of them; his father Johann was so jealous of his abilities that they were estranged.

Bernoulli made some outstanding discoveries in hydrodynamics, which deals with fluids in motion, and published them in 1738. A 'fluid' is anything that can flow. So it excludes solids, but includes gases like air and liquids like water. Thanks to Simon Stevin (**1586**) and Blaise Pascal (**1653**), we already knew that the pressure in a fluid standing still (the subject of hydrostatics) depends only on the depth; the deeper you go, the greater the pressure.

By using Newton's laws of motion (**1686**), Bernoulli found that pressure changed if the fluid was moving. The faster it went, the more the pressure within it dropped. Bernoulli's Principle is easily demonstrated with a vacuum cleaner. Reverse the connection so you are blowing air out. A ping-pong ball will hang suspended in the airflow, even if it is tilted over. Bernoulli says the air pressure is lower where the air is moving, so air from outside the stream is continually drawn in and keeps the ping-pong ball in place.

If air or water is flowing through a pipe that gets narrower, the fluid must speed up (otherwise the fluid will back up). So the pressure at the constriction will drop and water or air can be drawn into the stream from outside. Such a Venturi pump can drain water from the bottom of a boat or pull air out of enclosed spaces. The drop in pressure can be used to measure how fast the liquid (or the boat) is moving. In your car's carburettor, fast-moving air reduces the pressure above a tube containing

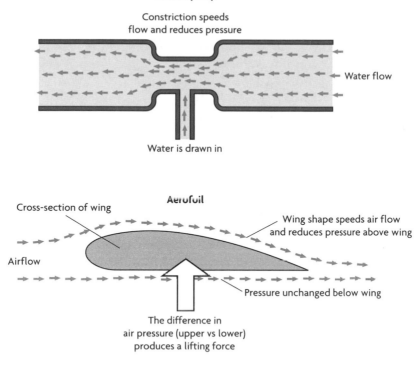

Venturi pump

Constriction speeds
flow and reduces pressure

Water flow

Water is drawn in

Aerofoil

Cross-section of wing

Wing shape speeds air flow
and reduces pressure above wing

Airflow

Pressure unchanged below wing

The difference in
air pressure (upper vs lower)
produces a lifting force

The Venturi pump and the 'aerofoil' are two powerful and useful demonstrations of Bernoulli's Principle that the pressure in a fluid drops when the fluid is in motion.

petrol. The petrol evaporates faster in the lower pressure and the vapour is pulled into the airstream heading for the cylinders.

Bernoulli's Principle also explains how aeroplanes stay in the air. The shape of the wings makes the air flowing over the top go faster, and so the pressure above the wing drops. Pressure under the wing is unaffected and the imbalance is like a force pushing up, able to support the weight of the plane. When you fly, thank Daniel Bernoulli.

How Old Is the Earth (and How Can We Know)?

The age of the Earth, or even whether it had an age, has always been an issue. Some of the ancient Greeks thought our planet, and the universe that enclosed it, had always existed, that its age was infinite. The Church had a different view, at least as far as the Earth was concerned. Adding up the ages of the Biblical patriarchs, and making a few other assumptions, suggested that the Earth had been created about 4000 years before the time of Jesus. That answer seemed to satisfy the many astronomers and other scientists who believed in the Bible, and that was most of them.

1740

The English astronomer Edmond Halley was one of those who thought otherwise. He argued that if the saltiness of the sea was the result of minerals being washed out of rocks by rain and other erosion and then carried to the ocean by rivers, then it would have taken very much longer than 6000 years to make a sea like the Mediterranean as salty as it was. French engineer Henri Gautier had a similar idea: he measured the quantity of rock fragments carried by rivers to determine the rate of erosion of the Earth's surface. That, too, suggested that the Earth was very old.

In 1740 leading French biologist Compte de Buffon (**1749**) joined the debate. He suggested that the Earth had once been a molten ball of rock flung off by the Sun, perhaps as the result of a collision with a comet. If so, it would have taken at least 75 000 years to get cold—perhaps much longer. Buffon made his guess after measuring the rate at which very hot balls of iron cooled. The Church responded to this challenge by condemning Buffon and burning his books. That by no means ended the argument, which was to rage on through the nineteenth century (**1862**) and beyond.

The Cassini Family Maps France

1740 While astronomers at the major national observatories in Paris, London and elsewhere (**1669**) had their eyes on the sky, their feet were on the ground. The big task of the day was navigation—finding the location of ships at sea. But the same methods could locate places on land with equal accuracy (or more accuracy, since the ground did not ·move during the observations). Celestially determined locations of key landmarks would be combined with the methodology of surveying to produce the detailed, reliable maps increasingly vital in peace and war.

The first country to attempt this immense task was France. It was begun by Giovanni (Jean) Cassini, the Italian-born first Director of the Paris Observatory, and continued by his descendants—four generations of Cassinis over a century—despite wars and other upheavals that regularly dried up resources. The starting point was a baseline grid laid out north–south through Paris, with astronomical observations to precisely locate each end of the baseline. Then a series of triangles was constructed, each side forming a baseline to establish the third point of a new triangle, and so on.

By the time the task was completed in 1740, France was encased in a network of 400 interlocking triangles. Now the detail could be added—towns, rivers, mountains and other landmarks—and France was properly mapped for the first time. A pattern was set that would soon spread across the English Channel, throughout Europe and around the world. The completed maps revealed some major errors in the commonly used maps of the day, which had usually been simply copied from others and never checked. Very early on, King Louis XIV remarked that the remapping of France had cost him much of his domain.

To make such a map, a basic scale was needed, one that linked latitude and longitude with distances across the ground. Early on, a key endeavour was the

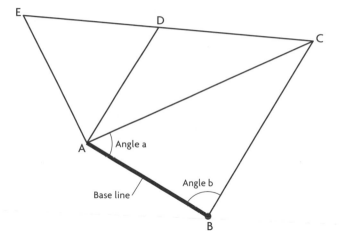

To create maps of the whole of France, the Cassini family set up a vast network of triangles by surveying. Starting with a precisely determined baseline AB they observed a prominent landmark at C and found the distances AC and BC by measuring the angles a and b. AC then became the baseline to locate another landmark at D; AD was used to position E and so on. In 1672 much the same method had been used to find the distance from Earth to Mars.

measurement of the length of a degree of latitude, the north–south distance between points say 50 and 51 degrees from the equator. That was the forerunner to the challenge of determining the shape of the Earth (**1736**), and later to the finding of the length of the metre when the metric system was introduced at the end of the century.

Reaumur and Celsius: Even Better Thermometers

Multifaceted French researcher René Reaumur, better known for his studies in biology (for example, he discovered the power of digestive juices), was another pushing the art and science of measuring temperature, building on the work of Isaac Newton and Daniel Fahrenheit (**1701**). The scale on Reaumur's alcohol thermometers, first made in 1730, had water freezing at 0 degrees and boiling at 80 degrees. These fixed points were set with great accuracy but the scale was never widely used and has long been obsolete, even in France.

The key name in temperature measurement nowadays is that of Anders Celsius of Sweden. Celsius devised his Centigrade scale in 1742, with water boiling at 0 degrees and ice melting at 100 degrees. This was soon inverted to provide the temperature scale now in almost universal use. Celsius' mercury-in-glass thermometers became widely used in laboratories and increasingly accurate. With their aid, Joseph Black of Scotland (**1761**) made crucial discoveries, identifying latent heat and specific heat. By the 1770s they could read temperature to one-tenth

THERMOMETER SCALES AND TYPICAL TEMPERATURES				
	Celsius	Fahrenheit	Reamur	Kelvin
Surface of the Sun	6000			
Iron boils	2750			
Iron melts	1525			
Gold melts	1064			1337
Surface of Venus	400			673
Lead melts	327			400
Pure water boils at normal pressure	100	212	80	373
Highest recorded air temperature	58			331
Normal human body temperature	37	98		310
Pure water freezes	0	32	0	273
Mercury freezes	−40			233
'Dry ice' turns to gas	−78			195
Lowest recorded air temperature	−88			185
Oxygen boils	−183			90
Hydrogen boils	−253			20
Helium boils	−269			4
Absolute zero	−273			0

Temperatures found on or near Earth; some given on the various scales used over the last 500 years.

of a degree. Celsius himself noted that water did not in fact always boil at 100 degrees. The boiling point of water and other liquids rose and fell in line with changes in air pressure.

Abraham Trembley: Plants Versus Animals

Since the days of the ancient Greeks, living things had been divided into plants and animals. Plants lived and grew; animals lived, grew and had sensations. They could also move by themselves. The third division of the natural world, minerals, could only grow and neither lived nor felt anything. All that seemed simple enough, but the division was not always easy to apply.

Some of the diverse living things that master microscopist Anthony van Leeuwenhoek (**1673**) had found in water looked like plants: they attached themselves to something, such as another plant, and had a stem and the appearance of branches or leaves. He called them 'polyps'. These caught the attention of Abraham Trembley, a young man serving as tutor to the sons of a nobleman in Holland. His 1744 report on the matter had a long title typical of the time: *Memoir on the Natural History of a Freshwater Polyp with Arms shaped like Horns.* Under the microscope, he saw a polyp apparently reach out to grasp an object and put it into what could have been its mouth. If the water was disturbed, the tiny organism appeared to create a foot to help it move.

Such independent movement was more fitting for an animal than a plant. Trembley wondered if he had found some missing link between the two. He also saw one polyp apparently become two, with a bud forming and breaking off. He experimented by cutting polyps in two with scissors, and then into many pieces. Each fragment became a whole polyp. Trembley knew his Greek legends and was reminded of the fearsome Hydra, with its capacity to grow two new heads if one was struck off. He gave the name to these creatures, clearly tiny animals rather than plants, and it stuck.

Around the same time, similar polyps were found living within beds of coral, moving their tiny tentacles, drawing in food. So corals too were animals, not plants. They were soon shown to produce the masses of limestone that make up a coral reef. So an animal can make a mineral. The living and non-living drew closer together.

The Leyden Jar: A Way To Store Electricity

In the mid-eighteenth century, electricity was a hot topic. Ewald von Kliest, the Bishop of Pomerania, in modern Poland, wondered if the newly discovered 'electric fluid' (**1733**) could be stored, say in a glass jar, as water **1745** can be. If so, it could be more easily studied. His experiments proved fruitful. A glass jar with a little water in it, held in the hand, could store enough charge to deliver stunning electric shocks.

His discovery was to be known to posterity, however, as the Leyden Jar, because it was made simultaneously and independently by Pieter van Musschenbroek from Leyden in Holland. Musschenbroek was canny enough to have a paper on the invention read in front of the French Academy of Science. Von Kliest merely told a friend. So the man from Holland got the credit.

The Leyden Jar later became known as a condenser, since the electrical fluid was condensed or concentrated in it. In modern terminology, it would be a capacitor. The water and the moist hand were soon replaced with other conductors, such as metal films inside and outside the jar. In time even the jar shape would prove unnecessary, the only requirement for the storing of charge being a layer of insulating material (a dielectric—like glass) between two conductors.

FARAD

The farad, named after the nineteenth-century physicist Michael Faraday (**1813**), is the unit of capacitance, that is, the ability of a capacitor (condenser) to store electrical charge. A capacitor with a capacitance of 1 farad can hold 1 coulomb of charge with a potential difference of 1 volt between its plates.

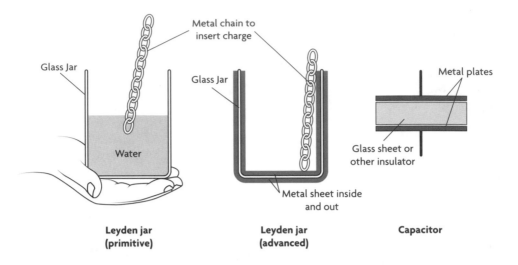

The diagram shows three stages in the evolution of the Leyden Jar as a way of storing electric charge. More modern capacitors use synthetic materials as the insulating layer and are able to store large amounts of charge by wrapping the insulator and conductor into a roll with many layers.

The improved 'jars' could hold massive amounts of charge, delivering powerful shocks and impressive sparks, and were soon part of the electrical amusements of the nobility as well as of the apparatus of the researchers. In fact the name Leyden Jar was first applied by the French priest and experimenter Jean Nollet (**1733**), who was soon using it in some spectacular demonstrations. In 1746 the King watched as a big charge from a Leyden Jar passed through 180 soldiers holding hands, and later along a kilometre-long line of 700 monks. According to reports, the 'subjects' all jumped into the air at once, proving that electricity moved very fast.

The Eighth Metal: Zinc Laid Bare

1746

Archaeologists have dug up objects more than 3000 years old made of the metal brass. We know that the ancient Romans had a recipe for making brass, but they did not know why it worked. They heated copper with a mineral called calamine. The resulting metal was quite different from copper, both easier to work and able to take a sharper edge, much like bronze, and equally resistant to rusting. It could be either cast or hammered into shape. Later, the alchemists tried heating calamine just with some carbon. When it cooled, they found a whitish metallic material clinging to the sides of the chimney (and therefore likely to be overlooked).

The 'last alchemist', Paracelsus (1520), was perhaps the first European to claim knowledge of 'zincum' as a metal with distinctive properties (so adding to the seven metals known to the Ancients); he perhaps learned of its existence from Indian metalworkers. Otherwise, zinc was not decisively identified until 1746 by the German

Andreas Marggraff (who also proved there was sugar in 'sugar beet'). Others may have found it around the same time, but Marggraff studied the metal methodically and is therefore commonly acknowledged as its (European) discoverer.

Zinc proved to be an important and useful metal. It reacted with acids more vigorously than iron to release hydrogen (**1766**) and was the source of some of the gas that filled the balloons carrying the intrepid Charles brothers aloft in **1783**. A little later it was paired with copper in some of the early electrical 'batteries', following their invention by the Italian Alessandro Volta (**1800**).

Later still, the technique of 'galvanising' coated sheet iron with a thin layer of zinc to provide protection against rust was developed. Being more chemically active, the zinc corrodes first. The ancient zinc ore calamine proved to be the compound zinc carbonate, and is still used as a lotion to relieve skin irritations and diseases.

Pierre Maupertuis: The 'Economical Universe'

French thinker Pierre Maupertuis was caught up in lots of things. In **1736** he led the arduous expedition to Lapland to help measure the shape of the Earth, partly to vindicate Isaac Newton, whose philosophy he followed. He studied families where various members had six fingers, looking for patterns of inheritance. He had provocative thoughts about human reproduction. And he seems to have anticipated ideas Charles Darwin held 200 years later. Most forms of life that have ever lived, he said, have died out. Those that are still with us have some special adaptation in their bodies or behaviour to increase the chance of living and having offspring. This sounds like Darwin's 'survival of the fittest' (**1859**).

Most intriguing was his idea about economy of effort in nature, his *Principle of Least Action*. French mathematician Pierre Fermat had used something like it to explain the laws of reflection and refraction of light more than a century earlier (**1621**): a beam of light always takes the quickest path between two points.

In 1746 Maupertuis came up with a more general idea. He argued that of the infinite variety of ways an object could move, the one it 'chose' was the one that needed the least amount of a quantity he called 'action'. This was the time taken for the trip multiplied by the energy of motion. He thought that the universe, as the handiwork of the Creator, was designed perfectly to function with least effort: 'Nature is thrifty in all its actions'. So there was a good deal of theology or at least philosophy in the idea.

But, as mathematicians would prove, many laws of nature follow inevitably from it. It is a powerful unifying principle. For example, a particle moving in accordance with Newton's laws of motion (**1686**) will keep action at a minimum, and vice versa. All other possible paths will involve more action than the one that it actually follows. Even the Uncertainty Principle (**1927**), which forms part of the bizarre world of twentieth-century quantum physics, is based on the idea that there is a smallest amount of action that nature allows us. We cannot reduce that to zero.

The Human Body: From Anatomy to Physiology

1747

A great deal of science is routine and methodical—the piece-by-piece, step-by-step gathering of information. This is very true of anatomy: laborious, time-consuming work by dozens of researchers in half a dozen countries, taking advantage of the increasing power of microscopes, revealed over the centuries the intricate structures, the architecture and plumbing of the human body. The names we remember are often those of the 'synthesisers', those able to pull together the work of many and present it coherently.

Such a man was Dutchman Bernard Albinus. His beautifully illustrated 1747 book *Tables of the Skeleton and Muscles of the Human Body* (English translation of title) was perhaps the pinnacle of visual anatomy, displaying the muscles and bones, nerves and organs of both the adult and the foetus with an unsurpassed clarity. Albinus was a product of the medical school at Leyden in Holland and the influence of Hermann Boerhaave, greatest of teachers (**1724**). He was followed, indeed surpassed, by the impossibly precocious Albrecht von Haller, who had composed vocabularies in Greek and Hebrew when only 10 years old. Von Haller's book *Elements of Physiology* came only a decade later, but already the wind had changed. Now the concern was with how the body worked, what all these marvellously detailed structures actually did.

Take, for example, the nervous system. By observation and experiment, von Haller had separated two types of nerves: sensory nerves that carry impressions to the brain, and motor nerves that relay the messages from the brain to activate the muscles. He concluded that the brain must be the seat of thought and sensation since all nerves start and finish there. Von Haller did not know how the nerves worked; he thought of them as tubes containing a special fluid. The experiments of Luigi Galvani (**1786**) would suggest that electricity was somehow involved.

Meanwhile, understanding of other basic body processes was growing. Digestion was proving to be more than just the physical grinding up of food: von Haller showed the importance of bile in the digestion of fats; France's René Reaumur discovered how the juices secreted by the stomach break down food; Italian Lazzaro Spallanzani (**1768**) found that digestion begins in the mouth with the action of saliva. By the 1780s, the act of breathing was seen as a way of getting oxygen into the blood and wastes like carbon dioxide out. Von Haller discovered that within the body, food and oxygen combined in a process similar to burning to release heat for warmth and energy for movement, but he was not yet sure where this happened.

James Lind: A Cure for Scurvy

1747

Every ship's captain knew of the threat of scurvy on long voyages. The disease brought physical weakness and depression, and caused livid spots on the skin and swollen and blackened gums. Untreated, the worst cases died. Yet it appeared that access to fresh food could clear the symptoms in a matter of weeks. Some more specific treatment or cure was needed; something to lessen the impact of scurvy on the crews and passengers of both naval and commercial ships.

In 1747 English naval doctor James Lind was aboard HMS *Salisbury* patrolling the English Channel during the War of the Austrian Succession. In spite of good food and 'sweet water', scurvy broke out. Within 10 weeks at sea, 80 men out of 350 were down with the disease. Lind knew that the juice of citrus fruits such as oranges, lemons and limes had long been suspected to be an effective remedy. So he designed an experiment to see how the 'citrus treatment' compared with other remedies.

Lind chose 12 members of the crew already suffering from scurvy, and grouped them in pairs. Each pair had a different addition to their normal rations of gruel, mutton broth and ship's biscuit. These included spoonfuls of vinegar, cups of seawater, a mash of garlic, mustard and horseradish, and a quart of cider. One pair had an 'elixir' of complex composition. The last pair had two oranges and one lemon each day. Only the pair drinking cider and the pair eating fruit showed any improvement at all, and the citrus pair by far the most. One of them was fit again for duty in six days.

The findings were very clear, and quickly accepted by the Admiralty. Juice of lemons and oranges (and limes) was the best cure for scurvy. Nevertheless, 40 years passed before an official order was made to supply lemon and lime juice to all Royal Navy ships. Within another year, scurvy had all but disappeared and the number of sailors hospitalised for any cause was halved. Another consequence was the adoption of the nickname 'limey' for English sailors, and ultimately for all English abroad.

A fundamental question remained. Why did the treatment work? What did the juice of lemons and oranges contain that prevented scurvy? A final answer to that was a long time coming. The answer, we know now, is vitamin C (**1906**).

Compte de Buffon:
The Natural World Summed Up

In 1749 the first volumes of a monumental work of science appeared: *Natural History, General and Particular* by Frenchman Georges Leclerc, later Compte de Buffon. These 44 volumes—eight published after his death—summarised all that was known about the natural world to that time. Brilliantly written for a popular audience, they were peppered with Buffon's own ideas and speculations.

1749

Like many of the time who became involved in science, Buffon had some family fortune and was trained initially in the law. His interests were broad, covering mathematics and Earth sciences as well as what would later be called biology and ecology. One of his great achievements was to convert the French Royal Garden (*Jardin du Roi*), of which he was the keeper from 1739, into a research centre and museum, extending it with plants collected from all around the world. He joined the

French Academy of Sciences at the age of 27, and was later patron of the young Jean Lamarck (**1809**).

Where others, including Linnaeus, saw a basic difference between plants and animals—for example, animals could move and experience sensations, plants could not—Buffon emphasised the similarities of all living things. He did note, as did others, that plants and animals were distributed around the world in ways that gave each continent its distinctive residents. Anticipating the ideas of Charles Darwin (**1871**) and others on human evolution, Buffon compared men with apes and wondered if they could have had a common ancestor.

One Galaxy or Many?

1750

The stars are not evenly spread across the sky. The most ancient observers noted the faint band of light, dubbed the Milky Way, which circles the heavens among the stars. In **1610** Galileo had shown that this was made up of stars too.

In 1750 the English astronomer Thomas Wright came up with an explanation. Stars did not fill space uniformly in all directions, but are grouped into a vast assembly shaped like a wheel. The view across the wheel reveals many more stars closely clustered than the view along the axle. The Milky Way therefore represented the rim of our star system, our galaxy (the name comes from the Greek word meaning 'milk'). William Herschel, the eminent discoverer of Uranus (**1781**), took up the idea and around 1785 drew images of our galaxy shaped like a lens.

But what lies outside our star system, vast as it is? What fills the rest of the universe, or is our galaxy alone in the cosmos? The great German philosopher Immanuel Kant probably did not know one end of a telescope from the other, but his mind was untrammelled. He visualised the whole of space filled with objects like our galaxy, elliptical congregations of stars, 'island universes' as he called them. He saw evidence for these among the many observed 'nebulas', fuzzy patches of light among the stars. These were often mistaken for comets; to avoid confusion the leading French comet hunter, Charles Messier, drew up a catalogue of about 100 such objects in 1781. These are still known as Messier objects: M42, for example, is the Great Nebula in the constellation Orion.

But there were many more that Messier did not list. With their big telescopes (**1789**) Herschel and his son John catalogued thousands. Some of these could have been elliptical, such as the one in the constellation Andromeda (Messier's M31), and may have been Kant's 'island universes', lying far beyond our own. But most astronomers thought not. Herschel was one such, declaring in 1791 that all the thousands of nebulae were simply collections of gas within our own galaxy, perhaps stars and planets in the process of forming. The matter rested there for nearly a century; the debate was resumed in **1847**.

Steam Power, Iron and Chemicals

EIGHTEENTH-CENTURY TECHNOLOGY

The start of the Industrial Revolution could be placed in 1705, the year that Thomas Newcommen of England took the steam-powered water pump fashioned by Thomas Savery a decade earlier and made it drive a piston and beam instead of simply sucking up water. The engine was still very wasteful of the coal burnt to make the steam; in each cycle, the hot cylinder was doused with cold water to condense the steam and make a partial vacuum inside the cylinder, with the external air pressure pushing the piston.

Scotsman James Watt had a solution: condense the steam separately, so the cylinder could always remain hot. He built a prototype in 1765; later versions converted the piston's back-and-forth motion into a smooth rotation. By 1776 he was making steam engines in partnership with the industrialist Matthew Bolton in Birmingham. Industry now had a reliable and abundant power source to replace wind and water, humans and horses.

And there were now more machines that needed such an energy source. The textile industry had developed successively better and faster machines to spin and weave linen, cotton and woollen cloth. They included the flying shuttle (1733), the spinning Jenny (1764), the spinning mule (1779) and the power loom (1785). Industrial wealth in Britain and later elsewhere was founded first on textiles.

On the cotton plantations of America, Eli Whitney's cotton gin (1793) boosted productivity by speeding the separation of fibre from seed and so made the slaves that grew the cotton an even more valuable commodity. Productivity in agriculture, which still supported most of the population in Europe, was boosted by the introduction of four-year crop rotation and improved methods of sheep breeding.

Other industries were scaling up: pottery, for example, imitating the Chinese porcelain; glassmaking and papermaking; and of course iron to make the engines and other machinery. With coke replacing charcoal in the iron smelters, iron became cheaper and of better quality. Wrought iron and crucible steel raised quality still further. The great iron master John Wilkinson could now more accurately bore out the cylinders of steam engines so that they were more reliable (he could also bore better barrels for cannons).

Wilkinson was already dreaming of iron ships. Certainly iron had new uses. The first bridge made totally of iron was built in England in 1779. Hydraulic lime cement, which could set underwater, was another major advance. Its inventor, John Smeaton, used it to build the third Eddystone lighthouse off the coast of England (completed in 1759).

To transport the raw materials and the products of industry, rivers were linked and then bypassed by a growing network of canals, beginning in England in 1759. These reigned supreme until the advent of railways in the 1830s. Communications still depended on horses and carriages, though the French engineer Claude Chappe pioneered a network of mechanical semaphore (signal) stations on hilltops 20 kilometres apart. These could relay a message from Paris to Marseilles in 20 minutes. The electric telegraph finally rendered that obsolete around 1850.

Growing industry increased the demand for chemicals, such as sulphuric acid, soda ash (1791) and chlorine for bleach (1774), and so drove the invention of new processes to make them. Around 1792 William Murdoch pioneered the use of coal gas for lighting and heating.

The flow of manufactured goods—cloth, paper, pottery, glassware, copper plated with silver, iron plated with tin—raised the standard of living, at least in wealthy homes. Other innovations during the 1700s included banknotes and bankers' cheques, a regular postal service, soda water, the water closet, the hot-air balloon, and if you were really adventurous, the parachute.

1751–1800

The World Stage

IN 1776 THE OUTPOSTS OF Britain along the eastern coast of North America came together and declared independence from the 'mother country', though another five years of war were needed to secure that independence. To the north, Britain had previously battled France for control of what is now Canada. This conflict was an outrider of the Seven Years War, the regular confrontation between the various European states for territory, power and resources. As with previous conflicts, the major gains were made by Prussia, the centrepiece of the Germany to come.

Thirteen years later, revolution came to France. After years of violence and terror, power passed to the soldier Napoleon Bonaparte, who knew the power of science and technology and promoted it. Confrontation between France and the rest of Europe was to last until 1815 (the Napoleonic Wars). During the struggle, France adopted the metric system of weights and measures, which then spread to most of Europe (except Britain) through French conquests.

Russia, under Catherine the Great, continued its push for modernisation, including sending students abroad to study science and technology. It continued to expand at the expense of a declining Turkey, gaining the Crimea. At the same time, Austria began to push the Ottomans from the Balkans. Looking west, Russia joined with Prussia and Austria to dismember Poland, birthplace of Copernicus three centuries before.

ARTS AND IDEAS

The Enlightenment, with its emphasis on materialism, tolerance, personal autonomy and intellectual freedom, reached its zenith during these five decades. In 1751 the first edition of the *Encyclopaedia of Arts and Sciences*, edited by Frenchman Dennis Diderot, appeared, summing up the principles of the movement. Those ideals contributed much to the two great political revolutions of the period: the American and the French. The Enlightenment viewed the science of the greats, such as Galileo and Newton, as the model for human inquiry, based on observation and experiment.

In music these were the years of classicism, typified by the Germans Joseph Haydn and Wolfgang Amadeus Mozart. New musical forms with complex structures appeared: the sonata, the string quartet and the symphony. The steady perfecting of the piano saw it replace the harpsichord and give rise to new types of music, including the 'art song' with piano accompaniment. Towards the end of the period, music became more dramatic, heralding Beethoven and the rise of romantic music in the next century.

Writing was abundant and diverse. Dr Samuel Johnson compiled the first dictionary of the English language. Frenchman Voltaire produced his novel *Candide*, satirising the Enlightenment philosophy that 'this is the best of all possible worlds', and soon after, *A Treatise on Tolerance*. Englishman Adam Smith published his *Inquiry into the Wealth of Nations*; German Immanuel Kant *The Critique of Pure Reason*; and Robert Burns his collections of Scottish ballads. Englishmen Samuel Taylor Coleridge and William Wordsworth published their *Lyrical Ballads*, including *The Rime of the Ancient Mariner* and *Tintern Abbey* respectively.

In art and decoration, the rococo was now the dominant style. It was more delicate and sensual than the baroque from which it had developed, reflecting the tastes and love of luxury of the French aristocracy and later of the other European courts. Painter Jean Watteau of France was the epitome of this. In England, the emphasis was on portraits; Joshua Reynolds and Thomas Gainsborough set the pace. Engraver William Hogarth continued to record the decadence of the upper classes and the social problems among the poor.

CHEMISTRY GROWS UP

While all areas of science advanced during this half-century, chemistry in particular came of age. Important gases such as carbon dioxide, nitrogen, hydrogen, oxygen and chlorine were first clearly identified, the popular but erroneous theory of phlogiston was finally dispatched through the genius of Antoine Lavoisier, and a new, rational language for chemistry was devised. Chemists established beyond all doubt the importance of precise measurement.

Elsewhere arguments continued. The 'neptunists' battled the 'plutonists' over the origin of rocks; in biology, the 'epigenists' contended with the 'animalculists' about the way animals developed after conception. At the same time, agreement slowly grew that fossils were the remains of plants and animals once living and gave vital insights into the history of our planet.

Electricity was still high on the agenda for the physicists. Lightning proved to be a giant electrical spark. We learnt the law that governs forces between electric charges and that electricity has a role in animals. By the end of the century, physicists could generate sustained electric currents.

Telescopes grew in size and sophistication and a new planet was discovered. Two major international enterprises to observe transits of Venus settled the size of the solar system. The Earth was weighed; speculation began on how the Earth and the Sun first came to be. A Yorkshire clockmaker provided the final answer to the challenge of finding longitude; a cataclysmic earthquake in Lisbon, Portugal, was methodically assessed, a first step in seismology.

The monumental achievements of Carl Linnaeus brought order to the vast numbers of known plants and animals. The notion that life could appear spontaneously from non-living matter was struck another blow, but somehow survived into the next century. We learnt how plants can restore 'injured air', a vital step towards understanding photosynthesis.

A 'germ theory' of disease was proposed for the second time, and once again rejected by the majority, even though a way was found to provide protection against smallpox. The first stirrings of 'occupational medicine' came with the observation that chimneysweeps suffered from a particular form of cancer.

Benjamin Franklin Flies a Kite

American inventor, author and statesman Benjamin Franklin was the first significant scientist to emerge from across the Atlantic. His 'kite' experiment ensured his fame today. According to the popular story, he flew a kite fitted with a metal point into a thunderstorm to prove that lighnting was really a gigantic electric spark. This conjecture was not original; Francis Wall in England had put it forward 50 years before. Franklin was not even the first to try the experiment but his is the name everyone knows.

1751

The proof came: again according to the usual account, electric charge surged down the damp kite string, charging a Leyden Jar (**1745**) and causing a spark to jump from a key on the string to his hand. It could have killed him, and it did kill at least one researcher who tried to repeat it. Franklin may never have done it at all, being apparently well aware of the dangers. Still, whether or not he ever did what he claimed, it soon seemed clear to most people that lightning was electrical in nature, similar to the sparks made by the popular electric machines (**1733**), if vastly larger in scale. Thunder was the crackle of the sparks as the heated air expanded, again immensely magnified.

Franklin had been studying electricity for years; in 1746 he had put forward a new idea of what electricity actually was. In place of the theory expounded by the Frenchmen Nollet and Dufay that there were two sorts of electricity (**1733**), Franklin pushed a 'one fluid' model. There was only one sort of charge, he claimed, but an object could have too much or not enough. In the first case, the object would be 'positively charged', in the second 'negatively charged'. The idea caught on, and was enlarged upon a decade later by the German Francis Aepinus. Electrical effects, he said, were the result of the one electrical 'fluid' (which was thought to have no weight) being unevenly distributed. In the long run it proved incorrect; Nollet and Dufay were closer to the mark. 'Positive' and 'negative' charges are two distinct forms of electricity (**1897 Thomson**).

Carl Linnaeus: The Catalogue of Life

1753

To Swedish naturalist Carolus Linnaeus (or Carl von Linne) we owe our modern system for naming and classifying living things. It is still in use—though with many modifications—250 years after he first popularised it. Though he drew on the work of many others (**1730**), Linnaeus may with justification be called the 'father of taxonomy'. The publication of his masterpiece of botany, *Species Plantarum,* in 1753 marks the formal date for the beginning of the 'new order' in naming plants and animals.

Fascinated by plants from an early age, and initially headed for the Church or medicine, Linnaeus was Professor of Botany at Uppsala University from 1741 until his death, nearly 40 years later. He travelled widely through Europe, especially in the north, collecting and classifying. His labours generated a string of books; the most famous, *Systema Naturae*, first appeared in 1735 and went through many editions.

Linnaeus elaborated his classification schema by systematically placing species and genera into larger collections of similar organisms called orders, classes and kingdoms. So the kingdom Animalia contained the class Vertebrata (all animals with backbones), which in turn contained the order Primates (meaning 'first'), which included the genus *Homo* and the species *sapiens*, so giving the name 'wise man' for human beings.

Other levels have since been added to this hierarchy to represent other degrees of affinity. Many of the names Linnaeus gave have long since been replaced with others based on different characteristics, but his binomial system and its hierarchical arrangement remains central to taxonomy today. In his early work Linnaeus thought that species were fixed and unchangeable. Later in life he was not so sure, noting

CLASSIFYING LIVING THINGS: SOME EXAMPLES				
	Human being	**House cat**	**House fly**	**Sunflower**
Kingdom	Animals			Plants
Phylum (Division)	Chordates (with spinal chord)		Arthropods (with jointed limbs)	Angiosperms (with flowers)
Class	Mammals		Insects (with six legs)	Dicotyledons (with two seed leaves)
Order	Primates ('most important animals')	Carnivores ('meat-eaters')	Diptera ('true flies')	Asterdales
Family	Hominidae (humans and great apes)	Felidae (all cats)	Muscidae (two-winged flies)	Asteraceae (daisies and sunflowers)
Genus and species	*Homo sapiens*	*Felis domestica*	*Musca domestica*	*Helianthus annuus*

This modern style of classification grew from the work of Linnaeus. To recall the order of classifications, remember 'Kings Play Chess On Fine Glass Surfaces.' The first letter of each word is the first letter of each category. The top groupings in the plant kingdom are often called divisions rather than phyla. This example shows one of several common ways to classify plants.

that hybrids can be formed between species and even genera. However, Charles Darwin's notion of open-ended evolution (**1859**) without divine guidance would have shocked his religious beliefs.

Linnaeus left an additional legacy through his students, 19 of whom went on voyages of discovery. These included Daniel Solander, who accompanied James Cook (**1769**) and brought back the first plant species from Australia and the South Pacific. ➤➤**1757**

Joseph Black and 'Fixed Air'

Joseph Black was a busy Glasgow doctor and medical teacher, whose patients included the philosopher David Hume and the nurse of the infant novelist Walter Scott. He was unmarried, a great frequenter of clubs,
played the flute well and was reputedly 'popular with the ladies'. He still found time for research and was regarded even by the great French chemist Antoine Lavoisier as one of the founding fathers of chemistry.

In 1754, in a thesis for his doctorate, he reported strongly heating the mineral magnesia alba (well known to doctors as an antacid and purgative) and similar minerals such as limestone or 'chalk'. Black observed that they lost much of their weight from the heating and surmised that some invisible 'air' been driven off by the heat. It was already known that when these substances were dropped into acids they bubbled and fizzed, apparently giving off the same gas.

Black collected some of this gas, which he called 'fixed air', perhaps because it had been 'fixed' inside the minerals and only set free by heating or by acid. He found the 'fixed air' quite unlike ordinary air. A candle would not burn in it, and a mouse put in a closed jar of the gas quickly died. It seemed very similar to the 'gas silvestre' that the Dutchman Jan von Helmont had identified 100 years earlier (**1644**) given off by burning wood. This was an important insight. If the same gas came from a mineral and from something once living, maybe the animate and inanimate parts of nature were linked.

Heating limestone had left 'quick lime'. Black found that if he left quick lime in the open around the laboratory it slowly turned back into limestone. This suggested that ordinary air had some 'fixed air' in it all the time and therefore air was not a single substance (an 'element') but a mixture. This led to a simple way of testing for the gas. If it was bubbled through water with quick lime dissolved in it ('lime water'), the water turned white as particles of limestone reformed.

Nowadays, Black's gas is called carbon dioxide. ➤➤**1761**

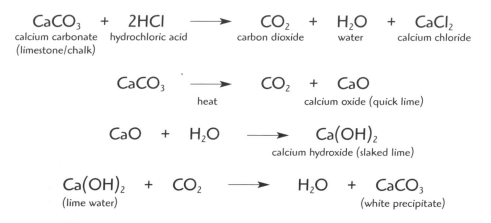

$$CaCO_3 + 2HCl \longrightarrow CO_2 + H_2O + CaCl_2$$

calcium carbonate   hydrochloric acid   carbon dioxide   water   calcium chloride
(limestone/chalk)

$$CaCO_3 \xrightarrow{\text{heat}} CO_2 + CaO$$

calcium oxide (quick lime)

$$CaO + H_2O \longrightarrow Ca(OH)_2$$

calcium hydroxide (slaked lime)

$$Ca(OH)_2 + CO_2 \longrightarrow H_2O + CaCO_3$$

(lime water)   (white precipitate)

These equations show in modern nomenclature how Joseph Black discovered that limestone and similar compounds release carbon dioxide when heated or acted on by acids. When carbon dioxide is passed through 'lime water', the solution turns milky due to the formation of calcium carbonate.

The Destruction of Lisbon

1755

On 1 November 1755 the Portuguese capital, Lisbon, was all but destroyed by perhaps the largest and most deadly earthquake in recorded history. One-third of the city's population, 90 000 people, died immediately or soon after. Damage was multiplied by a subsequent tidal wave and fire. Much of southern Portugal suffered severe damage and loss of life. It was a calamity beyond belief.

However, some good came from the catastrophe. The Spanish prime minister instigated a countrywide survey of the earthquake's impact. Every parish was required to report on how long the quake had lasted, how many aftershocks were felt and what sort of damage resulted. Thesex reports, the first of their kind, still exist and have allowed modern experts to reconstruct the earthquake. Estimates give it a reading of nine on the Richter scale (**1935**), as high as any earthquake since.

Parishes were also asked to report on any odd behaviour by animals, since folklore suggested that animals could tell if an earthquake was coming, and on any changes in the level of water in wells. Already people were thinking that perhaps earthquakes could be predicted by monitoring such signs.

The rebuilt Lisbon contained the world's first earthquake-resistant buildings. Wooden models of planned buildings were tested for their stability by being shaken by the heavy tread of marching soldiers.

What had caused the calamity? Popular thought, supported by the Church, made it an Act of God, representing perhaps His anger against heresy. Suspected heretics were reportedly hanged on sight in the days that followed. More-rational suggestions were also made. In 1760 the scientifically prolific English churchman John Michell (**1783**) argued in front of the Royal Society that the origin lay in the rubbing together of deeply buried layers of rock, perhaps under the influence of water. For others, the power of earthquakes, like that of volcanic eruptions, showed the immense energy latent within the Earth, linked to some source of internal heat.

Carl Linnaeus: The System of Nature

The Swedish naturalist Carl von Linne (Linnaeus in Latin) was a compulsive classifier, seeking a name and place in the natural order for everything. His *Systema Naturae*, which reached its tenth edition in 1757, **1757**
not only encompassed all known plants (*Vegetablia*) and animals (*Animalia*), but also covered minerals and gemstones (*Lapidae*), the four ancient Greek elements, the known celestial bodies, humankind, what he termed 'Wisdom', and the 'Scientific Method' (that is, the techniques of classification) and even God as the Creator of everything on the list (Linnaeus was keen to honour his Lutheran upbringing).

Like the classification of diseases, which the physician Linnaeus undertook on another occasion, the ordering of minerals was largely a waste of time. He had very little to go on, just colour and lustre and the shapes of the mineral crystals. He had no idea what was in them (their chemical composition); that sort of analysis was still decades away, though interestingly fellow Swedes such as Carl Scheele would play a major role in it. Even classification by cleavage (**1784**) and the useful Mohs scale of hardness (**1820**) were not yet available.

Linnaeus had to find names to describe all the known plants and animals, the numbers of which were already very large and rising rapidly. He mined not only Latin, which was still the language of science in northern Europe (England and France had taken to writing about science in the vernacular), but also ancient Greek. This was once again being taught and studied in universities after centuries in obscurity following the eleventh-century schism between the Latin and Greek wings of the Church.

He also laboured over precise definitions for the terms he used to describe the appearance and structures of plants and animals and where they lived. This was the sort of precision that the French chemist Antoine Lavoisier and friends would seek in naming chemicals 30 years later (**1787**). It also meshed with the growing realisation across science that measurement mattered, that accuracy in observing and recording was crucial and that a universally agreed 'language of science' was essential for progress.

John Dolland: Clearing the View to the Stars

1758

Scientists using microscopes and telescopes in the mid-1700s were frustrated by the imperfections of even the best instruments of the time.

One problem was the blurred coloured fringes that surrounded the image, making good observations difficult. This 'chromatic aberration' was the result of the different colours of light being bent by different amounts as they passed through the lenses. It was to defeat this problem that Isaac Newton had pioneered the 'reflecting' telescope, which uses a mirror rather than a lens to gather light (**1672**).

The hero of the hour in this matter was John Dolland, son of a Huguenot weaver in East London. He was a weaver for a time himself until he set up in business with his son in 1752 as a maker of optical instruments.

The solution seems to have been suggested first by one Chester Moor Hall, a 'gentleman' of Essex. He argued that the different parts of the human eye refract rays of light in such a way that they produce an image free from colour. Perhaps combining lenses composed of different materials, such as water and glass, which bent light by different amounts, might do the same. As a man of independent means, Hall was perhaps indifferent to fame; at least he took no trouble to communicate his invention to the world.

When he heard Hall's suggestion, Dolland had some doubts. It seemed to conflict with the ideas of Newton, but he began to experiment. He soon found success with the water–glass combination, and more practically with combinations of different glasses, such as 'crown glass' and 'flint glass'. By 1758 his 'achromatic' lenses were free of the troublesome coloured fringes. He gained the gratitude of observers, and also the Copley Medal from the Royal Society. He was a Fellow two years later and in 1761 he became optician to the king.

So his persistence turned out well for him and for science. His invention brought telescopes with lenses back into favour, as these made better use of the available light. The biologists were comfortable again with multi-lens ('compound') microscopes; within half a century these were able to magnify 1000 times. Both the macrocosm and the microcosm began to yield more secrets.

Caspar Wolff: Victory to the 'Epigenists'

The development of humans and other animals from conception to birth was debated for hundreds of years. English doctor William Harvey, best remembered for his insights into the circulation of blood (**1616**), argued for what he called 'epigenesis'. An ill-defined substance called 'germ material' was held in the egg produced by the female of the species. When acted on in some way by semen from the male, it began to take shape, forming its various limbs and organs ready for birth. This would be the same for all animals. '*Omnia ex ova*', Harvey proclaimed in Latin—'All things come from eggs'.

A contrary view came from the 'pre-formists'. They argued that the young were already complete in every detail in microscopic form in the egg ('encapsulated' was the jargon) and throughout gestation simply grew in size. There was no real evidence for this. The microscopes of the day could not see such a thing; indeed the human egg or 'ovum' had not even been sighted.

In **1673** Dutchman Anthony van Leeuwenhoek had discovered spermatozoa in semen. This produced a new theory that the pre-formed human is carried by the sperm, perhaps as a tiny adult 'homunculus' or 'manikin'. Leeuwenhoek had called the various tiny creatures he discovered 'animalcules', so the adherents to this view were the 'animalculists'. The 'ovists' held that the pre-formed human was in the egg, with the extremists claiming that every member of the human race had lain pre-formed within the eggs of the first woman, Eve.

Careful observation helped to settle the matter. Around 1759 the German researcher Caspar Wolff studied chick embryos as they developed inside fertilised hen eggs. Chicks were readily available and studies on them were easy to do, and therefore popular. It was clear that Harvey and the 'epigenists' were nearer the truth. There was no complete tiny chick present from the start, but rather some undifferentiated tissue that gradually developed the appearance and attributes of a chick as it grew in size. The same, he argued, would be true of humans, and studies of spontaneously aborted foetuses gradually provided the needed evidence.

But another question remained unanswered. What were the relative roles of the ovum and sperm? The sperm somehow stimulated the ovum to grow and develop. But was that all it did? That was a question for the nineteenth century.

Giovanni Arduino Organises the Rocks

For dwellers in northern Italy like Giovanni Arduino from Verona, the Alps were never far away. Anyone with an interest in rocks would be drawn towards them, with their massive cliff faces built from layer upon layer of rocks of many types. There was another more practical reason. Italy seemed less endowed with mineral resources than some other nations. Arduino was interested in mining, and therefore in ascertaining the sorts of rocks in which useful deposits might be found. This led him to try to classify the main groupings of rocks as they appeared in cliff faces and the walls of mine shafts, using the appearance of the rocks and the occurrence of fossils as markers.

Drawing on the century-old theory put forward by Nicholas Steno (**1667**) that rock strata had been laid down one after another with the oldest at the bottom, Arduino came up with a basic classification, which he first published in 1759. The lowest and oldest rocks were termed 'primary'; these had no fossils. Above them came 'secondary' strata, which were often tilted out of the horizontal and were commonly rich in fossils. The 'tertiary' layers above were fossil-rich but usually horizontal. The uppermost 'quaternary' rocks were the most recent and usually contained solidified lava from volcanoes or beds of sandstone or shale obviously laid down at the bottom of some ocean or lake.

It was a simple system, but widely accepted. Some of Arduino's terms are still used. Later researchers found reasons to subdivide the major groupings, giving names that were usually linked to locations where the strata formed prominent outcrops. Hence rocks of the 'Devonian' period, containing many fossils of primitive fish, were easy to find in Devon, England. The 'Silurian' rocks, the next layer down, had first been studied in a region in Britain once inhabited by a tribe called the Silures. The rocks of the much more recent 'Jurassic' period, forming major cliffs in the Jura Mountains in Switzerland, would prove rich in dinosaur fossils. And so on. Modern geologists recognise hundreds of such subdivisions.

Joseph Koelreuter: All About Hybrids

1761 In 1694 German naturalist Rudolf Camerarius had declared, somewhat controversially, that plants needed to exchange pollen to reproduce; in today's lingo, plants 'have sex'. Sixty years later, his countryman Joseph Koelreuter, director of the botanical gardens at Karlsruhe, took another step. He saw that pollination was aided by insects, some birds and even by wind. He was the first to claim that flowers make nectar purely to attract insects and birds, with pollination being the deliberate, if accidental, consequence. Another German, Christian Sprengel, a schoolmaster from Spandau, argued that everything about a flower—its shape, colour, perfume and so on—is designed to maximise the chance of pollen reaching its target.

Koelreuter was puzzled as to why plants made so much pollen. Millions, even billions, of pollen grains are released, with few seeming to do any good. Koelreuter decided that the pollen of each flower is distinct, able to fertilise only plants like the one that produced it or perhaps a very closely related plant. If the two plants were too different, there were no offspring, or the offspring were unable to reproduce. A striking example of this was the flower that English gardener Thomas Fairchild had created in 1719. He crossed a carnation with a Sweet William. The resulting sterile plant was dubbed 'Fairchild's Mule'.

This led Koelreuter to discover some rules about such hybrids. When pollen from one plant is used to fertilise a close relative (assuming this to be possible), the offspring have some characteristics of each parent, showing that both the ovule (the female reproductive organ) and the pollen help shape the next generation. He found that he could turn one variety of tobacco plant into another by continually cross-pollinating the offspring, generation by generation, with pollen taken only from the second variety.

He also demonstrated 'hybrid vigour': the offspring of cross-pollination between two varieties is generally stronger and more resistant to adverse conditions. A similar interest would lead the monk Gregor Mendel to repeat and extend this work a century later (**1865**). Sprengel showed that such cross-fertilisation goes on all the time; Charles Darwin (**1859**) was to show that cross-pollination, and the mixing of characteristics that it brings, is an essential driving force in evolution.

The Transits of Venus

On 6 June 1761, and again on 3 June 1769, the planet Venus passed across the face of the Sun as seen from Earth. Half a century earlier, the English astronomer Edmond Halley had proposed that careful measurement of the

1761

exact time taken for such a transit of Venus, viewed simultaneously from a number of widely spread locations, could be used to measure the distance from the Earth to the Sun, and so confirm the size of the solar system (**1672 Cassini**).

In the first major international scientific enterprises in history, networks of more than 60 stations were coordinated by the French Academy and the Royal Society to view the two transits. The first took place while England and France were at war! Observations were taken throughout Europe, at St Petersburg in Russia and at remote sites in St Helena, Lapland, India, Siberia, South Africa, Canada, Mexico and Mauritius. James Cook and his crew observed the **1769** transit from Tahiti. All these were major expeditions, in which the observers risked their health and even their lives in challenging environments.

It was an extraordinary achievement. Many of the observers returned excellent data, timing the duration of the seven-hour transit to within a couple of seconds. They also had to calculate the latitude and longitude of the observing sites, which was tough to do in the days before John Harrison's chronometer was widely available (**1765**).

The outcome was worth the effort. Combining the best results from the two transits, a figure of just less than 150 million kilometres was obtained for the Earth–Sun distance, the 'astronomical unit'. This is only 1 part in 100 away from the value accepted today.

Joseph Black and the Two Hidden Heats

1761

As thermometers (**1742**) became steadily more reliable and useful, fundamental discoveries about the nature of heat began to be made. Joseph Black, the Glasgow doctor already well known for finding 'fixed air', aka carbon dioxide (**1754**), was in the vanguard here. His two powerful insights were backed by observations made independently at much the same time by Johann Wilcke in Germany and Jean Deluc in Switzerland.

He noted, as others had done before, that when ice was melting or water was boiling its temperature did not seem to change, despite all the heat that was being absorbed. He gave the name 'latent heat' to this phenomenon. By careful experimentation he determined the amount of heat needed to, for example, boil a certain amount of water and make steam, and found that an equivalent amount of heat was released as that steam condensed back into water. This powerful insight explains many things in nature: why, for example the drying of sweat cools the skin, and how the condensation of water vapour inside a cloud liberates energy and can make the cloud grow into a 'thunderhead'.

He also noted that when different materials, such as a range of metals, were heated, their temperatures did not rise at the same rate. Each substance required a distinctive amount of heat (a 'specific heat') to raise its temperature by, say, one degree. These understandings enabled Black's friend James Watt to reduce the wastage of heat in his new steam engine; a separate condenser for the steam let the cylinder and piston remain at the one temperature. They also made clear the difference between heat and temperature or, in Black's terminology, between 'quantity of heat' and 'intensity of heat'.

Not everything became clear. Argument continued as to what heat actually was (**1798**).

Markus Plenciz: 'Every Disease Has its Organism'

1762

To the question 'Why do we get sick?', doctors over the millennia have supplied a wide range of answers. Some blamed 'an imbalance of the four humours in the body'; this led to ineffective and sometimes dangerous therapies such as blood-letting and purging. Others ascribed illnesses, especially widespread epidemics, to 'miasmas' of contaminated air; disease seemed more common in swampy areas or damp smelly houses. We still give one deadly disease the name 'malaria', which means 'bad air'.

Early users of the microscope, notably the Dutchman Anthony van Leeuwenhoek (**1673**) found tiny living things in many kinds of liquids: pond water, saliva, intestinal contents, plaque from between teeth and liquids in the process of 'going off' often swarmed with 'animalcules'. The first to propose these as the cause of disease seems to have been a doctor in Vienna, Markus Plenciz. Writing in 1762, he suggested that there were 'seeds' of various diseases in the air, organisms like the animalcules, but of many types. Each type, when brought into the body by breathing or swallowing, could multiply and cause a specific disease. 'Every disease has its organism', said Plenciz. This idea revived the 'seeds of disease' concept put forward 200 years before by Girolamo Fracastoro (**1546**).

Plenciz was ahead of his time and of the evidence, with little to support his speculations. He could not link any organism with a specific disease. But vindication did finally come in the form of the 'germ theory' of disease created a century later by Louis Pasteur and others (**1862**). Fittingly, Plenciz had been the first to use the word 'germ'.

Where Am I? (2): The Challenge of Longitude

In 1714 the British Parliament passed the Longitude Act. An attractive prize was offered to 'such person or persons as shall discover the Longitude'—£20 000 if the longitude of a ship at sea could be established within one degree, £40 000 if within half a degree. These were immense sums for the day, but the problem was urgent. Shipwrecks were disturbingly common, mostly because the sailors did not know exactly where they were. In 1707 one such disaster cost four navy ships and 2000 men.

1765

A ship's captain could find his latitude quite easily (**1731**) but reliably finding longitude east or west of his home port had so far proved impossible. The solution had long been known, at least in principle. The captain simply compared his local time set by the Sun with the time at the same moment back home. Each hour's difference in time meant 15 degrees difference in longitude.

But what time was it back home? There were two possible ways to find out, but both had problems. You could take a clock from the place you had left, but would it keep good enough time? Or you could use a 'clock' that would be visible from both places, say something in the sky. For example, the position of the Moon among the background stars is like the hour hand among the numbers on a clock. The Moon's motion is irregular, speeding up and slowing down for a variety of reasons, so calculating the required tables was a huge amount of work, even though Isaac Newton (**1687**) had showed how it could be done. Still, the Astronomer Royal Nevil Maskelyne produced his first *Nautical Almanac* in 1767 (it has come out every year since), which enabled the average captain to find his longitude to within a degree (about 100 kilometres).

That was good, but a Yorkshire carpenter turned clockmaker, John Harrison, did much better. He took the first route, as Christiaan Huygens had tried to do (**1656**). By 1765 he had created what many thought impossible—a 'chronometer' so reliable that it lost or gained less than a minute in a six-week voyage. This gave the longitude well within the acceptable error. Envious manoeuvrings among the members of the 'Board of Longitude' scandalously denied him his reward for years, but with the intervention of the king, 'Longitude Harrison' finally got the money and his place in history.

The Lunatics of Birmingham

1765 The need for scientists to get together to network and exchange ideas spawned not only the Royal Society of London (**1662**) and the French Academy of Science, but also the Lunar Society in Birmingham. This was a much more exclusive group, never numbering more than 14. Yet so influential and well placed were its members that the society has been called the 'think tank' (or even the 'revolutionary committee') of the Industrial Revolution.

The society began to meet in 1765, once a month on nights of the full moon. This arrangement eased the walk home through the dark Birmingham streets; William Murdock, one of the members, would later pioneer gas street lighting. They usually met at the Soho house of industrialist Matthew Boulton. The core membership also included: chemist and clergyman Joseph Priestley, discoverer of oxygen (**1774**); medical man Erasmus Darwin (**1794**), grandfather of Charles; James Watt, the inventor of the steam engine and partner in business with Boulton; master potter Josiah Wedgwood; William Withering (**1785**), the doctor who pioneer the use of digitalis to treat heart disease; the education reformer and telegraph pioneer Richard Edgeworth; and soap industry pioneer James Keir.

Rarely in human history would so much brainpower and entrepreneurial spirit have previously been gathered in one room. On the fringe of the core group were other leading figures: Americans Benjamin Franklin (**1751**) and President-to-be Thomas Jefferson; and master civil engineer John Smeaton, inventor of hydraulic cement and builder of the latest Eddystone Lighthouse.

These men shared concerns and passions. Politically liberal, and mostly dissenters in religion, they sympathised with the goals (if not the methods) of the French and American revolutions. But they also supported capitalism, private property and enterprise. The application of science to solve the problems of industry was the focus of their talk and action—issues like the building of canals and the need for accuracy in weights and measures.

The monthly meetings combined jolly fellowship with serious purpose; the members found the ambience exciting and optimistic and cheerfully called themselves 'lunatics'. They sincerely believed that they were helping to build a better world through knowledge linked to enterprise, where the material needs of all would be satisfied through enlightened industry. The society held its last meeting in 1813. By that time nearly all of the founding members were dead, but the spirit the society had captured remained very much alive.

Mikhail Lomonosov: The Unknown Russian

Mikhail Lomonosov was the first Russian scientist of real importance. Had he lived in Paris or London instead of St Petersburg, he may well have been one of the best known scientists of his day. He did enough to earn that

1765

place; his research papers show that he duplicated or even anticipated the work of many other researchers: Antoine Lavoisier in his rejection of phlogiston (**1775**) and his belief in the conservation of matter; John Dalton in his atomic theory (**1808**); Daniel Bernoulli on the kinetic theory of gases (**1738**) and Benjamin Thomson on heat as motion (**1798**). His insight that nature undergoes slow, continuous evolution, rather than either staying static or enduring regular upheaval, had much in common with the views of James Hutton (**1778**).

Lomonosov's impressive list of achievements includes a catalogue of 3000 minerals and the first observation of the freezing of the liquid metal mercury. He also demonstrated that soil, peat, coal, petroleum and amber have all originated in some way from living things. He maintained correspondence with some of his contemporaries and was an honorary member of the scientific academies in Sweden and Bologna. But most of his work was unknown outside Russia until after his death in 1765, aged 54.

Son of a fisherman from northern Russia, Lomonosov had been sent to Germany to study. On returning home, he rose to be Professor of Chemistry and later Rector at St Petersburg University. Eager to advance education in Russia, he helped found Moscow State University, later named after him. Lomonosov was prominent in the St Petersburg Academy of Science (founded by Peter the Great) at a time when it was dominated by foreigners.

The Russian writer Pushkin commented that Lomonosov 'embraced all the branches of learning'. Certainly he wrote poetry (in the German style, having lived there), and a grammar to reform the Russian literary language; this combined everyday speech with Old Slavonic used in the Church. His history of Russia was one of the first ever; the revival of the Russian art of mosaics owed much to him.

The transit of Venus in **1761**, four years before his death, linked Lomonosov to observers all over the planet. From his measurements he determined that Venus was roughly the same size as the Earth and that it seemed to have an atmosphere, the edge of the planet's disk against the Sun appearing blurred.

Henry Cavendish and His 'Flammable Air'

Henry Cavendish was an aristocrat and a man of means, 'the richest among the learned and the most learned among the rich'. His family were the Dukes of Devonshire and owned the great country house at Chatsworth. They were later to endow the famous Cavendish Laboratory at the University of Cambridge.

1766

Cavendish was an eccentric and reclusive bachelor, terrified of women and caring nothing for his appearance. He may have suffered from Asberger's Syndrome, a form of autism. He was, however, a first-rate experimenter in many fields, and a Fellow of the Royal Society when only 29, though much of his work was unpublished in his lifetime.

Cavendish did publish his work on gases. He was particularly interested in a gas that bubbled off pieces of iron or zinc when they were left in a bowl of acid. He was not the first to note it; previous observers included Robert Boyle, a hundred years before (**1661**) and Stephen Hales (**1727**). Cavendish was first to give it serious study (from around 1766) and so is usually called its 'discoverer'.

This 'air' had some unexpected properties. Firstly, it was very light; no less dense gas has ever been discovered. A balloon filled with it first lifted men far above the ground in **1783**. Secondly, unlike the other known gases, ordinary air and carbon dioxide, Cavendish's gas burnt fiercely. According to the conventional wisdom, which Cavendish accepted, things burnt because they were full of 'phlogiston' (**1723**). So well indeed did this 'flammable air' burn that Cavendish wondered if he had isolated pure phlogiston, which would have been a famous feat.

Cavendish worked on his new gas for 20 years. In 1781 he mixed flammable air with ordinary air in a vessel and exploded the mixture with a spark. On cooling, the inner walls of the vessel were covered with 'dew', which proved to be water. That led to the gas gaining the name we use today, 'hydrogen'. This combines two Greek words to mean 'water-maker'. ➤➤**1798**

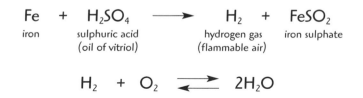

$$Fe + H_2SO_4 \longrightarrow H_2 + FeSO_2$$

iron   sulphuric acid   hydrogen gas   iron sulphate
       (oil of vitriol)   (flammable air)

$$H_2 + O_2 \rightleftarrows 2H_2O$$

Henry Cavendish first showed the reaction going to the right (burning hydrogen in oxygen makes water). Antoine Lavoisier made the reaction go to the left by heating water vapour over red hot iron, breaking the water into hydrogen and oxygen.

Lazzaro Spallanzani: Living from Non-living?

1768 Since the time of Aristotle, it had been commonly claimed that living things could arise spontaneously from non-living matter; maggots from rotting meat, mice from dirty hay, flies from sweat. The Italian priest Lazzaro Spallanzani was among those who helped put an end to such ideas.

Spallanzani built on the work of his countryman Francesco Redi, who had a century before (**1670 Swammerdam**) shown that if meat was kept away from flies, it stayed free of maggots. This proved that maggots were the larvae of flies. But the meat would still go bad and start to smell. In 1768 Spallanzani found that while a meat broth would quickly 'go off' if left in the open, it would stay fresh if first boiled and then kept in a sealed container. Even a flask open to the air kept the broth from rotting if a bend in the flask spout trapped particles of dust and stopped them falling in.

These observations convinced Spallanzani that what caused the rot to set in was in the air or on the dust particles, and that decay would not begin spontaneously. The 'what' were bacteria ('microbes'), which had already been seen in early microscopes (**1673**), though no one yet knew what they did.

Spallanzani was a man of wide interests, studying law and teaching logic and Greek at different times, but he was drawn to science early. He took an interest in the reproduction of animals such as mammals, showing that it required the fertilisation of an egg by sperm. He performed the first known artificial insemination, using a dog.

James Cook: Sailing for Science

The star of exploration late in the eighteenth century was the English navigator James Cook. In three voyages over a decade, he sailed tens of thousands of kilometres over the Pacific and Southern oceans. In 1769,

1769

after visiting Tahiti in the *Endeavour* to observe the transit of Venus (**1761**), he sailed west to discover the unknown coast of the continent he called 'New South Wales'. Two biologists went with him—Englishman Joseph Banks, later director of London's Royal Botanical Gardens and President of the Royal Society (**1778**), and Daniel Solander from Sweden, who had trained under Carl Linnaeus (**1753**).

His second and epic voyage (1772–75) in the *Resolution* took him further south to define the limits of the hypothesised 'Great South Land' and perhaps even to discover it. He sailed around it without sighting a coast, though passing within a few hundred kilometres on several occasions. He judged that if it existed, any continent would be cloaked and rimmed in ice and lie almost entirely within the Antarctic Circle, which he was the first to cross. The first sighting of the Antarctic coastline had to wait another 50 years. He was also first to circumnavigate the Earth from west to east. Cook had the benefit for the first time of John Harrison's (**1765 Longitude**) new chronometers to help fix longitude.

Cook's third voyage (1778–79), also in the *Resolution*, resulted in the discovery of Hawaii, but the real quest was the western end of the North-west Passage, a long-sought route between the Atlantic and the Pacific above North America but below the ice. He explored the coastlines of British Columbia and Alaska before passing through Bering Strait (Dutch explorer Vitus Bering had found in 1728 that North America and Siberia were not joined but instead were separated by a narrow stretch of water). Thick ice drove him back, ultimately, to Hawaii, where he was killed in a dispute with the native population.

Cook was a meticulous observer, outstandingly skilled in navigation and astronomy. His voyages drew together many of the themes of discovery of his time: the link to the transits of Venus (**1761**), the use of chronometers, the presence on board of trained scientists to document discoveries, the quest for the North-west Passage and the conquest of scurvy (**1747**)—care and diet kept his crews mostly free from that disease.

Antoine Lavoisier: Can Water Become Earth?

1770

Even in 1770, the old Greek notion of the four elements (earth, air, fire and water) had some attraction among chemists, along with the possibility of transmuting one into another. After all, the prolonged boiling of water caused it all to evaporate and leave a small residue. Was water being transmuted into earth?

Destined to be one of the greatest of researchers, Frenchman Antoine Lavoisier was the son of a wealthy lawyer and had practised law himself before turning to science. At the age of 27, but already a member of the French Academy of Sciences, Lavoisier tested the water-into-earth idea. He boiled very pure water in a glass vessel for 100 days, capturing and recondensing all the steam. A sediment did appear but careful weighing at all stages showed that it did not come from the water. It resulted rather from the disintegration of the inner wall of the vessel.

So water was not changed into earth. More importantly, Lavoisier had shown what could be proved by taking care with measurements. This approach would be the basis of the many contributions he was to make to chemistry in the next two decades, and indeed of all the chemistry to come. ➤➤**1773**

Fossils: Mementoes of a Lost World

1770

When large bones and teeth of strange creatures began to be dug up in the mines and quarries supporting the Industrial Revolution, they were initially thought to be from large specimens of creatures still living somewhere, including giant humans or, more bizarrely, from mythical creatures like dragons. One hundred years of digging and study was needed for an alternative view to take hold. Resistance came from the religious community, including many researchers; the suggestion that some of God's creations may have been rendered extinct was thought heretical, contrary to Scripture.

However, the evidence continued to mount relentlessly. In 1705 a giant fossil tooth found on the banks of the Hudson River in North America was thought to be from a giant human; actually it was from a mastodon, an extinct giant elephant. In 1719 an 'Almost Complete Skeleton of a Large Animal in Very Hard Stone', as the discoverer headed his report, later proved to come from a huge swimming reptile. In

1787 a 'giant bone' found in New Jersey in North America was identified almost certainly as being from what was later called a dinosaur, perhaps the first ever collected.

Slowly the picture took shape. It seemed hard to deny that great reptiles that could swim or fly, huge elephants and sloths and many other creatures once roamed the Earth but were now extinct.

One of the most impressive fossils was found in a chalk mine by the Meuse River in Holland around 1770. Thirty metres underground, workers uncovered a skull more than a metre long embedded in stone. Stopping their work, they called in Dr Hoffman, a local physician who was keenly interested in fossils and rewarded miners who alerted him to new finds. The bones were transported topside in triumph, but were soon appropriated by the owner of the land under which they had been found. The skull was later a prize of war, transferred to Paris by victorious French forces.

In 1799 a French fossil expert asserted that the skull came from a prehistoric swimming reptile, a view confirmed in 1808 by the even more knowledgeable George Cuvier (**1812**), also of France. It was later dubbed a Mosasaur, combining the Latin form of the name of the Meuse River with the Latin for 'lizard', and later still called *Mosasaurus hoffmani*, in honour of Dr Hoffman. It was the most powerful evidence yet that life on Earth had once been very different.

Joseph Priestley: Restoring 'Injured Air'

In 1771 the English radical churchman Joseph Priestley, almost as devoted to science as he was to God, lit a candle and put it under a jar. The candle burned for a while and then went out. No surprises there; lots of people had seen that happen. Likewise a mouse placed under the jar soon expired.

1771

Priestley pondered the implications of this. It seemed to him that the air in the jar had somehow been 'injured' by the burning candle and the breathing mouse so that now nothing could live or burn in it. If so, somewhere in nature there must be some way of undoing that damage; otherwise countless candles and mice and other agents since time began would have so degraded the air that no life would be possible on Earth. There was no evidence, he argued, that the air was any less wholesome now than it had ever been. How could that be? What was constantly restoring the quality of the air?

Priestley guessed that green plants were involved, that they were able to return to the air whatever it was that the burning candles and breathing mice had removed or poisoned. To test the idea, he placed a sprig of mint under the jar with the burning candle, which of course soon went out. Priestley gave the mint plenty of time, nearly four weeks, to do what he suspected it would do.

To avoid disturbing the jar, he had to relight the candle by using a 'burning glass'

> *In completing one discovery we never fail to get an imperfect knowledge of others of which we could have no idea before, so that we cannot solve one doubt without creating several new ones.*
>
> JOSEPH PRIESTLEY

to focus the heat of the Sun onto the candle wick. It burst into flame, and burnt as brightly as it ever had. Priestley was delighted; he had found how to 'restore' the injured air, and even more, had found that plants carry out this act of restoration all the time. It was a great moment.

Priestley was in correspondence with the American Benjamin Franklin (**1751**). On learning of his results, Franklin wrote, 'I hope this [rehabilitation of air by plants] will give some check to the rage of destroying trees that grow near houses, which has accompanied our late improvements in gardening from an opinion of their being unwholesome.' ➤➤**1774**

Daniel Rutherford and 'Noxious Air'

1772

The identification of carbon dioxide ('fixed air') by Joseph Black (**1754**) had a payoff for his young student Daniel Rutherford. Following up on his teacher's work more than a decade later, he kept a mouse in a closed container until it died. He then burnt first a candle and then some phosphorus until they would burn no more. He then passed the remaining gas through lime water to soak up the fixed air produced during the burning. When Rutherford tested the gas that remained, he found that nothing would burn in it, and a mouse quickly suffocated and died. He gave the gas the appropriate name of 'noxious air'.

Rutherford and Black used the still popular theory of phlogiston (**1723**) to explain what they had seen. They believed that something burning released this 'fiery stuff' which was absorbed by the surrounding air. Air that a mouse had breathed until it expired, or in which a candle had burnt until it went out, was clearly saturated with phlogiston and could absorb no more. So nothing could live or burn in it. As a result, 'noxious air' was also dubbed 'phlogisticated air'. Some decades later it was named nitrogen, because it was found in the compound known as nitre (or saltpetre), a component of gunpowder.

As others were doing, Rutherford went on to find the other major component of everyday air. He named it 'fire air', now called oxygen. He later had a distinguished career as Professor of Botany at Edinburgh University and Keeper of the Botanic Gardens. He was a maternal uncle of the novelist Walter Scott.

Joseph Lagrange Points the Way

The greatest mathematicians are very versatile, leaving behind them a trail of discovery across many areas of research. Joseph Lagrange was such a man, and a citizen of Europe, rather than just one country. Born in Italy, devoted to Latin as a boy, for 20 years active in the Berlin Academy of Science, he is usually regarded as being French. Certainly he ended his career in France, working on the introduction of the metric system and teaching at prestigious colleges. He was prominent in the French Academy of Science and was ultimately made a count and a member of the Legion of Honour by Napoleon.

The Lagrange Points are just one item in his immense catalogue of discoveries. They are not merely mathematical ideas but real locations of importance in space. Lagrange found them while working on the 'three body' problem. Isaac Newton's laws of motion and gravity, set out in his *Principia* (**1686–87**), made it possible to figure out how a small object like a satellite will behave as it is pulled in its orbit by the gravity of the Earth. But the Earth is also orbiting the Sun, so the satellite is being

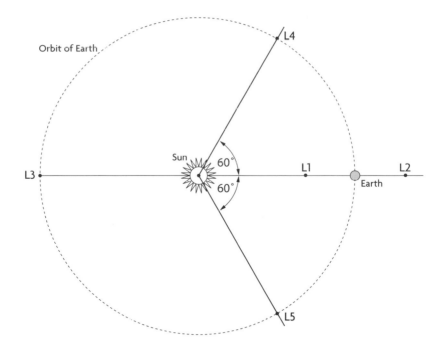

The Lagrange Points were discovered by French mathematician Joseph Lagrange. A satellite placed at any of the five positions will hold its position relative to the Earth and the Sun, so we always know where it is. The L2 point is ideal for positioning instruments to see into deep space without interference from the Sun, since it is always in the Earth's shadow. The successor to the Hubble Space Telescope may be placed there.

pulled simultaneously and in different directions by the Earth and the Sun. This makes it a 'three body' problem, and it can be terrifyingly complex to solve if the pulls of the Sun and Earth are about the same. In most cases the satellite would follow a very erratic and unpredictable path.

Lagrange wanted to know if there were any spots in space where a satellite would hold its position relative to the Earth and the Sun so we would always know where to find it. His calculations, which won him a prize from the academy in 1772, turned up five such 'Lagrange Points', named L1 to L5. Three of them lie on the line running through the Earth and the Sun. For instance, L2 is on the far side of the Earth from the Sun.

Two others, L4 and L5, sit on the Earth's orbit, 60 degrees ahead and 60 degrees behind the Earth as it moves through space. Nowadays, space enterprise is interested in these points. L2 is always in shade, so it is a good place to park sensitive instruments to look into deep space (**1992**). L4 and L5 are likely places to assemble all the hardware for a really big space mission. Lagrange would surely never have envisioned such things.

Antoine Lavoisier: Why Do Things Burn?

1773
The 30-something Antoine Lavoisier, a rising star among French chemists, was very keen to find out why things burn. He had his doubts about the usually accepted explanation, that something burning was releasing phlogiston, the 'essence of fire'. When a piece of lead was heated in air and formed a coloured crust known as a 'calx', it too was losing phlogiston. The theory had been popular since German Georg Stahl wrote it up in **1723**, since it seemed to explain so much.

But Lavoisier knew there had been problems with the theory from the start, and these got worse as chemistry became more precise, more concerned with measurement. How could it be that a candle burning in air (and losing phlogiston) got lighter, while a piece of lead being heated (and losing phlogiston as it formed a calx) got heavier? Were there two sorts of phlogiston, one of which weighed less than nothing?

To find an answer to the riddle, Lavoisier heated lead in a closed container so nothing could get in or out. The lead itself got heavier as the calx formed, but the whole container weighed the same before and after. It seemed that the weight gained by the lead had been lost by the air in the container. So the calx formed not because the lead had lost phlogiston, but because it had taken in something from the air.

Likewise, the candle 'burned away' and lost weight, but again the total weight of the container and contents did not change. The missing matter was still there, presumably having combined with something in the air to make an invisible gas. This 'something' in air allowed things to burn and let metals form a calx. Perhaps, too, it was extracted from iron ore when it was heated with charcoal to release the iron. The old theory said the charcoal gave the iron ore phlogiston; now Lavoisier could suggest the charcoal pulled his unknown substance from the ore.

Here, then, was a viable alternative to phlogiston as an explanation for combustion, calcining and smelting. One step more was needed—to find the 'agent of combustion', the 'something' in air that made it all happen. Lavoisier would have to wait only a year for the answer. ➤➤**1775**

Carl Scheele Finds Chlorine

In 1774 the dogged Swedish pharmacist Carl Scheele, in one of his countless experiments, added the material we now call manganese dioxide to what we now know as hydrochloric acid, but which he called 'marine acid'. He got a surprise. The lively reaction delivered a pungent greenish yellow gas, now identified as chlorine. The name comes, like 'chlorophyll', from the ancient Greek for 'green'.

1774

$$4HCl \;+\; MnO_2 \;\longrightarrow\; Cl_2 \;+\; MnCl_4 \;+\; 2H_2O$$

hydrochloric acid   manganese dioxide   chlorine   manganese chloride   water
(marine acid)   (pyrolusite)   (dephlogisticated marine acid air)

In the terminology of the time, Scheele dubbed his discovery 'dephlogisticated marine acid air'. It is worth explaining why. Fifty years earlier, Englishman Stephen Hales had reported heating a mixture of sea salt and sulphuric acid, releasing a pungent but colourless gas Hales called 'marine acid air' because it had come from salt. When dissolved in water, the gas made 'marine acid' (hydrochloric acid). The gas Scheele liberated from marine acid was clearly different, so it needed a new name. He thought it lacked the 'principle of fire' that made things burn (**1723**), hence it was 'dephlogisticated'. All this showed how urgent was the need to tidy up the language of chemistry.

Even without phlogiston, chlorine was obviously quite active. It was never found free in nature, so the gas had a high 'affinity' for other substances (again, that was the contemporary terminology), uniting with them in compounds that were hard to break open. Early on, chlorine was found to be a powerful bleach, removing colour from plant materials and so whitening linen and cotton cloth. This was a vital new technology for the rapidly growing textile industry. One of the first to note this was the French chemist Claude Bertollet, who would work with Lavoisier and others on the 'new language' of chemistry published in **1787**.

Since chlorine gas is hazardous (indeed poisonous, as its use in World War I gas warfare was to demonstrate), a safer method was soon developed in the form of 'bleaching powder'. This was called 'chloride of lime', and was made by reacting chlorine with 'slaked lime', that is, limestone heated to make 'quick lime' and mixed with water (**1754**). This compound also found use as a disinfectant (**1847 Semmelweis**); today's swimming pools are kept algae-free by chemicals that release chlorine. ➦**1777**

Joseph Priestley: The Discovery of Oxygen

1774

The 'pneumatic revolution' begun by English clergyman Stephen Hales (**1727**), was turning chemistry into a science; it was continued by another English churchman, Joseph Priestley. Hales had been an orthodox believer; Priestley was a Unitarian. His religious views and radical politics brought him trouble. His house and laboratory in Birmingham were sacked by an angry mob in 1791.

Priestley improved Hales' 'pneumatic trough' by using mercury rather than water. He could then collect gases that dissolved in water, gases such as Joseph Black's 'fixed air' (**1754**). Priestley found that absorbing carbon dioxide, as we now call it, in water produced a fizzy liquid with a sour taste. This was the prototype of 'soda water', so called because the gas came from heating a mineral called soda (aka sodium carbonate), and all the soft drinks since. Priestley moved on to discover other gases, adding (in modern terminology) ammonia, nitrous oxide, hydrogen chloride and sulphur dioxide to the existing list of air, carbon dioxide and hydrogen.

Priestley's greatest discovery came in 1774 from the mercury he was using in his trough, or at least from the red crust or 'calx' that formed on mercury when it was heated gently in air. Priestley heated some of this 'red calx' with a large 'burning glass' to focus the Sun's heat. The result was remarkable. The original mercury reappeared as tiny shining globules, and a gas was released, ready to be collected. This gas had unexpected properties: things burned in it much more easily and brightly than in ordinary air. A glowing splinter of wood burst into flames. Mice thrived in it, and Priestley himself reported being much revived when he took a whiff.

Priestley was a believer, not only in God but also in 'phlogiston', the 'fiery stuff' released by materials as they burnt. Most of his fellow chemists believed likewise. Priestley assumed that his new 'air' was so depleted in phlogiston that it sucked it eagerly from what was burning, so increasing the fire. So he called his gas 'dephlogisticated air'. Thanks to the Frenchman Antoine Lavoisier (**1775**), we call it oxygen today.

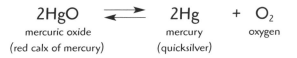

$$2HgO \;\; \rightleftharpoons \;\; 2Hg \;\; + \;\; O_2$$

| mercuric oxide | mercury | oxygen |
| (red calx of mercury) | (quicksilver) | |

This equation summarises what Joseph Priestley and Antoine Lavoisier found. Gentle heating drives the reaction to the left, making 'red calx' from mercury. Stronger heating makes the reaction run to the right, releasing oxygen and liquid mercury.

Nicholas Desmarest: Volcanoes Leave Their Mark

The Giants Causeway on the northern coast of Ireland is a major tourist attraction. Huge columns of dark stone line the cliffs; many others have broken off to form an immense road. Such landmarks are found in many places, including in the Auvergne region of France, where they drew the attention of Nicholas Desmarest who was Inspector-General for all the manufacturing industries in France. He travelled extensively for his work, usually on foot in order to indulge his other interest, namely the study of rocks. Desmarest became convinced that the huge hexagonal columns he saw in the Auvergne, like those in the Giants Causeway, had formed from the hardening of molten rock poured out by now-extinct volcanoes.

1774

He studied the matter methodically and in 1774 published an essay accompanied by a geological map. This showed the sites of the ancient volcanoes and how the lava they had poured out had been weathered since it hardened, producing many different landscapes. In some places, the array of columns has been likened to a giant pipe organ. He also pointed out, perhaps for the first time, how much of the landscape is the result of erosion. For example, over time a stream will cut its way down through layers of rock to form a valley; many tributaries of a large river can form an interconnected network of valleys. This was clearly a very slow process— more evidence to add to the debate over the age of the Earth (**1740**).

The rock in the columns, called basalt, looked quite uniform to the unaided eye, lacking the obvious crystals of various minerals that can be seen in a piece of granite. Clearly, once poured out onto the ground the lava had cooled very quickly, and crystals had not had time to grow. If granite also formed from molten rock—and the 'neptunists' and 'plutonists' were arguing about that (**1775**)—then it must have formed deep underground, where the heat would be trapped and crystals given the leisure to form. The deeper underground, the bigger the crystals.

Percival Pott: Chimneysweeps and Cancer

London surgeon Percival Pott was among the most eminent in his profession, even though, in the manner of the day, he gained his licence to practise by joining the Guild of Barbers. In 50 years at St Bartholomew's Hospital he did much to improve surgical methods. For example, he did away with 'cauterisation' of wounds with hot irons, which was routinely done after surgery to stop bleeding. It is worth remembering that the two great inventions that made modern surgery possible, anaesthetics and antiseptic practice, were still a century in the future.

1775

Pott's major contribution to knowledge formed the starting point for occupational medicine. In 1775 he found that chimneysweeps were particularly susceptible to cancers of the scrotum, and guessed that the constant exposure to soot was the cause. For the first time, something in the work environment was suspected of causing a serious disease. Many more examples would follow. Tobacco, associated with recreation rather than work, was already under suspicion. In 1761 John Hill had established an association between the taking of snuff (powdered tobacco) and malignant (and often fatal) nose polyps.

Pott left other legacies through his students. Scots-born John Hunter was even more eminent than his master, becoming surgeon to the king and ultimately Surgeon-General. He elevated surgery from a mere technical trade to a profession equal to any other sphere of medicine. His studies of teeth were the beginning of modern dentistry, the teeth of the general population then being in appalling condition by modern standards. He gave teeth their modern names, such as molar, incisor and cuspid. Hunter also advanced understanding of inflammation and sexually transmitted disease, and wrote a well-known book on gunshot wounds, based on his first-hand observations as army surgeon during the Seven Years War against France.

One of Hunter's students became better known still—Edward Jenner, pioneer of vaccination against smallpox (**1796**).

Antoine Lavoisier: Replacing Phlogiston with Oxygen

1775

Late in 1774 the English scientist–cleric Joseph Priestley visited the now celebrated Antoine Lavoisier in Paris. He told the Frenchman of his recent discovery (**1774**) of a gas that made things burn much more brightly than in ordinary air. Lavoisier was excited; perhaps Priestley's new gas was the 'agent of combustion' Lavoisier needed to do away with phlogiston for ever (**1773**). He quickly repeated Priestley's work, taking it further in his methodical way. Sadly, when he came to publish his findings, he made no reference to Priestley, though at other times he acknowledged how much he owed to the Englishman.

Lavoisier 'calcined' mercury: that is, he heated it with focused sunlight for many weeks and watched a red calx form. The container he used sat upside down in water, and the water level rose in the vessel as the experiment went on, showing that something was being absorbed from the air within. The process seemed to stop when about one-fifth of the air had been used up. Lavoisier found that the remaining unused air would support neither combustion nor life. He called it 'azote', Greek for 'no life'. It was Daniel Rutherford's 'phlogisticated air' (**1772**), later called nitrogen.

Lavoisier now collected up all the bits of calx and heated them again, more strongly this time. Just as Priestley had found, the calx decomposed, leaving behind bright beads of mercury and releasing a gas. As well as could be determined, the same amount of gas came out as had been absorbed into the mercury calx during the first heating. He found, as others did, that the gas had remarkable properties: fires burnt more brightly in it and it made animals like mice very lively. He named it 'vital air' at first; others called it 'fire air' or 'pure air' or even 'empyrean air', for much the same reasons.

In 1778 Lavoisier settled on the name oxygen in the belief that it was found in all acids (the name means 'acid-maker' in Greek). He had some basis for the idea; sulphur, phosphorus and carbon all burnt in oxygen, releasing gases or ashes (oxides) that made acids when dissolved in water. Plausible enough but incorrect. As was later shown, not all acids contain oxygen, and that not all oxides are acidic. But the name stuck.

Though he had named it, Lavoisier was clearly not the first person to deliberately make oxygen. Both Priestley and the Swede Carl Scheele has done so before him, Scheele making the first discovery but not the first announcement (**1777**). Two more Englishmen, Stephen Hales (**1727**) and John Mayow (**1674**), knew that something like it existed in everyday air.

Here, then, was the truth about the formation of a calx, and by extension about all forms of combustion: they involved not the release of phlogiston, but a chemical reaction with oxygen from the air. The phlogistonists had been right up to a point. Combustion, calcination and the other reactions do involve the release of an agent by one substance and its combination with another. They just had the wrong agent. If we replace 'phlogiston' with 'absence of oxygen', we see the connection. Substances burn in air not because it lacks phlogiston but because it contains oxygen. They will not burn in pure nitrogen because it has no oxygen, not because it is saturated with phlogiston. Lavoisier had made phlogiston unnecessary. What is more, he had collect oxygen in a jar: it was clearly a real substance. No one had ever done that with phlogiston.

Lavoisier made a powerful case, but he did not instantly win everyone over. The dichard phlogistonists like Joseph Priestley and Henry Cavendish clung to their views, but the doctrine of phlogiston died with them. ➤➤**1785**

Abraham Werner: Do Rocks Come from Water?

In the late eighteenth century, two big debates racked the area of science we would later call geology. One was about the way in which the surface **1775** of the Earth had been shaped (**1778**). The other concerned the formation of rocks. The Biblical story of a great flood that had covered the whole Earth in ancient times was an enduringly popular one, including among early scientists. They saw in it a possible explanation for the current appearance of the Earth and the existence of rocks. Leading them was German Abraham Werner, founder of the theory called 'neptunism', named after the ancient Greek god of the oceans.

In 1775, when only 25, Werner became a teacher at the mining academy in Freiberg. Through his efforts this became one of the leading schools in the country. Werner, too, became renowned for his methodical studies of rocks and minerals; he was perhaps the most notable researcher of his time in this field. His fundamental notion that rock layers are formed one after another in a time sequence was crucial for geologists who came after him, even though da Vinci (**1517**) and Steno (**1667**) had thought of it earlier.

Werner and his followers visualised a primeval chaotic ocean enveloping the globe, from which the various types of rocks had settled out one by one in layers. First to precipitate, when the waters were still deep, were the 'crystalline rocks', like granite; these had the most obvious patches of various minerals. They were usually found deepest down and so must have formed first. Later, as the waters receded, the various sedimentary rocks, such as shale, sandstone, conglomerate and limestone, formed from the precipitation of silt, sand, rock fragments and tiny shells.

Werner and the other neptunists were ultimately both right and wrong. Their theory about sedimentary rocks was confirmed and accepted by others, but a rival school of thought, 'plutonism' (named after the ancient god of the underworld), led by James Hutton of Scotland (**1778**), maintained that the crystalline rocks and others like basalt had once been molten, hardening as they cooled. In the case of those rocks, the plutonists were right.

Carl Scheele: Missing Out on the Credit

1777

Busy Swedish pharmacist Carl Scheele is a good example of how you can miss out on getting the credit in science by not getting into print fast enough. His laboratory notebooks clearly show that he found what we now call oxygen several years before the man usually named as the discoverer, English parson Joseph Priestley (**1774**). He used the same methods too, heating compounds such as red calx of mercury to drive off the loosely bound oxygen. Thanks to his publisher's negligence, his findings were not known to the world until 1777. Similar tardiness yielded primacy in the discovery of nitrogen to the Scotsman Daniel Rutherford (**1772**).

Scheele was certainly prolific. In addition to oxygen and nitrogen, he identified the elements chlorine (**1774**) and molybdenum, and helped compatriot chemists discover cobalt, tungsten, nickel and manganese, among others. A useful new tool was the blowpipe, which sent a stream of air into a flame to make it hotter as it played on a piece of mineral. The colour of the flame, the vapours formed and the deposits left behind all gave clues as to whether some unknown element was hidden in the sample.

Scheele named a dozen or more new acids associated with living things, such as acetic acid from vinegar, citric acid from lemons, malic acid from apples and lactic and uric acids from animals. He also found some nasty gases, including the highly poisonous hydrogen cyanide and hydrogen sulphide ('rotten egg gas'). Scheele routinely did 'taste and sniff' tests on anything he found; this may well have contributed to his early death at the age of 44, as it probably did for the Englishman Humphrey Davy a few decades later (**1802**).

Joseph Banks: Botanist to the World

Wealthy landowner and ardent naturalist Joseph Banks became President of the Royal Society of London in 1778; it was a powerful post that he was to fill in rather autocratic fashion for a record 42 years until his death.

1778

Banks was a remarkable individual, combining the passionate in-the-field researcher with the man of affairs, the patron of science with the scientific administrator.

Banks first came to notice in his 20s as chief naturalist on board HMS *Endeavour*, captained by James Cook (**1769**) and sent by the Admiralty to observe the transit of Venus from Tahiti. Three observing sites were set up in case one was obscured by cloud; Banks manned one of these.

From there, the expedition discovered and explored the east coast of 'New South Wales'. Banks and his assistant Daniel Solander, who had studied with Carl Linnaeus (**1753**), brought back to England the first plant and animal specimens from the new continent. After classification, the vast haul yielded 110 new genera and 1300 new species among the plants alone. Banks' own name is commemorated in the distinctive *Banksia* genus of Australian plants.

Banks continued to collaborate with Solander, and with other collectors worldwide, throughout his life, especially once he took over management of the botanical gardens at Kew, west of London. These were still the property of the royal family. The gardens prospered enormously under his care. The number of species under cultivation increased 20-fold, with specimens gathered from around the world at Banks' request by missionaries, diplomats, naval officers and tradesmen. His network was immense.

Banks had a hand in many things: in the choice of Botany Bay in Australia as the site for a penal colony; in commissioning HMS *Bounty* under William Bligh to take breadfruit from Tahiti to feed slaves in the West Indies; in financial and moral support for William Smith, then working on the first ever geological maps (**1799**); and in the founding in **1799** of the Royal Institution by Benjamin Thomson.

Unlike some of his contemporaries, such as the members of the Lunar Society in Birmingham (**1765**), Banks had no strong political opinions, and was able to stay on good terms with both the king and Benjamin Franklin during the American Revolution. Later, he was an enduring link between English and French scientists when their two nations were at war. He certainly left a very strong mark on the science of his day.

Jean Deluc and James Hutton:
A History of Catastrophe?

1778

Pondering why the surface of the Earth looks as it does, with its massive mountains, deep valleys and vast plains, many thinkers in the mid-eighteenth century imagined the life of our planet as short and violent, a series of catastrophes. Prominent among these was Jean Deluc of Switzerland, who in his younger days had walked regularly in the Alps and become an expert in their natural history. He made sturdy portable barometers and pioneered measuring the heights of mountain peaks from the reduction in air pressure. He also found that water had its maximum density at 4°C, which is why ice floats.

Deluc came to England in 1773 when his family business failed. For 44 comfortable years he was 'reader' to Queen Charlotte, and was elected to the Royal Society. Travels in many lands to study their rocks led him to champion catastrophism. He saw ample evidence that the Earth has been devastated at intervals by cataclysmic floods, earthquakes and eruptions. The Bible, in which many or even most scientists believed, set the life span of the Earth at only 6000 years, so there seemed little time for anything else. Expounded in 1778 in a book by Deluc, catastrophism remained influential into the nineteenth century under the patronage of Deluc's supporter, Georges Cuvier of France (**1812**).

Scotsman James Hutton had a very different view, which he called 'uniformitarianism'. To Hutton, the processes that formed rocks and moulded the surface of the Earth in ancient times were (a) exceedingly slow and (b) very similar to those still going on today. So the Earth had to be very old, perhaps infinitely old, but certainly much older than the age deduced from the scriptures. Given enough time, the most immense changes, such as the raising of mountain ranges or the levelling of plains, could be accomplished very gradually.

Hutton was quite a contrast to Deluc. His early interest was in chemistry, which led to medicine, but he never seriously practised as a doctor. He also had trained in law. He invented a process to make 'sal ammoniac' (ammonium chloride), a significant industrial chemical, and this gave him a steady income. He later took to farming in Norfolk and outside Edinburgh, finding great pleasure in rural society and time to develop his ideas. Later again, he settled in Edinburgh, in a house crammed with fossils and chemical apparatus, and made friends with James Watt and Joseph Black.

Insightful as he was, Hutton's books, such as *The Theory of the Earth*, published in 1785, were so densely written and obscure that few could make sense of them, though Hutton was apparently much more intelligible face to face. Fortunately he had disciples who could communicate the ideas much more clearly; these included John Playfair, who published *Illustrations of the Huttonian Theory of the Earth* in 1802, and Charles Lyell (**1829**), whose 1830s writings had such an impact on the young Charles Darwin.

Looking back, it seems that Hutton, Deluc and their various supporters claimed too much for their theories. Both sides had some of the truth. We now know that much of the Earth's appearance is a result of very slow changes over long periods of time (and that was certainly the dominant view in the nineteenth century). But catastrophic events have also been influential, including immense volcanic eruptions and asteroid strikes (**1980**). The same is true of the neptunism/plutonism debate in which Hutton was active (**1775 Werner**). Rocks are formed both underwater (sedimentary rocks) and from hardening magma and lava (igneous rocks).

Jan Ingenhousz: How Plants Purify Air

We have several reasons to remember Dutch doctor Jan Ingenhousz. One is as a champion of inoculation against smallpox (**1796**). Invited by a family friend who happened to be President of the Royal Society and physician to King George III, Ingenhousz came to London in 1765, set up a successful practice and met scientific luminaries such as Joseph Priestley.

1779

The king sent Ingenhousz to Vienna to inoculate the family of the Austrian Empress Maria Theresa (which included the young and later tragic Marie Antoinette). There he became Court Physician, with time and means to experiment. He returned to London in 1779 and published a book with the impressive title *Experiments upon Vegetables: Discovering their great Power of Purifying the Common Air in the Sunshine and of Injuring it in the Shade and at Night*. The title sums up what he had found in a decade of research in Vienna.

Priestley had shown (**1771**) that air made unfit for breathing or unable to sustain a flame could be restored by the presence of a green plant. Everyone knew that plants needed light to live and grow; a plant left too long in the dark withered and died. Was light also needed for a plant to clean up the air? Ingenhousz found that it was. Not only was a plant deprived of light unable to cleanse air degraded by burning candles or breathing mice, the plant itself 'injured' the air. It 'breathed' as an animal did.

Ingenhousz noted something more. If the stem and leaves of a plant were placed in water, small bubbles formed on the leaves and green parts of the stem, but only in daylight. In the dark the bubbles disappeared. So by day plants release a gas that might be called 'pure air', but at night they do not. The bubbles came off slowly. Given enough time, a sample of the gas could be collected and tested. It caused a glowing splint of wood to burst into flame. So the gas was 'fire air', identified by Priestley in **1774** and named oxygen in 1778 (**1775**). This was the gas now known to be needed for breathing and burning. And plants made it.

Two hundred years of research were needed to clarify this process, including finding and naming the green pigment, chlorophyll (**1818**) that makes it happen.

Horace de Saussure: Understanding the Alps

1779

Horace de Saussure was in the happy position of being able to combine his science with his love of the outdoors. As Professor of Philosophy at the University of Geneva for 25 years, he had the Swiss Alps at his doorstep. Like his countryman Jean Deluc (**1778**), he explored them extensively, studying the rocks and plants, making maps and weather observations, and publishing the first of his scientific journals in 1779 under the title *Voyages in the Alps*. Saussure was a serious mountain climber, and was the third to reach the summit of Mont Blanc, the highest peak in the Alps.

He made his greatest contribution in 'geology' (he may have been the first to use this new name for the subject). He saw the Alps as a key to understanding the history of the whole Earth, and became expert in the composition and structure of its rocks. This led him to support the 'neptunist' point of view (**1775**), which argued that all rocks formed under water, and to oppose the 'plutonists'.

Being outdoors so much, he took a great interest in the weather, making thousands of readings of temperature, pressure and other variables, many of them at high altitude. To determine the amount of moisture in the air, he invented the 'hair hygrometer'. This used strands of human hair that stretched or shortened depending on how moist or dry the air was. He had instruments, mostly of his own design, to measure everything, from the blueness of the sky to the temperature deep in the soil. He found that the temperature 10 metres underground lagged behind the surface temperature by six months, and that the temperature at the bottom of a deep lake never changed.

William Herschel Finds a New Planet

1781

Since ancient times, only five planets (other than Earth) had been known: Mercury, Venus, Mars, Jupiter and Saturn. With the Sun and Moon, that made seven heavenly bodies. Seven being a powerful number, no more were expected to be found. That did not stop William Herschel, German-born but living in England, adding to the number in 1781. Using one of the best telescopes of the time from his observatory in Bath in the west of England, he found a faint star-like object that could not be a star, since it moved very slightly from night to night against the background of the other stars.

Herschel called his new planet George's Star (in Latin) to honour the king, much as Galileo had initially named the moons of Jupiter after the Medicis when he found them in **1610**. It soon became known as Uranus, and is so known today. In Roman mythology, Uranus was the father of Saturn, as Saturn had been the father of Jupiter.

Measurements showed that the new planet took nearly three times as long to orbit the Sun than did the most distant known planet, Saturn. It was therefore, according to Kepler's Laws (**1609**), nearly twice as far away. This one discovery doubled the known size of the solar system. There was some debate as to whether it was a comet rather than a planet, but that was settled once its orbit was found to be almost circular rather than elongated.

Herschel had devoted most of his time up until then to music, as an organist and composer. Now he was famous, and astronomy took over his life. Appointed the King's Astronomer (not the same post as Astronomer Royal), he continued to study the sky for 40 years from Slough, west of London, where he built the biggest telescope ever made (**1789**). His sister Caroline was likewise talented in both music and astronomy. She was her brother's assistant in both fields for many years, and made significant discoveries in her own right, including a number of new comets. She was arguably the first woman to figure prominently in the history of science.
➤➤1789

Watching the Weather

As the Enlightenment glided on, natural phenomena like the weather became less mysterious and frightening. A century before, Edmond Halley (**1676**) had shown the close link between changes in air pressure, as measured with a barometer, and major weather events. Storms were more likely when the pressure was falling; a 'rising barometer' usually meant the weather would improve. This suggested that the weather could be explained and even predicted.

1783

The idea was strengthened by an insight by the American Benjamin Franklin (**1751**). By following newspaper reports, he noted a tendency for major storms to move from west to east, at least at the latitudes of Europe and the American colonies. If we knew where a storm was at a particular time (marked by a region of low pressure), we might be able to estimate where it would be one to two days ahead and give warning.

But first we would need to gain an overview of the state of the weather by drawing together observations taken simultaneously over a large area. Throughout the seventeenth and eighteenth centuries, various individual scientific bodies and local rulers had organised networks of observers, but their efforts were mostly short-lived.

In the 1780s the Mannheim Meteorological Society in Germany put together an extensive network, enabling researchers to draw a 'weather map' for the whole of Europe for every day in 1783. This was a massive achievement, but the time taken for data collection meant that the maps were not completed until months or even years after the event. It was not until the introduction of the electric telegraph in the mid-nineteenth century that weather maps could be assembled fast enough to be the basis of weather forecasting. By the 1860s weather maps showing air pressure variations over large areas were being published in the major daily newspapers (**1869 Galton**).

Jacques Charles: Science and Ballooning

1783

Frenchman Jacques Charles featured in both the study of gases and the daredevil pursuit of ballooning. With little scientific education, the 30-something Charles was working in a Paris government office when the American ambassador, Benjamin Franklin, paid a visit. Charles was inspired through learning of Franklin's scientific experiments (**1751**) to teach himself some science. Within a few years he was giving public lectures; soon after he was in the Academy of Science with a rising reputation.

Early in 1783, the Montgolfier brothers sent a hot air balloon aloft carrying some animals. The Academy asked Charles to examine this new mode of travel. He thought that for safer and surer ballooning, the balloon needed to be filled with the very light 'flammable air' (hydrogen)—discovered by the Englishman Henry Cavendish (**1766**)—rather than air heated over a fire.

By August that year, Charles had filled his first balloon with hydrogen made by

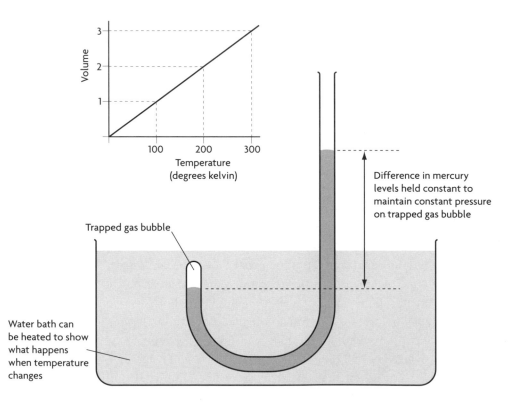

*With this apparatus, which is very similar to that used by Robert Boyle in **1659**, Jacques Charles showed that if the pressure on a trapped bubble of gas is held constant, the volume of the bubble increases directly as the temperature rises. Charles' Law was confirmed by a number of other investigators.*

pouring sulphuric acid over scrap iron (later zinc). It could lift only 10 kilograms as it rose from a field in Paris where the Eiffel Tower stands today. The onlookers included Benjamin Franklin. By year's end, Charles, his brother and other brave souls were taking untethered flights over Paris hanging beneath a gas-filled balloon.

It soon became apparent that as balloons rose, they expanded in the declining air pressure; Boyle's Law predicted as much (**1659**). Hot-air ballooning showed that heating a gas also expanded it. In 1787 Charles explored the relationship using a J-shaped tube, as Robert Boyle had done in his study of gases. Charles placed the tube in a water bath so that he could vary the temperature of the gas. By keeping a constant difference between mercury levels in the two arms of the tube, he held the pressure of the trapped gas bubble steady. The result was clear. The volume of the trapped gas rose and fell in step with changing temperature. This direct relationship is commonly, and perhaps unjustly, called Charles' Law.

Charles was not the first to announce the finding (his fellow French chemist and balloonist Joseph Gay-Lussac did that in 1802); nor was he first to discover the relationship. That honour belonged to another Frenchman, Guilliame Amontons back in **1702**. Charles' apparatus was understandably better than Amontons'. His more precise measurements later helped others locate the temperature at which a gas, shrinking in volume as it chilled, should come to occupy no space at all. This temperature would be dubbed 'absolute zero' (**1848**).

John Michell Imagines a 'Black Hole'

Little-known John Michell passed his later years quietly as a country parson in Yorkshire, England, yet he left his mark on many areas of science. He had been a Professor of Geology at Cambridge as a younger man and had even been elected to the Royal Society.

1783

Among many outcomes of a productive life, Michell invented the torsion balance, which could measure very small forces through the use of a twisted thread. He did little with it himself, but his friend Henry Cavendish used it to measure force between electric charges (he was beaten into print in **1785** by the Frenchman Charles Coulomb, who had invented the same apparatus). Cavendish also used the balance to measure the force of gravity and so weigh the Earth (**1798**).

Michell was the first to suggest (**1755**) that earthquakes are caused by movements and tensions between underground rock layers. He also proposed a method for working out the actual brightness of the stars and developed highly efficient ways to make artificial magnets.

Perhaps most startlingly, Michell was first to visualise an object or region of space from which light could not escape. In 1783 he reasoned that anything falling from infinity towards a star say 500 times the mass of the Sun would arrive travelling faster than light. So even a ray of light, with its immense speed, could not escape from the star and reach infinity; rather it would inevitably slow to a stop and fall back into the star. So no light could leave such a 'dark star'. The reasoning, based on Newton's ideas, was fallacious, but black holes would in time be shown to exist (**1916**).

René-Just Hauy: Master of Crystals

1784   The major contribution made by the French priest René-Just Hauy to the study of crystals began, so the story goes, with an accident. While looking at a fine collection of crystals of various minerals owned by a friend, he dropped one. On his hands and knees clearing up the mess, he noted that the fragments, though of different sizes, all had very much the same shape.

Hauy had already done some work in botany and attended lectures in mineralogy. He began an intense study of many types of crystals and found a common feature. Whatever different crystals of the same mineral might look like externally, they all share some form of inner structure. It was as if the particles in the crystals were arranged in a distinctive way relative to one another. This produced lines of weakness along which the mineral would 'cleave' in regular pieces with sides set at fixed angles to one another. 'Cleavage' became one means of identifying minerals.

Here was an early example of what would be recognised as a truth in chemistry in the nineteenth century. The properties of a substance depend not only on what elements or compounds are in it, but on how the atoms and molecules of those chemicals are arranged. Structure and organisation can be as important as composition. The growth of organic chemistry would provide ample evidence of that (**1828**).

Hauy wrote up his ideas in an influential book in 1784, shortly after his election to the French Academy of Science. At the request of the leading figures of the academy, he gave lectures on his new ideas and was soon the acknowledged expert.

Hauy began life as the son of a poor weaver and owed his education to priests who took a liking to his singing voice. He taught science at a church-run college for 20 years before retiring to concentrate on his research. He suffered at the hands of the revolutionaries, as did many priests who refused to take an oath of allegiance, but survived to become Professor of Mineralogy at the Paris Museum of Natural

History. He died in poverty, but his reputation was such that Napoleon had made him an honorary canon at Nôtre Dame Cathedral and one of the first members of the Legion of Honour.

William Withering: A Cure for the 'Dropsy'

1785

The members of the exclusive Lunar Society of Birmingham (**1765**) were a mixed crowd, but all were eminent in their fields. William Withering was a doctor from Edinburgh. He joined the 'lunatics' when he came to Birmingham General Hospital in 1779. His fame rests mostly on a single discovery. It seems he was treating a patient with dropsy (heart failure due to a build-up of fluid). The patient improved remarkably after taking a traditional herbal remedy. Withering's insight was that some particular substance in the mixture had done the job, and he found it in the leaves of a plant called the foxglove.

The active ingredient is now known as digitalis, after the plant's scientific name. It works by both slowing and strengthening the heartbeat, improving circulation and so clearing the congestion. Withering worked on his treatment for 10 years, doing clinical trials and finding side-effects. In 1785 he published his findings in *An Account of the Foxglove and Some of its Medical Uses.*

Herbal remedies were, of course, an established part of medicine at the time. Some were of long standing, others were new to Europe, such as the 'miraculous bark' from South America that cured malaria and later yielded the pure compound quinine. Finding new cures and treatments was one of the main reasons plants had been studied, at least in the age before Linnaeus (**1753**). Now for perhaps the first time, we knew why a traditional remedy worked, and how to use it to our best advantage.

Antoine Lavoisier: The End of the 'Four Elements'

1785

By 1785 the fall of the venerable 'four-element' theory (earth, air, fire, water) was all but complete. The idea of 'earth' as an element was already long gone. So many different 'earths' had been discovered, with such widely varying compositions, that the notion of a single 'essence of earth' seemed not only unsupportable but pointless.

The notion of 'air' as an element had suffered under the close scrutiny of the Swede Carl Scheele, the Englishman Joseph Priestley and the Scots Daniel Rutherford and Joseph Black. Their ideas had been summed up, expanded and confirmed by the Frenchman Antoine Lavoisier. Those who accepted their evidence (and not everyone did) now had little doubt that air, at least the everyday kind, was a mixture of gases, not a simple substance. Most of it was 'dead air' or 'noxious air' that would neither burn nor allow burning, and in which animals suffocated. A smaller proportion, about 20 per cent, was 'vital air' or 'eminently breathable air'; this allowed things to burn and animals to live. Lavoisier had named these gases azote and oxygen; the first we now call nitrogen. In addition, air contained small amounts of 'fixed air' (carbon dioxide) and water as a vapour. These very different components could be quite simply separated. No 'essence of air' could explain that variety.

Water was next to be denied element status. Englishman Henry Cavendish had already created water by burning his 'flammable air', called hydrogen ('water maker') by Lavoisier. Burning hydrogen combined it chemically with oxygen, making water—a compound. In 1785 Lavoisier illustrated the true nature of water in a striking manner in front of a large audience. He boiled water and passed the vapour over red-hot iron, which decomposed the water into its constituent gases. The hydrogen and oxygen were collected, mixed in a large glass jar and exploded with a spark. The original water reappeared.

That left just 'fire'. Lavoisier had quite convincingly replaced the notion of an 'essence of fire', aka 'phlogiston', used to explain burning, with a theory that involved oxygen (**1775**). That ought to have been the end of it. But Lavoisier himself muddied the water by insisting that there remained an 'essence of heat', a weightless fluid present in larger amounts in hot bodies than in cold ones. He put this 'caloric' high on his list of elements (**1789**). Half a century of experiment would be needed to discredit the idea, though it was first challenged as early as **1798** (by his wife's second husband!). ➤➤**1789**

Charles Coulomb and His Law

1785

By now it was clear from many observations (**1733**) that electric charges attracted and repelled each other. But what law did those forces follow? One hundred years earlier the same question had been asked about the force of gravity; Isaac Newton had come up with the definitive answer (**1687**).

American Benjamin Franklin and Englishman Joseph Priestley found a hint: if a hollow metal container was electrified, all the charge sat on the outside, and no effect was felt within. According to Newton, there would be no force of gravity inside a hollow sphere. This suggested that electric charges obeyed the same sort of law that masses did, an 'inverse square' law. It looked as if doubling the distance between two charges reduced the force of attraction (or repulsion) between them to a quarter.

The first to try to prove this was the reclusive Englishman Henry Cavendish (**1766**), but he was, as usual, slow to publish his findings. As a result, the French military engineer Charles Coulomb was first into print, in 1785. The law of attraction and repulsion between electric charges is therefore known as Coulomb's Law. Had it been Cavendish's Law, it would have said the same thing.

COULOMB'S LAW

The force between any two electrical charges is directly proportional to the size of the charges multiplied together, and inversely proportional to the square of the distance between them. The law is expressed by the following equation:

$$F = \frac{k\, q_1 \times q_2}{d^2}$$

F is the force in metres

q_1 and q_2 are the two charges in coulombs

d is the distance between them in metres

k is a constant

The value of k in this equation (and a similar one for the force between two magnetic poles) is controlled by the ease with which electric and magnetic fields are set up in a vacuum. These are the numbers James Clerk Maxwell used to calculate the speed of light (**1871**).

COULOMB

The coulomb is the unit of electrical charge: 1 coulomb of charge is carried by a current of 1 ampere flowing for one second. It is also equal to the total charge of six hundred thousand million million million electrons.

The charges that could be collected for testing were small, so the forces between them were weak. To gain the needed sensitivity, Coulomb invented the torsion balance. Rather than using the force of gravity, as an ordinary balance does, the torsion balance used the forces embedded in a twisted metal wire. Cavendish used the same equipment that was independently invented by his friend, the remarkable John Michell (**1783**), and later measured the force of gravity with it (**1798**). As with gravity, the forces depended on more than distance. The size of the electric charges involved was also a factor. Coulomb's precedence saw the unit of electric charge being called the coulomb (it could have been the cavendish!).

Though remembered today mostly for his work on electricity and magnetism (he also found that the force between magnets is inverse square), Coulomb was versatile and productive, prominent from 1785 in the French Academy and active in many fields, including some outside science. He took little part in the stormy political events of his time, but under Napoleon served as Inspector General of Public Instruction. He played a major role in setting up the network of lycées (secondary schools) across France.

Luigi Galvani: Animal Electricity

1786 By the late eighteenth century, powerful and sophisticated machines were available to separate and accumulate large amounts of 'static' electricity through friction. These build-ups of charge, stored in devices such as the Leyden Jar (**1745**), could generate impressive shocks and sparks—causing much entertainment—but not a sustained or continuous flow of charge.

Credit for making the first electric current goes to the Italian anatomist Luigi Galvani, though he did not know he was doing it. His interest was in what he named 'animal electricity'. Various animals, such as the torpedo fish and the 'electric eel' were known to deliver a shock, sometimes very severe, similar to that from a Leyden Jar, and therefore assumed to be due to an electrical discharge.

Galvani came at the subject from the other end. He noted that the leg muscles of a frog he was dissecting twitched when an 'electric machine' was running in the room. He directed the electric discharge into the tissues with a copper probe and they twitched even more. The big surprise came when he hung up the frog legs, bound in copper strips, on an iron rail. They continued to twitch even though the electric machine was not running.

Galvani had a ready explanation. Though the frog was clearly dead, the muscles still contained an 'electric fluid', a 'vital force' that was liberated by the touch of copper and iron. The frog legs had been 'galvanised' into action. This term was proposed by Galvani's compatriot and contemporary Alessandro Volta, and widely used. Ironically, it was Volta who later challenged Galvani's explanation, showing that the frog's legs had nothing to do with the effect (**1800**). Galvani had accidentally made the first 'galvanic cell', which used chemical reactions to generate an electric current.

Galvani's electrical life force may have been an illusion but his name lives on; not only in 'galvanise' but in the 'galvanometer', an instrument developed later to measure small electrical currents, and 'galvanising' as applied to iron and steel to protect them from corrosion.

A New Language for Chemistry

1787 Although rational chemistry was beginning to take shape, thanks to the work of Antoine Lavoisier and others, the language of chemistry remained in a mess. People still spoke of compounds by fanciful titles that had come down from the days of alchemy: 'flowers of zinc', 'sugar of lead', 'butter of arsenic', 'spirits of salts', 'oil of vitriol'. Furthermore, many chemists over the centuries had invented their own terminology, which conflicted with that of others.

Antoine Lavoisier may have been the leading chemist of his day, but it was his countryman Guyton de Morveau who most clearly saw the urgency of the need for reform. Lavoisier helped, as did others, including Claude Bertollet, and the new grammar and vocabulary of chemistry was published in 1787 as *The Method of Chemical Nomenclature*. The titles of compounds now reflected mostly what they contained. The suffixes of names were important. To call a compound hydrogen sulphide meant it contained sulphur and hydrogen and nothing else. Hydrogen sulphite contained some oxygen as well, hydrogen sulphate even more.

Names could include numbers as words to indicate how much of an element was present. Carbon dioxide contained twice as much oxygen (compared with the amount of carbon), as carbon monoxide. It was all very ordered and logical. There were other subtleties. Mercuric oxide, previously called 'red calx of mercury', had a different ratio of mercury to oxygen from that in mercurous oxide.

Some common names have persisted, of course. Hydrogen sulphate is usually called sulphuric acid (but no longer 'oil of vitriol'). Since water has long been proven to be a compound of oxygen and hydrogen, it should properly be called 'hydrogen oxide', or even more precisely, 'oxygen dihydride', but no one does.

Antoine Lavoisier: Listing the Elements

Frenchman Antoine Lavoisier's *Elementary Treatise on Chemistry*, published in 1789, summed up his (and the world's) new understanding of chemistry, based on 30 years of precise experiment. It also contained a very important list: the 31 substances that he considered 'elements', as defined by the 'sceptical chemist' Robert Boyle in **1661**—substances unable to be broken down into anything simpler. Since Boyle's time, chemists had succumbed to the lure of the erroneous phlogiston theory; Lavoisier was now returning them to the more productive path first explored by Boyle a century earlier.

Lavoisier's list is surprisingly accurate; only two substances on the list are totally misplaced. Lavoisier thought both light and heat (which he called 'caloric') were elementary substances; during the nineteenth century both of them would be proven to be forms of energy (**1798**). Heading the list were the three gases discovered over the previous 20 years: oxygen, hydrogen and 'azote' (nitrogen). Lavoisier did not include chlorine, because his mistaken ideas about acids containing oxygen (**1775**) required chlorine, which was found in hydrochloric acid, to be a compound. Then there were some common solids: sulphur, phosphorus and charcoal (carbon). Most of the rest were metals; seven known from ancient times and ten discovered since.

Lavoisier also listed substances he thought were elements but which would within a few decades be prised apart chemically to reveal the true elements within them. The element calcium would be found in his 'lime' (or chalk), magnesium in 'magnesia', barium in 'barytes', aluminium in 'argill' or 'earth of alum', and silicon in 'silex' (or silica as we would say today), the compound in sand. Lavoisier thought these were elements, not only because he could not break them down by heat or chemical action, but also because he knew of no element from which minerals with their properties might be formed. It was a reasonable point of view.

Surprisingly, Lavoisier left a couple of 'earths' off his list of elements: soda (which would prove to be the source of sodium) and potash (ditto potassium). Perhaps he really thought these were compounds; certainly they were well known and had useful practical applications, such as in making glass and soap. He listed the three remaining substances as 'radicals': the 'muriatic' radical associated with the acid of that name, the 'fluoric' radical and the 'boracic' radical. Again, these were not really elements, but elements were hidden within them: chlorine, fluorine and boron respectively.

The French Revolution began the year Lavoisier published his revolutionary book. Five years later he was still only 51 and at the height of his powers of discovery. He was at heart a genuine reformer, wanting to improve the lot of ordinary people as he had improved chemistry. But he had served the old regime as a member of the hated fermée, which collected taxes on behalf of the king. In revenge, the revolutionaries brought him to the guillotine. 'The regime has no need of wise men', they declared. 'It is in need of justice.' The mathematician Joseph Lagrange (**1772**) put it differently: 'It took only a second to sever that head. A century may not be enough time to produce another like it'.

William Herschel: Bigger Is Better

1789   By the time William Herschel was born in Hannover in modern Germany in 1738, telescopes had been sweeping the night sky for nearly 130 years. They had advanced mightily in size and power since Galileo's **1610** primitive 30-power refractor. With larger, better lenses arranged in more sophisticated ways, 'refracting' telescopes were now gathering more light, detecting much fainter and more distant objects and seeing more detail in nearby ones. The troublesome coloured fringes that had frustrated early observers were a bother no more, thanks to John Dolland's achromatic lenses of **1758**. Refractors were the first choice of most astronomers, since they wasted less light.

But refracting telescopes were limited in size. A lens was supported only around its edge; a very large one would collapse under its own weight. 'Reflecting telescopes' that used mirrors supported from behind clearly had an advantage. There was no obvious size limit. And size mattered. If you double the 'aperture' of a telescope, you can see twice as far and make out details half as big.

Herschel led this drive for larger apertures. He had emigrated to England in 1772, perhaps because the king of England had ancestors from Hannover. It is not clear why he switched from his early employment in music to astronomy, but by 1776 he was building his own telescopes.

He started with one 2 metres long, with which he found the new planet Uranus in **1781**. With financial support from the king, he doubled the size, then doubled it again. In 1789 he finished with a monster 12 metres long, carrying a mirror more than a metre across. Constructed in Slough, west of London, the scaffold and turntable used to support the telescope and turn it to point around the sky were bigger than a house. Nothing would rival it in size for more than half a century.

To be honest, the beast was hard to use, and Herschel made many of his major discoveries, such as new moons around Saturn and Uranus, with smaller instruments. Nonetheless, he had pointed observational astronomy into the future. In time, telescopes that dwarfed Herschel's creations would be built. ➤➤**1800**

The Sad Tale of Nicholas Leblanc

Late in the eighteenth century, growing demand for various chemicals of importance to industry was outpacing supply. Chief among these were the 'alkalis', vital in making glass, soap, textiles and paper. There were two **1791** main forms: 'potash', produced by steeping wood ashes in water; and 'soda ash' (aka 'washing soda'), made by boiling down various seaweeds such as Scottish kelp and 'barilla' from the Canary islands. Supplies were unreliable, quality was variable. More abundant and reliable sources were urgently needed.

The French, who were at war with much of Europe, were unable to import soda ash. In 1783 King Louis XVI had his Academy of Science offer a prize to anyone who could 'make the alkali' (that is, soda ash) by decomposing sea salt by the 'simplest and most economic method'.

The starting point was already known. A century and a half before, German chemist Johann Glauber had heated sulphuric acid with sodium chloride (sea salt) to make hydrochloric acid. This left a residue (Glauber's Salt), also called 'salt cake'. Various people thought soda ash (sodium carbonate) could be extracted from the salt cake but no one could do it cheaply and easily enough to meet demand.

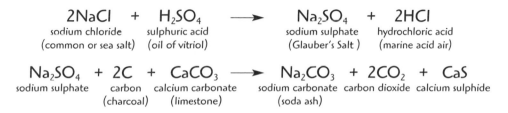

$$2NaCl + H_2SO_4 \longrightarrow Na_2SO_4 + 2HCl$$

sodium chloride   sulphuric acid   sodium sulphate   hydrochloric acid
(common or sea salt)   (oil of vitriol)   (Glauber's Salt)   (marine acid air)

$$Na_2SO_4 + 2C + CaCO_3 \longrightarrow Na_2CO_3 + 2CO_2 + CaS$$

sodium sulphate   carbon   calcium carbonate   sodium carbonate   carbon dioxide   calcium sulphide
           (charcoal)   (limestone)   (soda ash)

These two equations reveal how Nicholas Leblanc converted common salt into soda ash for use in industry, generating hydrochloric acid and calcium sulphide as troublesome wastes along the way.

No one, that is, until Nicholas Leblanc, a doctor employed the Duke of Orleans, came up with his 'process' in 1791. He roasted the salt cake with charcoal (carbon) and chalk (calcium carbonate). Off came carbon dioxide, leaving a mass of sodium carbonate and calcium sulphide, a waste which could be separated from the product by washing. The problem appeared solved. Leblanc's patron built a big plant outside Paris and production began. Demand was high because of war.

Unfortunately, it did not quite work out for Leblanc. The plant was soon appropriated by the revolutionaries, the Duke was beheaded, Leblanc was forced to reveal the secret of his process and others quickly put it to work. He was never compensated for the takeover of the plant, and by the time he was able to resume production his competitors had too much of a head start. The Academy of Science had been dissolved so he never got the prize. Little wonder that, bankrupt and despairing, he finally shot himself.

His process, too, ultimately met its end. Environmentally unfriendly, and producing various nasty wastes, it was superseded in 1861 by the Solvay method. At least his descendants finally got some reward. In 1855 Napoleon III made a payment to them in lieu of the prize. A statue of Leblanc now stands in Paris.

Erasmus Darwin: Grandfather of Genius

1794 Mention the name Darwin and most people immediately think of Charles Darwin and *The Origin of Species* (**1859**), not Erasmus Darwin and *Zoonomia*. Charles Darwin's grandfather was, however, a very significant scientist in his time; his book, published in 1794, was highly influential. In *Zoonomia* we find ideas that anticipate those of the great proponents of evolution who followed, in particular Jean Lamarck (**1809**).

Erasmus Darwin clearly believed that plants and animals changed over long periods of time, that species were not immutable, that the relationships between plants and animals expressed in tables of classification actually reflected the history of life on Earth. He saw that living things were adapted to life in different environments: the beaks of birds were hard or soft, long or short, narrow or broad, depending on where they lived and what foods they found to eat.

Like Lamarck, Erasmus Darwin saw these adaptations as a consequence of the never-ending struggle to survive. By such struggle, some creatures gained some new features to help their survival and passed that improvement on to their offspring. His grandson would have a different view of the process, but the outcome, greater fitness to survive, was the same. Over a long enough span of time, such a process could have produced life on Earth in all its variety from a common origin.

Erasmus Darwin had many talents. He practised medicine, wrote poetry admired by Coleridge and Wordsworth, and invented a canal lift for barges, a mechanical speaking machine and a carriage that would not turn over, among other things. He did not patent any of them for fear of damaging his medical reputation. His scientific interests were broad, too. He suggested that the Moon had once been part of the Earth and had been torn out long ago, leaving the chasm now filled with the Pacific Ocean.

Darwin corresponded with Joseph Banks (**1778**) and Carolus Linnaeus (**1753**) and translated Linnaeus' books into English. He was a founding member of the influential Lunar Society of Birmingham in England (**1765**). There he met the master potter Josiah Wedgwood; both his son Robert and his grandson Charles later married Wedgwoods. Like many of his fellow 'lunatics', Darwin was politically radical, believing in democracy, education reform and free love. Such views brought him under suspicion during the Napoleonic Wars and his public reputation never really recovered.

Edward Jenner: The Battle Against Smallpox

1796

In the eighteenth century smallpox was still a dreaded disease, able to blind, maim, disfigure or even kill those who contracted it. No one knew what caused it, and treatments were 'rule of thumb' and generally ineffectual. The toll was vastly greater among indigenous populations of Africa, America and Oceania, where smallpox was unknown until the coming of European invaders.

Since the start of the century, inoculation had been practised in Europe as a way of conferring some sort of immunity to the disease and preventing its spread. Smallpox sufferers rarely got the disease twice, suggesting that one infection generated resistance to a second attack. In inoculation, pus from a pustule on a smallpox sufferer was rubbed into a small incision on the skin. The patient was thereby given a mild dose of the disease, from which they would hopefully recover and be immune thereafter. The drawback was that until the disease had run its course, patients were infectious and could pass the scourge on to others.

A better technique was put forward by English country doctor Edward Jenner, who practised in Gloucestershire. He drew on a local folk tradition that milkmaids had attractive, unblemished skin because they had contracted a mild infection called cowpox from their daily work and this protected them from the much more serious smallpox. In 1796 he tested the idea, using the same method of inoculation, but with pus from a cowpox pustule. The patient, eight-year-old James Phipps, quickly

> *The joy I felt at the prospect before me of being the instrument destined to take away from the world one of its greatest calamities [smallpox] was so excessive that I found myself in a kind of reverie.*
>
> EDWARD JENNER

recovered. Six weeks later Jenner gave him the standard smallpox inoculation and the boy was completely unaffected. Cowpox had conferred immunity against smallpox.

Jenner called his new method 'vaccination', from the Latin *vacca* meaning 'cow'. It was the start of a new era in public medicine and was to lead 200 years later to the total global eradication of smallpox. From the work of Louis Pasteur (**1886**), and others in the next century, ways were found to confer immunity to other deadly diseases. Jenner's discovery also raised questions about how such immunity was actually generated. How indeed does the body constantly fight off infection on a daily basis? Those questions had to wait about 100 years for an answer (**1882 Metchnikoff; 1891**).

Pierre Laplace:
How the Solar System Came To Be

1796   The achievements of Pierre Laplace were so far reaching and profound that he has been dubbed the Newton of France. As with most mathematicians, his best work was done when he was young, between the ages of 20 and 40, and the last 40 years of his life were spent setting it all down in a series of powerful and influential books.

His *System of the World*, published in 1796, contained one of his most imaginative ideas, a proposal as to how the solar system had formed, producing the Sun at the centre surrounded by orbiting planets, some with moons, and the comets. As Laplace saw it (and the philosopher Immanuel Kant at about the same time), the solar system began as a vast cloud ('nebula') of gas and dust. All the particles pulled on one another through gravity, and the cloud began to shrink. Random movements set it turning; as it shrank it turned faster, like a skater pulling in her arms. The faster rotation made the cloud flatten and bulge out in the centre.

Within the cloud, particles of gas and dust came together randomly to form clumps. These rapidly grew through gravity and collision with other clumps to become ultimately the size of planets or even bigger. At the centre, the largest accumulation became the Sun; smaller accumulations further out became the orbiting planets, the satellites and the comets.

The vision explains most of the features of our solar system; for example, why all the planets orbit the Sun in much the same plane in the same direction, with the Sun itself rotating in the same direction. Over the years this 'nebular hypothesis' has been challenged on various grounds, and many details debated, but it is today the generally accepted explanation as to why the solar system, including our Earth, looks as it does.

In his book, Laplace raised the terrifying possibility that the Earth might one day be struck by a comet. It sounds a remote chance, but given enough time, he says, it is certain to happen. The consequences—the destruction and loss of life—would be dire. Such ideas are still around (**1980 Alvarez**).

Alessandro Volta: The First Electric Battery

A decade after Italian Luigi Galvani accidentally generated the first continuous electric current with bits of copper and iron and some frog legs (**1786**), his compatriot Alessandro Volta worked out what had really happened and so began a major technological revolution. Volta told the world of his discovery in 1800 in a letter read before the Royal Society of London by its then president, the biologist Joseph Banks.

One account of Volta's discovery goes as follows. Investigating Galvani's 'animal electricity' and its ability to cause physical sensations in the body, he connected a copper coin and an iron coin together with a wire and put them into his mouth, one above, one below the tongue. He felt a slight tingling and a salty taste in his mouth, which he attributed to the release of animal electricity. He soon found that he did not need his tongue. A pad of cloth or cardboard soaked in brine or dilute acid and placed between the two coins seemed also to stimulate electrical effects. Connecting the discs in turn to an electroscope showed that one was releasing positive electricity, the other negative.

Here then was the first 'Voltaic cell', able to generate a steady flow of electric current. It was now clear that Galvani's frog legs were superfluous and there was no 'animal electricity' involved. The essence of the effect, as Volta the physicist saw it, was contact between two different metals (such as copper and iron) moistened with a little salty water (the body fluids in the frog legs had provided that).

Volta soon found he could multiply the effect by placing cells side by side or on top of each other and connecting the copper plate in one cell to the iron plate in the next. This made a 'Voltaic pile', or, in another analogy, an electric 'battery', similar to placing a number of identical cannons side by side and firing them together.

It is hard to overstate the importance of Volta's discovery (**1800**). It is fitting that his name also lives on in the volt, the unit of the electrical pressure that drives a current through a conductor (**1826**).

Henry Cavendish: How Heavy Is the Earth?

English researcher Henry Cavendish was aristocratic, wealthy and decidedly eccentric. He made many important discoveries, including the gas hydrogen (**1766**) and the law of force between electric charges, usually associated with Charles Coulomb (**1785**). His last achievement was the most grandiose. He weighed the Earth.

To do it, he relied on Isaac Newton (**1687**), who had shown that the force of gravity between two objects depends on their masses as well as on the distance between them. Using the very sensitive torsion balance invented by his friend John Michell (**1783**), Cavendish was able to measure the force of attraction between large and small balls of lead. The force was so weak and the measurements therefore so sensitive that Cavendish set up the apparatus in a sealed room and read the scale from afar using a telescope. He then compared that force with the weight of the small ball, which represented the force between the small ball and the whole Earth. That let him calculate the weight of the Earth. The answer came out at six million million tonnes, which is close to the number commonly accepted today.

Cavendish was not the first into this field, though he was first to use this method. Twenty years before, Nevil Maskelyne, the Astronomer Royal and author of the first *Nautical Almanac* (**1765**), had used a weight on a string to attempt the same task. You would expect the weight to hang straight down, but the presence of a very large mass nearby, such as a mountain, would tug it ever so slightly to one side, out of the vertical. Remarkably, Maskelyne could measure that tiny shift and so figure out the gravitational pull of the mountain. An estimate of the mass of the Earth was the next step; his answer was in the same ballpark as Cavendish's.

Benjamin Thomson (Count Rumford): The Attack on 'Caloric'

1798

American-born adventurer Benjamin Thomson had a multifaceted life: it included teaching, military service for the British in the War of Independence (and probably some spying), and high office in the government of Bavaria, where he became a count. He was a practical man, with a string of inventions to his name, mostly to do with fireplaces, chimneys and applications of heat, such as a double boiler and a coffee dripper. He promoted the use of James Watt's improved steam engine. In between, his life rose and fell, and he spent little time in his homeland. He married twice, once to a rich American widow and later to Marie Lavoisier, widow of the great chemist. They separated after four years.

Thomson was the only important figure in science to emerge from North America between Benjamin Franklin (**1751**) and Joseph Henry (**1831**), but his reputation in the sciences rests on one major observation. As the Bavarian Minister for War, he was responsible for the manufacture of cannons. For each barrel, a solid cylinder of brass was cast and the centre bored out with a drill driven by horsepower. Thomson soon saw that the cannons got hot during the boring process, especially if the bit was

It frequently happens that in the ordinary affairs and occupations of life, opportunities present themselves of contemplating some of the most curious operations of nature.

BENJAMIN THOMSON (COUNT RUMFORD)

blunt, and suspected that some important insight into the nature of heat might come from what many people would have already observed.

As he told the Royal Society in 1798, he had a special cylinder cast and placed in a vat of water. The blunt horse-driven drill was applied and the surrounding water soon became warm. After two and half hours it was ready to boil. Clearly an immense amount of heat—equal, Rumford calculated, to nine large wax candles burning together—had been generated without fire. It seemed that as long as the boring went on, the water continued to boil. The supply of heat seemed inexhaustible.

In the accepted theory of the day, heat was an invisible substance, a probably weightless fluid called 'caloric'. This notion had been made popular by his wife's first husband (Lavoisier); Thompson reputedly commented that he would 'drive caloric from the stage', just as Lavoisier had done with the theory of phlogiston (**1775**). Two things seemed clear: the inexhaustible supply of heat made it most unlikely that heat was a 'substance', since it would surely have run out after a time; and the need for the drill to be turning to generate heat suggested a close relationship with motion.

Caloric was not at once dethroned. The influence of Lavoisier left many people clinging to the older idea for decades. One who embraced the new was the young Englishman Humphry Davy (**1801**), who found that two pieces of ice could be made to melt by being rubbed together. Here was another example of motion making heat. Perhaps Thomson was saying 'thank you for your support' to Davy when he appointed him a lecturer at the Royal Institution he had just established in London, so launching Davy's stellar career.

The Royal Institution was Thomson's other enduring legacy. He made some powerful friends during his several stays there, including Joseph Banks, the great naturalist, then President of the Royal Society (**1778**). They gathered a consortium of aristocrats and backers for the venture, though Thomson himself bankrolled the institution through its earlier years.

In its programs, the institution was to combine research with fostering public awareness of the impact of science on everyday life. Thomas Young of the 'double slit experiment' (**1801**) was briefly Professor of Natural Philosophy. Other scientists of distinction associated with the institution over the next 200 years included Michael Faraday (**1813**), Nobel Prize-winner Lord Rayleigh (**1894**), James Dewar (**1911**) and the Braggs, father and son, also Nobelists (**1915**). After a time, the work at the institution became too theoretical for Thomson's taste, and he lost interest.

The Ultimate Triumph of 'Strata' Smith

William Smith was the son of an Oxfordshire blacksmith who died when his son was young. After a minimal formal education, Smith became apprenticed to a surveyor and proved very adept. The work took him all over England and allowed many opportunities to observe the outcropping of rocks,

1799

something that had interested him since his youth. He went down mine shafts where he could easily observe and record the successive layers of rocks of different types.

Even more striking evidence came from hillsides cut open by the building of canals, such as the Somerset Coal Canal, on which he worked for six years. Across England, and in other countries, mining and canal building was boosting the study of rocks, which was beginning to be called 'geology'.

That rock layers occurred in a certain order was well known. Nicholas Steno (**1667**) and Giovanni Arduino (**1759**) (and even Leonardo da Vinci before them) had established the Principle of Superposition: that in any series of rock strata, the oldest rocks are at the bottom and the youngest uppermost. Smith added something very important. He noted that the different rock layers contained distinctive collections of fossils, stony lumps that often looked like shellfish or bones. Those groupings could be used to identify and link layers of rock that might appear in cliff faces kilometres apart. A rock layer outcropping in Somerset could be proved identical with one in Yorkshire.

In 1799 Smith started on his first geological map. He coloured in the various rock strata on a map of the Bath district, where he was living at the time. Soon after, he did the same on a small map of England, the first step towards his great geological map, published in 1815. This was the first large-scale map, and the most detailed so far produced, that used fossils to identify the strata.

That map was an immense labour and in time would bring him fame as 'Strata' Smith. He had first to struggle to raise the money to have it published (Joseph Banks was one of his backers) and then to endure both bankruptcy and the disdain of professional geologists at his lack of education. But he triumphed in the end and in 1831 was awarded the highest honour of the Geological Society of London, the first Wollaston Medal. His place in scientific history was now secure.

William Herschel: Heat Is Light

1800　William Herschel, German-born but resident in England from the age of 29, made profound contributions to astronomy, including the discovery of the planet Uranus (**1781**) and the building of the first big telescopes (**1789**). He studied the Sun through various coloured filters to reduce the amount of light, and noted that different colours allowed different amounts of heat to pass through. One guess was that the colours themselves had different amounts of heat associated with them.

To test this he set up a simple experiment, easily reproduced. A prism broke up sunlight into its various colours from violet to red. He placed a thermometer with a blackened bulb into each coloured region in turn and compared its temperature with the reading on a pair of other thermometers placed well beyond the end of the spectrum.

The results were surprising. Each area of colour was hotter than the controls, and the amount of heat detected rose along the spectrum from violet towards red. So Herschel placed the test thermometer just beyond the red, where no colour was visible. This registered the highest temperature of all.

It seemed that sunlight contained some form of 'light' that could not be seen. Herschel called these 'calorific rays', from a Latin word meaning 'heat'. Later experiments showed that these calorific rays could be reflected by a mirror, refracted by a prism and focused by a lens just like ordinary light. His rays, later called infrared radiation (**1802**) or radiant heat, were indeed like visible light, and kin to many other forms of radiation, such as ultraviolet light (**1802**), radio waves (**1888**), X-rays (**1895**) and gamma rays (**1899**), still to be discovered.

The Beginning of the 'Electric Century'

Alessandro Volta's 'piles' or 'batteries', developed in **1796** and disclosed in 1800, caused a sensation. Physicists everywhere were soon seeking new effects from, and possible applications of, the flow of electric current they provided. Discoveries flowed quickly, especially as bigger and bigger piles were built, delivering more electrical potential or 'voltage'.

1800

Almost at once England's William Nicholson and Anthony Carlisle found a remarkable effect when the wires from each end of a pile were dipped into salty or acidified water. The first observations may have been accidental, but repeated experiments showed tiny bubbles of gas collecting around each wire. These proved to be oxygen and hydrogen. The electric current had decomposed the water into its constituent gases, an effect later called 'electrolysis'. Englishman Humphry Davy, working at the Royal Institution in London, would discover new elements in this way (**1807**).

Davy became one of the great pioneers in the new electrical science. He found the money to build a huge pile with 2000 pairs of plates. When he connected the ends of the pile to a pair of carbon rods, slightly separated, the intense current and heat liberated incandescent particles of carbon. Invented in 1808, this was the 'electric arc', not the first case of electricity providing light (**1702**) but certainly the first useful one.

The impact of Volta's **1796** discovery on every aspect of life would be profound, including powerful new technologies as well as new tools for studying matter. Electricity would soon be making magnetic fields (**1820**), so powering electromagnets and, soon after, the electric telegraph and then the telephone.

By **1831** the first electric motors would be turning electrical current into mechanical power as an alternative to the steam engine. That, plus the invention of the electric light globe (**1882**) would herald the large-scale electrification of advanced societies. Efforts to pass electric current through gases (**1854**) would lead to X-rays (**1895**), neon signs and ultimately to electronics and television. By the end of the 'electric century', we would glimpse the fundamental unit of electric charge (**1897**).

1801–1850

The World Stage

IN 1815, 25 YEARS OF conflict between France and other powers (Britain, Prussia, Austria, Russia) ended on the battlefield of Waterloo. The subsequent Congress of Vienna established a conservative political system that discouraged war for several decades. But Europe was not peaceful. In 1830, and particularly in 1848, uprisings in many European capitals clamoured for political and social reforms, including constitutional government. Several rulers were overthrown.

Britain, now the 'workshop of the world', led Europe in political and social change, including the reform of parliament, the extension of the right to vote (1832), the abolition of slavery (1834) and the expansion of free trade (1846). Such progress began to remedy many inequalities and injustices, including the terrible living conditions in cities bloated by the growth of industry.

Bit by bit, the map of Europe was redrawn. In 1834 German states formed a customs union under Prussian leadership, foreshadowing the political union to come. Greece gained independence from a steadily declining Turkey (1829), Poland became a province of Russia (1832), Belgium and Luxembourg were separated from the Netherlands (1839).

European states were accumulating overseas empires, notably Britain and France in Africa, Asia and the Pacific. The Spanish domination of the Americas was beginning to weaken. In 1804 the 'Louisiana Purchase' of French-held territory greatly expanded the domain of the United States north and west. Conflicts with Mexico, freed from Spanish control, later added Texas and California (the latter the location of a major gold rush in 1849). In Ireland, a virulent fungus disease in the potato crop in 1845–46 caused widespread famine and led to major emigration, especially to the 'New World'.

ARTS AND IDEAS

With the overthrow of the French monarchy, the rococo style of art and decoration favoured by the old regimes was replaced by 'classical' styles modelled on Greek and Roman forms—as revealed by the new excavations at Pompeii—and by something more realistic, commenting on social conditions and the irresponsibility of governments. In England, Turner painted mostly landscapes but he was alert to the impact of the Industrial Revolution, particularly the advent of railways.

Music moved from the 'classical' era to the 'romantic', the transition typified by Germans Beethoven and Schubert. Leading composers mid-century included Berlioz in France, Chopin in Poland, Glinka in Russia, and the Germans Mendelssohn, Schumann and von Weber. Italians Rossini, Donizetti, Bellini and Verdi took to writing opera, along with German-born Meyerbeer. Star 'solo performers' such as the violinist Paganini and the pianist Liszt became prominent. Increasingly, performances were for the wider (paying) public rather than for royal and noble patrons and their courts.

The German brothers Grimm collected folktales and fairy stories, as did the Dane Hans Christian Andersen. Mary Shelley's 1817 novel *Frankenstein* drew on the 'animal electricity' experiments of Luigi Galvani. The 'great lover' Casanova published his memoirs; the first edition of *Webster's Dictionary* appeared. The publication of the *Book of Mormon* initiated a new church. German philosopher Friedrich Engels exposed *The Condition of the Working Classes in England* in 1845 and joined with his countryman Karl Marx to issue *The Communist Manifesto*.

The major literary forms had plenty of players: among poets, the Englishmen Wordsworth, Keats, Shelley and Byron; Americans Longfellow and Poe; Germans Goethe and Heine; and the Russian Pushkin. Leading novelists included Austen, Dickens and the Bronte sisters in England; Scott in Scotland; Sand and Stendhal in France; and Hawthorne and Cooper in America. As printing technology improved, daily and weekly newspapers became cheaper and more diverse.

ATOMS, CELLS AND BONES

Various lines of scientific inquiry, previously separate, began to interlink and cross-fertilise. Scientists learnt to identify chemical elements from the light they gave out. Electricity played a role in finding new chemical elements; suspicion grew that all chemical reactions were fundamentally electrical in some way. A better understanding of the properties of light helped us find the secret of fermentation, during which living things carry out chemical reactions. The distinction between living and non-living blurred, giving rise to what was later called biochemistry.

New words were carefully crafted to allow 'scientists' (itself a new word, along with geology and biology) to describe the ever-growing list of new phenomena. Major new theories and concepts gave sense and coherence to observations or pointed to the fundamental underlying unity within nature. The 'atomic theory' underpinned advances in chemistry, the 'cell theory' showed the common structure of all living things, the doctrine of 'uniformitarianism' demonstrated that the past can be explained by the present.

Electricity was linked to magnetism and both of them to light. The various forms of energy proved interchangeable at 'fixed rates of exchange', though the total amount of energy could not be changed. We found proof that the Earth definitely turns, and discovered how far away the stars were, why Saturn has rings, that we cannot get colder than 'absolute zero', that ultraviolet light exists, that 'coal tar' is not merely a waste, and that the solar system has eight major planets and many small ones.

The complex history of our planet grew steadily clearer. Bones of 'dinosaurs' and other creatures pointed to a past very different from the present. We began to accept that some life forms had become extinct (and tried to explain why that might have happened), and evidence grew of major shifts in climate—'ice ages'—in past eras.

As for the leading figures, it is hard to go past England's Michael Faraday, experimenter beyond compare, though both Louis Pasteur and Charles Darwin began their careers in this half-century. The Royal Institution, which Faraday headed for many years, was a new sort of place; people there were paid merely to do research and to tell the public about it.

John Dalton: Manchester's Favourite Son

1801 Englishman John Dalton's name is famous in science, and rightly so. The Quaker son of a Cumberland weaver, Dalton passed most of his life as a schoolmaster in Manchester; research was relegated to his spare time. His lifelong interest was in the weather and he reputedly logged some 200 000 observations in his daily journal. Sadly, his records, carefully preserved for over a century, were destroyed by bombing during World War II. His lifelong disability was colour-blindness, first diagnosed at the age of 26 when he brought his mother some scarlet stockings, thinking they were blue.

From his preoccupation with the weather came an interest in gases like air and water vapour. In 1801 he made his first contribution to science, his Law of Partial Pressures: when two gases are mixed, the pressure of the mixture is the sum of the pressures of each considered separately. He also noted that the pressure of the vapour over a liquid in a closed container increases when the liquid gets hotter. Others would show that a liquid boils when that vapour pressure equals the pressure of the surrounding air.

From his interest in air, there was a natural progression to his 'atomic theory', published in 1808, on which his fame mostly rests. He saw this as a concise and natural explanation for the way substances combine chemically. The fame this generated led to membership of the Royal Society and the French Academy, and to a government pension, but he kept on teaching.

Dalton never married, and rarely strayed far from Manchester. He died from a stroke in 1844. Four hundred thousand people filed by his coffin as it lay in state, and his funeral cortege stretched for 2 kilometres. At his request, his body was autopsied to determine the cause of his colour-blindness, sometimes called Daltonism. No defect was found with his eyes, but 150 years later one eyeball preserved at the Royal Institution yielded DNA to show an inherited lack of sensitivity to green light. ➤➤1808

The Short but Productive Life of Humphry Davy

Humphry Davy was the son of an impoverished woodcarver from Cornwall in England who died when his son was only 16. Nevertheless, by talent and diligence Davy became a leading figure in early nineteenth-century science, active in many fields. To support his family, Davy became an apprentice surgeon and apothecary; he was turned on to chemistry by reading Antoine Lavoisier's textbook (1789).

1801

Davy first made a reputation before he was 20, experimenting on himself (a dangerous practice that he kept up throughout his life) with the gas nitrous oxide. Breathing it intoxicated him—hence its name 'laughing gas'—and numbed pain. Davy argued that the gas could be used as an anaesthetic in minor surgery and dentistry. That suggestion was not taken up for over 40 years, by which time getting 'high' on laughing gas had become an essential activity at fashionable parties.

Davy's own entry to fashionable society came through his appointment in 1801 as a public lecturer at the recently founded Royal Institution in London (1798). Here Davy both researched and communicated science, quickly becoming Professor of Chemistry and then Director of the institution. He established the tradition, which persisted for over 100 years, of demonstrating new findings in science in front of audiences of men and women in evening dress; such events were soon part of the social round.

Davy made many important discoveries, including in electrochemistry (1807) and in correcting Antoine Lavoisier's contention (1775) that all acids contain oxygen. He was also very inventive; his miner's safety lamp, devised around 1815, is the main reason most people have heard of him. But he was not really a diligent and careful experimenter, and he was succeeded and indeed surpassed, both as scientist and communicator, by his one-time assistant Michael Faraday (**1813**). He reputedly regarded the brilliant Faraday, who was from humble beginnings like himself, as his 'greatest discovery', though he twice opposed Faraday's election to the Royal Society.

Davy himself received many honours. A Fellow of the Royal Society at only 25 and later its President for three years, he was knighted in 1812, three days before his marriage to a rich widow. He reputedly enjoyed his fame and lived well. Davy died when only 41, arguably from the accumulated effects of self-experimentation over many years. ➤➤**1807**

Giuseppe Piazza: Ceres Fills a Gap

1801

In 1801 the number of planets in our solar system grew by one. So what? William Herschel had found a new planet, Uranus, only 20 years before (**1781**). But this new planet was special. It was the first of many (in fact, thousands ultimately) found in much the same region of the Sun's territory. It was to prove very small compared with the other nine planets. And its existence had been predicted; Herschel had just happened upon Uranus.

The Titus–Bode Law was first created in 1766 by the German Johann Titus of Wittenberg, However, it was so assiduously popularised by Johann Bode, director of the Berlin Observatory, that it is commonly called simply Bode's Law. It is based on a number series, as shown in the table opposite. To Titus and Bode these numbers reflected the distances of the various planets from the Sun (at least the planets known at the time) in astronomical units (AU) (the distance of the Earth from the Sun). The numbers lined up impressively, and the fame of Titus–Bode increased again when Uranus was found 19.2 AU out from the Sun (the predicted distance was 19.6).

Clearly something was missing between Mars and Jupiter, perhaps a planet yet to be discovered. Astronomers started looking; Italian Giuseppe Piazza found a bright speck called Ceres in 1801. He was not really looking for a new planet, but Ceres was right in the slot at 2.8 AU. Others were found not far away, beginning with Pallas in

THE SPACED-OUT PLANETS		
Titus–Bode Law	Actual distance (in astronomical units)	Planet
0.4	0.39	Mercury
0.7	0.72	Venus
1.0	1.00	Earth
1.6	1.52	Mars
2.8		
5.2	5.20	Jupiter
10.0	9.54	Saturn
19.6	19.19	Uranus
38.8	30.0	Neptune
77.2	39.5	Pluto

The Titus–Bode Law begins with the following series of numbers: 0, 3, 6, 12, 24, 48, 96, 192, 384, 768. After the second number, each number doubles the one before.

If you add 4 to each number and divide by 10, you get the numbers in the table above. These are compared with the actual spacings of the planets in our solar system, given as astronomical units (the distance of the Earth from the Sun).

1802. Within a few decades so many were known that they began to look like the remains of a giant planet that perhaps once orbited between Jupiter and Mars and had since been torn apart. Certainly the 'planets' were small. None was more than a dot in the best telescopes of the day, and Herschel thought they were only a few hundred kilometres in diameter. He called them 'asteroids' because they looked like stars, and they are still so called today.

The Titus–Bode Law had started the hunt, but there is no reason why it should be true. Even the asteroids do not really fit the 'law', since their distances range from 2.1 to 3.5 AU, though the average is about right. The 'law' fell down badly with Neptune and Pluto and is now generally recognised as just a remarkable coincidence.

Thomas Young: Light Is a Wave

The 'double slit' experiment reported by English researcher Thomas Young in 1801 is one of the most famous ever done, and it continues to reverberate 200 years later. It was not entirely original; the Italian Jesuit Francesco Grimaldi had done something similar in **1665**. For him, as for Young, the findings made a strong case that light was best explained as a form of wave motion, like ripples on a lake, rather than as a string of particles, another theory popular at the time and which had enjoyed the support of Isaac Newton.

1801

Anyone can do the experiment. Set up a bright source of light and a screen onto which the light falls. Between the two place an opaque barrier with just two narrow

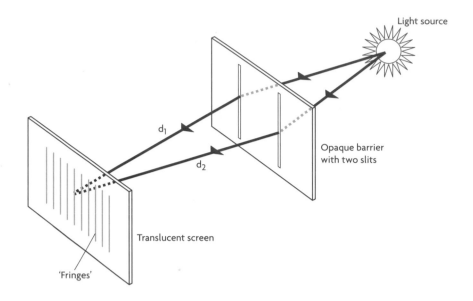

Rays of light passing through the two slits in the opaque screen interfere with each other when they reach the viewing screen, and produce a pattern of bright and dark 'fringes'.

If the path lengths d_1 and d_2 differ by half a wavelength (or one and a half, two and a half and so on), the two light rays cancel each other out (the crest of one wave matches the trough of another) and the screen is dark. If the difference is a whole number of wavelengths, the two waves reinforce each other and produce a bright fringe.

By this experiment Thomas Young proved the wave nature of light and measured its wavelength.

slits (they can be just two small holes) to let light through. Cover up one slit; the screen will show just a narrow bright streak produced by light passing though the other slit. Uncover both and something very different appears. A pattern of light and dark streaks (sometimes called 'fringes') will stretch across the screen. Young and others argued that the light-as-particles theory could not explain this, but the light-as-waves model could, very easily.

The explanation runs like this. Light arriving at a spot on the screen will have travelled slightly different distances from each of the slits. The two light waves will interact (or 'interfere'), and the outcome will depend on whether the crest of one coincides with the crest of the other (that is, they arrive 'in phase') or whether a crest of one arrives with a trough of the other, or something in between. Light waves arriving in phase will reinforce each other, and you will see a bright spot or line. A crest and a trough together will cancel out, and the screen will be black. The 'in betweens' will make various shades of grey.

So the 'double slit' is indeed powerful support for light-as-waves. But there is more. The two light waves will come in phase if the distances they have travelled (their 'path lengths') differ by exactly one or more wave lengths (the distance from one crest to another). So the spacing of the fringes and the other dimensions of the apparatus allow us to calculate the wave length of the light. This turns out to be very small, less than one millionth of a metre.

Like Grimaldi before him, Young saw the colours of the rainbow in the fringes. Young's explanation was simple; the different colours of light have different wave lengths, with violet light waves about half as long as red ones. The light and dark fringes of the various colours therefore fall in slightly different places, spreading the colours out in a spectrum as a prism does (**1672**).

Johann Ritter: Light Beyond the Violet

Johann Ritter at the University of Jena in Poland was engrossed in developing the technology of batteries to make and store electricity when news came through that English astronomer William Herschel had found **1802**
a new form of light (**1800**). Noting that Herschel's 'calorific rays' were invisible and lay outside the red end of the spectrum of sunlight (**1672**), Ritter wondered if some other unknown radiation was hidden in the dark beyond the violet end as well.

Herschel had used a thermometer to find his rays because they were associated with heat. Ritter employed the chemical silver chloride, which turned black when exposed to light (from that would grow the technology of photography). As Herschel had done, he made a spectrum from red to violet using a prism, and then put a dab of silver chloride under each colour. All the spots went dark, but spots near the violet end of the spectrum darkened much more quickly and completely than those near the red end.

He then dabbed some silver chloride in the dark region beyond the visible violet. Sure enough, it went black, and faster than with any of the visible colours. So there was some form of light out there, undetectable to the human eye but still able to make things happen.

Ritter called his new light 'chemical rays', since that was how they were found. We now call them 'ultraviolet', and Herschel's 'calorific rays' are known as 'infra-red'. These names have their origin in Thomas Young's **1801** work with the 'double slit' experiment. Young found that violet light had a shorter wavelength and therefore a higher frequency than red light. This meant that Ritter's rays had a higher frequency still, and were therefore 'above the violet'. By the same argument, Herschel's rays were 'beneath the red'.

Over the next 200 years we would learn how influential ultraviolet (UV) 'light' is. UV in sunlight stimulates the production of vitamin D (**1906**) in the skin but also causes sunburn and skin cancer, and can cause genetic mutations in living things, as X-rays do (**1926**). Its energy can break down chemical compounds, fading colours and weakening materials. Hence the fear in **1973** when there was a threat that we would be exposed to more of it.

Joseph Proust:
The Great Debate About 'Proportions'

1804

Late in the eighteenth century, French chemist Joseph Proust was living in Spain, well away from the excesses and dangers of the French Revolution. Between 1797 and 1804 he carefully heated metals like iron, copper, lead, tin and mercury so that they combined with oxygen, then calculated how much oxygen had been taken in. He found that the ratio of the amounts of oxygen and metal in the oxide was always the same, or perhaps one of only two or three fixed numbers. So a compound of oxygen and iron might contain 27 per cent oxygen or 48 per cent oxygen, but not some amount in between. There were just two forms of copper oxide: 18 per cent oxygen or 27 per cent oxygen were the only options.

At the time, these findings were dubbed Proust's Laws. We now call them the Laws of Constant Proportions and Multiple Proportions, and they seem simple and obvious. Yet Proust had to fight to get his ideas accepted by colleagues not yet in tune with the new ideas in chemistry. Proust's countryman Claude Bertollet, an eminent researcher and colleague of the great Lavoisier (**1787**), was among the sceptics. He thought the amount of each element in a compound might depend on how it was made, and claimed to have evidence of such variations.

The debate between Proust and Bertollet was vigorous, but support for Proust steadily grew. Many other researchers repeated his work. The German Jeremiah Richter had already shown that acids and alkalies always reacted together in fixed proportions, which meant that the resulting salts had fixed composition. Minerals collected from nature had the same composition as the same compounds manufactured in the laboratory. Some compounds could be produced in a variety of ways, but the composition still came out the same, contrary to Bertollet's expectations.

By 1805 the debate was just about over; Proust had won and chemistry took a great step forward. There were two other consequences. The concept of a 'compound',

---

LAWS OF CHEMICAL COMBINATION

John Dalton used these as powerful evidence for his atomic theory (1808).

Law of definite proportions or constant composition: A given compound always contains the same elements in the same proportion by weight.

Law of multiple proportions: When two elements combine to make more than one compound, the different weights of one element that combine with a fixed weight of the other element are always in a simple ratio.

dating back to Robert Boyle in **1661**, became more precise; constant composition was another way to distinguish compounds from mixtures. And the soon-to-be announced 'atomic theory' of John Dalton (**1808**) gained vital supporting evidence.

Luke Howard and Francis Beaufort: Winds and Clouds

Weather watchers had long had most of the instruments and systems they needed to describe the weather: rain gauges, barometers for air pressure (**1643**), thermometers for temperature (**1742**), hygrometers for humidity (**1779**). Some things they lacked; one was a way to assess the strength of the wind, since the wind gauge (the 'anemometer') had not yet been invented. That need was met by an English admiral, Francis Beaufort. His 12-point 'Beaufort scale', invented in 1806, described the effects of various wind speeds on the state of the sea and 'men of war' sailing on it; the scale was later extended to include descriptions of effects on land.

1806

For example, Beaufort's Force 2 wind was a light breeze (around 10 kilometres per hour in a modern measure) felt on the face and making leaves rustle in the trees. A Force 6 'strong breeze' (around 40 kilometres per hour) made small trees sway and caused some spray on the sea surface. A Force 10 'whole gale' (100 kilometres per hour) uprooted trees and turned the sea white with foam. Nowadays, the scale has only eight divisions but it still covers the gamut from 'calm' to 'hurricane'.

A few years earlier, another Englishman, London pharmacist and amateur scientist Luke Howard, had done a similar job, providing a vocabulary to describe the various types of clouds. He chose five words derived from Latin to be used in various combinations: cumulus (for 'a heap'), stratus (for 'a layer'), cirrus (for 'a hair'), nimbus (for 'a shower') and alto (for 'high'). So 'cumulonimbus' clouds are heaped up and bring showers (thunder clouds); 'altostratus' are high up and flat. Only about 10 of the possible combinations are used. Other observers, such as the biologist Jean Lamarck, had worked on a similar scheme, but Howard's is the one that has survived.

Humphry Davy: Electricity Reveals New Elements

Englishman Humphry Davy (**1801**) is well known for inventing the miner's safety lamp and the electric arc light, heading up London's Royal Institution and 'discovering' Michael Faraday. His fame in science rests mostly on his pioneering studies in 'electrolysis'. He made up this word to describe the decomposition of various liquids by the newly discovered electric current. Davy was one of the first to follow up the discovery by his countrymen Nicholson and

1807

Carlisle in **1800** that running electricity through water containing a little acid generated bubbles of hydrogen and oxygen. The water had been broken down into the two elements from which it is compounded.

Davy's first insight (a profound one, as history later showed) was that breaking down the water into its elements involved very similar processes to making the electric current using a 'voltaic pile' (**1796**). Chemical reactions can drive electric currents and vice versa. The new study was thus rightly called electrochemistry. More fundamentally, the willingness of substances to react chemically (which had been called their 'affinity') is ultimately electrical in nature.

Davy wondered if electrolysis could break down certain substances, like common salt, lime and soda, which had long been suspected of being compounds, not elements, but had resisted attack by heat and ordinary chemicals. Electrolysing solutions of these in water produced just hydrogen and oxygen. Davy's bright idea was to melt the substances before passing the current through.

Success came in 1807. From molten salt he obtained a new metal he called sodium; molten potash yielded potassium. Calcium, strontium, magnesium and barium soon followed in similar fashion. These mostly lightweight metals proved very active, reacting vigorously, sometimes violently, with air and water. That willingness to react was the reason the metals had been so hard to extract from their compounds.

John Dalton: Atoms Are the Answer

For nearly 200 years the ancient vision that matter was made up of minute hard particles in constant motion enticed European physicists. Atoms explained the behaviour of gases (**1661**) and they had encouraged Isaac Newton to propose a 'corpuscular' theory of light (**1672**). English schoolmaster John Dalton gained lasting fame by applying the same idea to chemistry.

Dalton came to such ideas through his lifelong interest in the weather and in gases

DALTON

The dalton is the measure of the mass of atoms and molecules against the mass of the most common carbon atom, taken as 12.

like everyday air (**1801**). Given that air is mostly oxygen and nitrogen, this led to experiments with the compounds of those two elements. Only three 'nitrogen oxides' were known: one of these contained equal amounts of oxygen and nitrogen by weight; in another the ratio was 1 to 1.7; a third had the ratio 1 to 3.4. It seemed that nature allowed no other ratios.

Joseph Proust in France had found the same rules with other substances (**1804**); and they were reflected in his newly developed Laws of Constant and Multiple Proportions. But Dalton went a stage further: he asked 'Why is it so?' He had already contemplated the idea of 'atoms', preferring the original term used by the Greek Democritus 2000 years before to the more fashionable 'corpuscles'. He saw that it provided a ready explanation for the constant composition of the nitrogen oxides: the particles of these compounds would be made up of small numbers of atoms of the two elements, with the atoms of a given element all weighing the same. In the lightest of the nitrogen oxides, one nitrogen atom would be linked with one oxygen atom; in another, two nitrogen atoms would join three of oxygen; in a third, one nitrogen atom would be alongside two of oxygen.

This scheme, plus the actual weights of the various compounds, implied that the ratio of the weights of nitrogen and oxygen atoms was seven to eight. Similar calculations could be done on dozens of other compounds, so generating a table of the distinctive 'atomic weight' for each element.

Dalton published his ideas in 1808 in *A New System of Chemical Philosophy*. There was initial controversy, some of which Dalton brought on himself by sticking with some erroneous ideas alongside obviously correct ones. For example, he thought that, with a few exceptions, compounds contained only one atom of each element. He clung to the resulting odd values for atomic weights in the face of other people's evidence. In fact Dalton would not have accepted the calculation we have done above.

He was stubborn in other matters too. He later refused to accept the new system of chemical symbols that the Swede Jöns Berzelius (**1814**) proposed in place of his own clumsy nomenclature. He remained an ardent supporter of the 'caloric' theory of heat even when evidence against it was growing (**1798**). And he was not really a careful experimenter: he was content with rough and ready measurements when better results and techniques were available.

Still, his 'atomic theory' changed the course of science, especially as supporting evidence from Berzelius and others began to pile up. Dalton had argued a case for the reality of atoms and built a new take-off platform for modern chemistry. One consequence of the new philosophy was a death blow (if one were needed) to the faint, lingering hopes of alchemists that one element could be transmuted into another; lead into gold, for example. Dalton's atoms were indivisible and immutable. A century would pass before that conviction was challenged, with profound consequences (**1919**).

Joseph Gay-Lussac: When Gases Combine

1808

During the French Revolution and its aftermath science became mixed up with politics. The great Lavoisier perished on the guillotine (**1789**) because of his political associations; Joseph Proust (**1804**) had escaped to Spain to avoid the troubles. Under Napoleon, who well understood the benefits of science for the national economy and defence, bright, able scientists had opportunities for advancement and reward.

Joseph Gay-Lussac was one whose life was greatly affected. The arrest of his father in 1792 saw the young Gay-Lussac sent to Paris to finish his education. There he fell in with some of the leading scientific figures of his day, including Claude Bertollet, who had worked with Lavoisier, and Pierre Laplace (**1796**). His career blossomed from that time and he became active in many fields. He was a leading figure in the growth of the chemical industry in France, contributing to improved production of sulphuric acid, gunpowder and oxalic acid. He was also adventurous, once ascending to over 6 kilometres in a balloon to see if the Earth's magnetic field weakened with altitude (it did not).

In pure science, his achievements concerned gases. He began in 1802 by discovering the relationship between the space taken up by a given amount of gas, held at constant pressure, and its temperature. Gay-Lussac found (as his countryman and fellow balloon enthusiast Jacques Charles had done 25 years earlier (**1783**) but had not published) that for each degree of fall in temperature, the volume of a gas shrank by a fixed amount. After careful experiment, he found the size of that decrease, so helping Lord Kelvin define absolute zero (**1848**), a temperature at which the gas would (theoretically) shrink to nothing.

Gay-Lussac also measured the volumes of gases involved in chemical reactions. The figures proved remarkably simple: for example, if temperature and pressure were held constant, 2 litres of hydrogen would combine with 1 litre of oxygen to make 2 litres of water vapour. (Englishmen Nicholson and Carlisle had found the same thing but in reverse in **1800** by breaking water into hydrogen and oxygen with an electric current.) Similar simple numbers applied to other reactions, such as between nitrogen and hydrogen to make ammonia.

GAY-LUSSAC'S LAW OF COMBINING GAS VOLUMES

When gases react, they do so in volumes that bear a simple ratio to each other and to the volumes of the products if they are gases, provided all volumes are measured at the same temperature and pressure.

From this, Gay-Lussac expounded his Law of Combining Gas Volumes in 1808: 'When gases react, their volumes bear a simple relationship to each other and to the volume of the product, if a gas'. It took only until **1811** for someone to tie this to John Dalton's atomic theory, also announced in **1808**.

Etienne Malus: Light from a Window

From his laboratory bench in Paris, Etienne Malus could see the sunlight reflected from the great glass windows of the Luxembourg Palace opposite. Malus was very interested in light, including its peculiar property called 'polarisation'. This phenomenon had first been noted and explained 120 years before by the Dutchman Christiaan Huygens. Polarisation was possible, said Huygens, because light is a series of waves in an invisible, undetectable medium called the aether, not unlike ripples passing over the surface of water. In water waves, the ripples are always on top. With light, it is more complex.

1808

Suppose you tie one end of a long rope to a post and hold the other end, pulling the rope taut. A quick movement of your hand will send a ripple along the rope, vertically, horizontally or at any angle in between depending on the direction your hand moves. Now pass the rope between two vertical posts. Only vertical ripples are now possible; the posts block side to side ones, making the ripples on the rope 'polarised'. Malus knew that light could be similarly polarised by a crystal of the mineral calcite, which allows light to pass through only one 'plane of polarisation'. Two crystals used together could detect if light was already polarised. Polarising filters today use synthetic materials, such as 'polaroid'.

By looking though a pair of calcite crystals at the palace windows, Malus found that at least some of the incoming light was polarised. So light can be polarised by

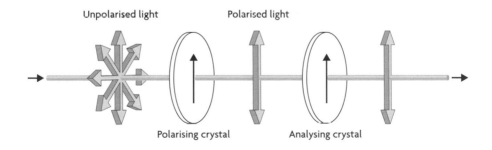

Unpolarised light   Polarised light

Polarising crystal   Analysing crystal

A ray of light is travelling from left to right. Before the first crystal, it is 'unpolarised', that is, vibrating from side to side but in all directions. The arrangement of atoms in the polarising crystal allows through only light vibrating in one direction, the 'plane of polarisation'. The second crystal will let the polarised light through only if its atoms are lined up parallel to those in the first. So rotating the second crystal will show that the incoming light is polarised and find its plane of polarisation.

reflection from ordinary glass and does not necessarily need fancy crystals. Further studies of light polarisation led to better understanding of what light is and to Louis Pasteur's discovery that nature is 'one-handed' (**1847**).

Scots physicist David Brewster later found that the degree of polarisation depended on the angle at which the light is reflected. Some light is reflected off a window; some goes into the glass and is bent (refracted). The greatest amount of light is polarised when the reflected and refracted rays are at 90 degrees to each other. This is Brewster's Angle.

Malus did not long outlive his discovery; he died aged 37, probably from tuberculosis contracted years before as a soldier in Egypt. He survived long enough to be elevated to the Academy of Sciences, and to receive the Rumford Medal from the Royal Society of London. England and France were at war at the time; science often transcends politics.

Jean Lamarck: The Reality of Evolution

1809

Frenchman Jean Lamarck's place in the history of science hangs mostly on his controversial theory of biological evolution. Animals and plants change over time because capacities they acquire during life, and which help them to survive, are passed on to their offspring. Wading birds have long legs because they have to catch fish. Their ancestors stretched their legs so they could wade in deeper water; those longer legs were inherited by the next generation. Disuse of limbs and organs made them whither, said Lamarck. Snakes have no legs because their ancestors did not need them. The examples could be multiplied many times over.

To give a 'why' to evolution, Lamarck visualised nature in a constant state of 'advancement towards perfection'. Organisms are continually moving up the ladder of life from simple to complex, gaining new abilities through the struggle for survival and passing those on. The hierarchy is crowned by the human race; we have already climbed up all the steps of nature.

Lamarck wrote his ideas up in his 1809 book *Philosophy of Zoology*. As an explanation of many things seen in nature, 'inheritance of acquired characteristics' was ingenious and comprehensive, strongly based on his own observations (**1815**). Yet a further century of observation and experiment has proved him wrong. Evolution as he proposed it does not happen.

Why, then, is he so remembered? Any student of biology 'knows' Lamarck claimed that the giraffe's long neck is the result of generations of stretching to reach higher leaves on trees. In fact, that example was invented by his strong critic English geologist Charles Lyell (**1829**) to ridicule Lamarck's whole scheme. By quoting extensively from Lamarck in his *Principles of Geology*, Lyell made Lamarck's views well known.

Lamarck's ideas gained very little support even in his homeland. For most French scientists, evolution did not happen at all. From the eminent Georges Cuvier (**1812**) down through the ranks, they thought that some species might be wiped out catastrophically and others might then be created by God anew; otherwise, being made perfect, species remained unchanged. Charles Darwin largely rejected Lamarck's explanations, but at least the Frenchman was willing to contemplate biological evolution. As Darwin acknowledged in *The Origin of Species* (**1859**), Lamarck had raised the possibility of 'all changes in the organic, as well as the inorganic, world being the result of law and not of miraculous intervention'. ➤➤**1815**

Mary Anning and the Dinosaurs

The remarkable Mary Anning started young in the still-young science of palaeontology. She was only 12 years old when she made her first significant fossil find. In 1811 the first complete skeleton of what was later

1811

called an ichthyosaur ('fish-lizard') was exposed by erosion from the cliffs near her home in Lyme Regis in the west of England. Her brother had found the head of the beast the previous year, but had thought it was a crocodile.

The early death of their father from tuberculosis drove the young Annings to collect fossils for a living, with ready sales to tourists, other collectors and, increasingly, to scientists. Poverty-racked Anning found a patron in a wealthy local collector who sold his own collection to keep her going. She made more important discoveries before her own early death at 47 from breast cancer. These included the first known plesiosaur (another swimming reptile) in 1821 and later the first pterosaur (a reptile which could fly), found outside Germany.

Fossil creatures similar to those Anning had found were now being dug up all over Europe, the remains of large, bizarre creatures unlike anything then living, though perhaps related. It became respectable to believe in extinction, that many creatures once plentiful were now nowhere to be found. Another English collector, William Buckland, found the bones of a creature he called Megalosaurus or 'giant lizard'. Other fossickers found bones and fragments of creatures with powerful legs, bony plated bodies, and tails massive enough to be weapons, but also of smaller, more delicate animals the size of large birds.

In 1841 English anatomist Richard Owen proposed a collective name for these creatures: dinosaurs, meaning 'terrible lizards'. They represented, he thought, a long-vanished family of reptiles—cold-blooded, egg-laying creatures, some able to swim or fly as well as run. Their teeth showed that some ate vegetation, others were carnivores. Some were truly massive. One found in 1849 had an upper arm a metre in circumference.

The English celebrity painter John Martin provided illustrations for popular books about the beasts, showing them (not really accurately) looking like dragons. Soon after, Richard Owen supervised the reconstruction of scenes of 'prehistoric life' in London's Crystal Palace Park. Public fascination with dinosaurs was aroused—and it has never faded.

Charles Bell: The Architecture of the Brain

1811

Scots-born physician Charles Bell could have made a career as an artist. His many drawings and watercolours of dissections were a vital record before the invention of photography. They were highly prized and skilfully executed, especially the famous illustrations of soldiers wounded at the Battle of Waterloo, where Bell served as an army doctor. According to one report, he operated on the wounded 'until his clothes were stiff with blood and his arms powerless with the exertion of using the knife'.

Bell's name lives on today in terms like Bell's Palsy, the degeneration of a facial nerve that causes one side of the face to drop. This underlines his life's work, which was to better understand how the nervous system operates, and how nerves connect identified centres in the brain to sense organs and muscles. In this he was very influential. His 1811 book, *An Idea of a New Anatomy of the Brain*, was called by some the 'Magna Carta of neurology'.

Bell distinguished the various areas of the brain and their functions, especially the cerebellum—the most primitive region and closest to the spinal cord, which looks after basic functions like breathing and heart-beat—and the cerebrum at the top and front of the brain, the centre of sensation and voluntary activity. He also noted that the spinal cord inside the backbone contains two sets of nerves, one connecting to the back of the brain (the 'dorsal nerves'), the other to the front of the brain (the 'ventral nerves').

With others, he discovered that the dorsal nerves are 'sensory nerves', carrying impressions to the brain from the eyes, ears, skin and so on, while the ventral nerves are 'motor nerves', transmitting messages from the brain to the muscles and organs, so causing movement and other changes. This established the basic division of labour within the nervous system, and provided a platform for more detailed exploration through to the present day.

Genial and kindly, and something of a dandy in his attire, Bell was much sought after as a teacher of anatomy. He took over the anatomy school established by the great John Hunter (**1775 Percival Pott**), which later became part of the new University of London. Bell was appointed a professor, but he did not care for the working conditions, and retired to private practice. He later quit London altogether ('a good place to live in but not to die in') and returned to Scotland and the fly-fishing of which he was so fond.

Amadeo Avogadro and His Very Big Number

1811

The year **1808** had been a major one for the maturing of chemistry as a science, where measurement mattered, not just observation. The French chemist Joseph Gay-Lussac had expounded his Law of Combining Gas

Volumes. This stated that the ratios between volumes of gases combining together to make a new compound were whole numbers or simple fractions. Three litres of hydrogen would react with 1 litre of nitrogen to make 2 litres of ammonia. Exactly.

In the same year, the English schoolmaster John Dalton had published his 'atomic theory', intended to explain why the proportions of the various elements in a compound are always the same, no matter how the compound is formed. The answer, he said, was that ammonia came in particles and each of these contained only small numbers of smaller particles or 'atoms' representing the nitrogen and hydrogen.

All the numbers being so simple, some more profound connection had to exist. In 1811 the Italian chemist Amadeo Avogadro set down his 'hypothesis' (as it is still commonly called). Paraphrasing a little, we can state it thus: take, say, a litre each of any two gases, no matter what, and if the pressure and temperature are the same, each litre will contain exactly the same number of particles. This leads to the definition of Avogadro's Constant.

Avogadro's Hypothesis was a startling suggestion, quite out of line with the thinking of the time. Luminaries like Dalton, Gay-Lussac and Berzelius never accepted it. For one thing, it required the particles of nitrogen and hydrogen (for whom Avogadro invented the term 'molecules') to each contain two atoms. This seemed impossible; surely two atoms of nitrogen, being alike, would repel each other and never bind into a molecule?

Scorned and ignored, Avogadro lingered in obscurity for 50 years. When both he and his rivals were dead, a colleague revived his hypothesis and it was soon the conventional wisdom. Freed from fixed ideas, chemists could see that there was no better way to explain the accumulation of experimental observations. Equal volumes of all gases under the same conditions really do contain equal numbers of particles.

And just how many particles is that? The numbers were not known with any certainty for a century after Avogadro but they are stupendous, because molecules are exceedingly small. For example, under 'standard conditions', 2 grams of hydrogen or 28 grams of nitrogen (a 'mole' of each) will fill about 22 litres. In that volume, we will find an 'Avogadro number' of molecules, some six hundred thousand million million million of them. Really.

Georges Cuvier: Extinctions Really Happened

1812

The prodigious career of Frenchman Georges Cuvier earned him the title 'father of palaeontology'. Certainly he did more than anyone else to systematise the study of fossil bones, and to visualise the creatures that had left them. So complete was his understanding of 'functional anatomy', the way in which the form of any part of an animal is related to its function, that he could reconstruct creatures with striking accuracy from only a few fragments. His book on the fossil bones of quadrupeds, first published in 1812, was a classic, but fossils of all kinds attracted his attention.

Cuvier's most important contribution was probably legitimising the concept of extinction. Many others before him had argued that fossils were the remains of creatures that had vanished from the Earth. The idea was controversial; having made all creatures, some critics argued, why would God allow some to die out?

Mammoths were a good example. Bones of these immense, elephant-like creatures had been found in many places. In Italy they were thought to be the remains of the beasts used by Hannibal in his invasion 2000 years before. The US President Thomas Jefferson, a corresponding member of the Lunar Society (**1765**), thought mammoths might still be living in the wilds of North America. Cuvier's response was to carefully compare the bones of today's elephants and fossilised mammoths. They proved sufficiently different in many details that they had to be put into different species. So the mammoths had gone, but the elephants remained.

And why had these creatures vanished? Cuvier though that the Earth had gone through periodic upheavals or 'revolutions' that had wiped out whole families of living things, profoundly changing the pattern of life on Earth. He did not use the term 'catastrophe' because others linked it with events like the Biblical Flood, which was apparently an 'act of God'. Cuvier's revolutions had natural causes. In promoting his view of things, Cuvier had to battle the 'uniformitarians' such as Charles Lyell (**1829**), just as his predecessors, like Jean Deluc (**1778**), had had to contend with James Hutton. Lyell and friends won the debate then, but the possibility of sudden violent change has been revived recently, with talk of strikes by asteroids (**1980 Alvarez**) and immense volcanic eruptions as major influences on the pattern of life.

A man who could remember, almost word for word, everything in the 19 000 books in his library, as Georges Cuvier was reported to be able to do, must have had a remarkable mind. The Frenchman's talents and achievements went far beyond scientific research. He had an almost full-time job as Secretary of the National Institute, which had been set up following the revolution to replace the Royal Academy. Under Napoleon, he was very active in spreading higher education, especially in the territories that had been annexed by France. He died, loaded with the highest honours of the French state, at the age of 63.

Michael Faraday:
Life Begins at the Royal Institution

Humphry Davy, first director of London's Royal Institution, was asked to name his greatest discovery. He reputedly said 'Michael Faraday'. If he did say this, he may well have been right. Talented as he was, Davy was surpassed in most measures by his protégé. The apprentice bookbinder had an ever-inquiring mind and read every scientific book that came into the shop. In the well-known story, Faraday was given tickets to attend lectures by Davy. He later wrote up his notes, added illustrations and forwarded them with a covering letter asking for a job. Davy was impressed and took him on as assistant and secretary in 1812. Soon after, the pair set out on a grand tour of Europe, together with Davy's snobbish wife who expected the low-born Faraday to eat with the servants.

1813

Enriched by the experience, Faraday returned to work at the Royal Institution. By 1821, married and 20 years old, he was Superintendent of the Laboratory. He succeeded Davy when the latter retired in 1825. During these years, he liquefied the gas chlorine for the first time and discovered the important hydrocarbon benzene by distilling whale oil.

Faraday's greatest achievements were in electricity and magnetism from **1831** on, following an earlier stumble (**1820 Orsted**). Under his direction, the Royal Institution did much to promote public awareness of science. His Christmas lectures for children, first given in 1826, continue to this day. Faraday's own lectures on 'The Natural History of a Candle', given late in life, were classics.

Faraday was never happy with the 'action at a distance' concept that Isaac Newton had proposed for gravity (**1687**). Like René Descartes (**1644**), he could not comprehend forces acting through 'empty space'. Descartes had filled space with 'subtle matter' to carry the forces mechanically. Faraday filled it with 'fields', zones of influence where a mass, a charge or a magnetic pole feels a force. Faraday argued persuasively for the reality of these fields, embodied in 'lines of force'. Theoreticians like James Clerk Maxwell (**1871**) put them into mathematical form (Faraday was never much good at that).

Two last legacies: Faraday gave this advice to the young physicist William Crookes: 'Work, finish, publish'. To the rest of us he said: 'Nothing is too wonderful to be true'. ➜**1831**

Jöns Berzelius: The Language of Chemistry

Englishman John Dalton (**1808**) is usually credited with originating and promoting the 'atomic theory' to explain chemical reactions, but he had a lot of help. Major support came from Swedish chemist Jöns Berzelius. Initially a physician and surgeon, Berzelius turned to chemistry when he had to

1814

> *I have seldom experienced a moment of such pure*
> *and deep happiness as when the glowing stick thrust into it*
> *lighted up and illuminated with unaccustomed brilliancy*
> *my windowless laboratory.*
>
> JÖNS BERZELIUS (WHEN HE FIRST COLLECTED OXYGEN IN A LABORATORY EXERCISE)

teach it. A long and very influential career followed.

While writing a textbook for his students, Berzelius rediscovered the Law of Constant Proportions (aka Proust's Law; **1804**). This led him to construct tables of 'atomic weights' (the relative weights of atoms of various elements), more carefully than Dalton had done. Berzelius found that most atomic weights were not quite whole numbers. This was a blow for a new idea, originated by English chemist William Prout, that the atoms of all elements are constructed from atoms of hydrogen, the lightest and simplest element.

The idea had a lot of appeal. It reflected a common desire, as old as the ancient Greeks, to reduce all matter to one fundamental substance, so increasing order and symmetry in the natural world. Unfortunately it did not fit the facts. As the careful Berzelius had found, an atom of oxygen weighs not 16 times as much as a hydrogen atom but 15.9 times. An oxygen atom made up of 16 hydrogen atoms was feasible; one made of 15.9 hydrogen atoms was not. Ironically, once we knew the true structure of atoms (**1919**), Prout's conjecture proved not so far from the truth.

Berzelius also provided a new language for chemistry, a way to represent elements and compounds in the chemical 'equations' that described how they reacted. Antoine Lavoisier and Claude Bertollet had shown how to do this in words (**1787**); now Berzelius devised symbols. These quickly replaced a clumsier system devised by Dalton, much to Dalton's annoyance, and remain in use today with little change.

Berzelius chose one or two letters, usually the initials of the name of the element in English or Latin, to represent one atom of the element; H for hydrogen, O for oxygen, N for nitrogen, C for carbon, S for sulphur, Fe for iron (ferrum), Au for gold (aurum) and so on. The elements in a compound were shown by letters and numbers: H_2O indicates that a molecule of water contains two atoms of hydrogen and one of oxygen; H_2SO_4 indicates that two atoms of hydrogen, one of sulphur and four of oxygen make up a molecule of sulphuric acid.

These chemical words can be formed into sentences that concisely state the participants in and outcomes of chemical reactions. For example, $FeS + H_2SO_4 = FeSO_4 + H_2S$ is a precise and compact way of saying that iron sulphide reacts with sulphuric acid to make iron sulphate and hydrogen sulphide, aka 'rotten egg gas'.
➤➤**1848**

Joseph Fraunhofer:
The Messages Hidden in Light

Spectroscopy means roughly 'seeing the colours (of light)', and it is a powerful tool for science today. Its origins lie, like so much else, with Isaac Newton. In **1672** he had reported that passing sunlight through a glass prism broke it up into a continuous spectrum of colours—the colours of the rainbow from red to violet.

1814

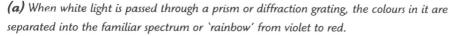

(a)

Violet   Blue   Green Yellow   Orange   Red

Wavelength 400 nm   700 nm

(b)

Violet   Blue   Green Yellow   Orange   Red

δ   γ   β   α

(c)

Violet Violet   Blue–green   Red

δ   γ   β   α

(d)

Violet Blue   Green   Dark red

δ   γ   β   α

(a) When white light is passed through a prism or diffraction grating, the colours in it are separated into the familiar spectrum or 'rainbow' from violet to red.

(b) If the white light is first passed through hydrogen gas, not all the colours emerge. Narrow black bands or 'lines' appear against the rest of the spectrum, as if those colours had been removed. This is an 'absorption spectrum'. Here four such lines are visible, labelled δ, γ, β and α.

(c) If hydrogen is heated until it glows in a dark room, most of its 'emission spectrum' is black. But four bright lines appear—one red, one blue–green and two violet—in the same positions as the dark lines in the absorption spectrum. This pattern of lines is characteristic of hydrogen and can be used to identify it. In **1913** Niels Bohr explained how these lines are generated.

(d) This spectrum shows the effect of 'red shift' *(1912)*. If the source of the hydrogen light (say, a star) is moving away, the distinctive lines are shifted towards the red end of the spectrum. α becomes a darker red, β becomes green rather than blue–green, γ shifts into the blue. The amount of red shift is a measure of how fast the object is moving away.

Nearly 100 years later, in 1751, the Scot Thomas Melville used a prism again, but this time to look at the light from a flame into which some crystals of salt had been sprinkled. He saw colours but not the continuous spread. Against a black background just some bright sharp lines of pure colour appeared; he realised that the colours on show depended on which salt was in the flame.

Around 1802 Englishman William Wollaston went back to Newton's work but with much better equipment. He saw the continuous spread of colours in the spectrum of sunlight, but something more as well. The background of colours was crossed by dozens of thin black lines, which seemed to be always in the same places in the spectrum, as if some very precise colours were missing. Wollaston did not do much with his findings, and did not see any connection with what Melville had found, even if he knew of it.

Around 1814 the methodical German Joseph Fraunhofer got busy. Fraunhofer made optical equipment, and initially broke sunlight into colours just to prove how good his prisms were. He carefully charted the black lines in the solar spectrum, increasing to over 500 the number of what were soon being called 'Fraunhofer lines'.

In about 1821 he invented the diffraction grating to speed up the work. This gadget was the offspring of the 'double slit' experiment Thomas Young had reported in **1801**. Young had shown that passing light through two slits caused it to break up into colours. Fraunhofer multiplied the effect by passing the light through hundreds of parallel lines scratched on a piece of glass, making what amounted to hundreds of slits. He also found that the closer the scratches were, the more the colours were spread out and the more precisely he could measure them.

The bright lines Melville had seen could be studied with gratings too. Soon the 'emission spectra' of dozens of hot gases and vaporised solids were known, and researchers were following up Englishman John Herschel's suggestion around 1823 that the spectra could be used as 'fingerprints' to identify the elements present in a sample. As an astronomer, Herschel would have been delighted with where that led (**1868 Lockyer and Janssen**).

Jean Lamarck: The Other Side of the Ledger

1815

Frenchman Jean Lamarck has the reputation of 'the man who got it wrong', at least as far as evolution was concerned (**1809**), whereas Charles Darwin 'got it right' (**1859**). But as with most scientists, even his best known ideas were only part of the man. One enduring legacy is the term 'biology' as an umbrella term for all the separate studies of various living things (botany, zoology, anatomy and so on). Lamarck did not invent the word, but he made it his own and the scientific world did too.

Though he came late to science after time in a seminary, the army, banking and medicine, Lamarck was certainly diligent and painstaking in his research in botany and other subjects. Though his reputation faded later, he secured a place in the French Academy of Science when aged only 34 under the patronage of the eminent Compte de Buffon (**1749**), who had helped him publish his first major book, *The Flowers of France*.

In 1793 Lamarck played a major role in the reorganisation of the Royal Botanical Gardens (*Jardin de Plantes*) into the Museum of Natural History. He was placed in charge of the collection of insects and worms, a less than prestigious position. The great ones, such as Carl Linnaeus (**1753**), had thought them not worthy of study. Knowing almost nothing to begin with, Lamarck became the expert on 'invertebrates' (animals without backbones), a term he coined. He reorganised the huge but chaotic collections of specimens and established systems to classify them that are still influential today.

In his big book on invertebrates published in 1815, Lamarck separated the major groups, distinguishing the six-legged insects from the eight-legged arachnids (such as spiders), the crustaceans (such as crabs and lobsters) and the annelids (the ringed or segmented worms). He quickly saw that many invertebrates differ from one another by only tiny degrees and this drove him to develop his theory of 'inheritance of acquired characteristics' to explain why it was so.

Lamarck came from the lesser nobility (he carried the title 'Chevalier') but always struggled for recognition and financial security. He ended his life in poverty, obscurity and blindness aged 65. Married at least three times, he was cared for by two devoted daughters. Placed in a rented grave for five years, his remains are now lost.

Pierre Pelletier and Joseph Coventou: Alkaloids Galore

Herbal remedies are nothing new. Every culture has used extracts from various plants to try to prevent, treat or even cure disease. Some worked better than others, perhaps because whatever was in them was more powerful or better suited to the situation. Better use of such remedies depended on knowing more about their 'active' ingredients (**1785 Withering**).

Two French pharmacists working out of the back of a Paris shop tackled this problem in a methodical way. Pierre Pelletier and Joseph Coventou suspected that the active ingredients in many medically useful plants were from a category of chemicals called 'alkaloids'. These were often bitter tasting, soluble in water and slightly alkaline, so that a little dilute acid would convert them into salts that could be crystallised and purified. They took their cue from an older fellow pharmacist, Friedrich Serturner, who lived just over the border on territory the French had conquered under Napoleon. Around 1805, Serturner had found what gave opium its long-known sleep-inducing properties. It was an alkaloid he called 'morphine' (from Morpheus, the ancient god of sleep).

Following the same path, Pelletier and Covenou extracted a string of alkaloids from plants. These included caffeine, emetine from the nausea-inducing plant ipecauanha, the poison strychnine and, most importantly, quinine, the anti-malarial found in the bark of the South American cinchona tree. This was so valuable in malaria-ravaged areas that it was called the 'Jesus bark'. Finding and purifying these chemicals meant they could be administered more precisely and efficiently, and ultimately produced synthetically, though that was still a long time off.

By carefully burning the alkaloids and collecting all the products, the pair discovered that they contained nitrogen, and were therefore somehow related to proteins. Nowadays some 3000 different alkaloids are known. They include pain-killers like codeine; medically useful drugs such as atropine, used to dilate the pupil in the eye; and mind-altering chemicals like heroin and LSD, the latter originally extracted from a fungus on grain.

The pair used similar methods to find some vegetable-based chemicals that were not alkaloids. The most significant of these was chlorophyll, the green pigment that seemed to be involved in plants absorbing light and releasing oxygen (**1779**). They gave it that name, meaning 'green leaf' in Greek.

Augustin Fresnel: Light in the Darkness

1819   In 1819 the French Academy of Sciences offered a prize for the best explanation of the diffraction of light; this odd behaviour apparently sees light bend around corners, rather than travel only in straight lines. This had first been seen long before (**1665**) and explained by Christiaan Huygens (**1690**) and others, by assuming that light is a form of wave motion. Now it was time to pin down the details by the use of mathematics, to predict just how bright or dark it would be at various points behind some opaque object

Among the entrants for the prize was 31-year-old Augustin Fresnel. He had been a government engineer and more recently had studied the various properties of light, such as polarisation, as described by his countryman Etienne Malus (**1808**). His analysis produced something quite unexpected. If a circular disc cast a shadow, the central point of that shadow, where you would expect the darkness to be deepest, would be marked by a bright spot. This seemed absurd, as crazy as Thomas Young's suggestion (**1801**) that two bright rays of light could fall on the same spot and produce darkness.

The claim was soon put to the test. Fresnel was proved right and won the prize. He also won some support for the wave theory of light, which most French scientists rejected in favour of the 'corpuscular theory' that saw light as streams of particles. He was later elected to the French Academy and awarded the Rumford Medal of the Royal Society of London at the age of 39, on his death-bed.

Fresnel has a particular place in the history of lighthouses. He served as secretary of the French Lighthouse Commission until failing health forced his resignation. However, his 'Fresnel lens' became widely used in lighthouses in France and the United States. A Fresnel lens is a beautiful object, as effective as a regular lens in sending into the life-saving beam every scrap of light from the weak whale-oil lamps of the time. But it is cheaper to make and much lighter for hoisting to the top of a lighthouse. Essentially an array of prisms to bend and direct the light, the lens looks as if rings of glass have been cut away. Some handheld magnifying glasses today operate on the same principle.

Friedrich Mohs: How Hard Is that Mineral?

Geologists out in the field trying to identify which particular mineral they were looking at did not have much to go on, mostly just general appearance, such as colour and lustre (shininess). Following the work of René-Just Hauy (**1784**), geologists could use cleavage, the shape of pieces a mineral breaks into when struck with a hammer. But chemical analysis to see what elements were present, say with the use of the blowpipe (**1777**) had to wait until they got back to the laboratory.

1820

Around 1820 the German geologist Friedrich Mohs provided another tool—the scale of 'hardness' that still bears his name. The beauty of this scale was that it could easily be used to identify minerals in the field.

The Mohs scale rates minerals from one to 10 on their hardness. A mineral with a hardness of five will scratch anything rated four or less but is itself scratched by something rated six or more. Level one hardness is typified by the mineral talc, so soft it is easily crushed into 'talcum powder'. Gypsum or 'Plaster of Paris' is the archetypal hardness two. Calcite, a crystal of limestone, is typical of hardness three. Feldspar, found as pink or white crystals in granite, characterises hardness six; quartz (silica), the main ingredient in sand and sandstone, has hardness seven.

Near the top end is corundum (aluminium oxide), at nine. Fragments of this, which is chemically the same as ruby and sapphire, are stuck to emery paper for shaping metal. Hardest of all, ranked at 10, is diamond. Modern tests show that diamond is really four times harder than corundum, 15 times harder than quartz and 1500 times harder than talc.

Hans Orsted: Magnetism Meets Electricity

To the Dane Hans Orsted is commonly ascribed the fame of first showing a clear connection between electricity and magnetism, both long studied (**1600**) but not thought closely linked. Some people had claimed that an electric discharge, such as a bolt of lightning, could affect a compass needle, or even make a piece of iron magnetic, but no one seems to have studied the matter methodically.

1820

Orsted's observation was simplicity itself. Bring a compass needle near a wire carrying an electric current, and the needle moves, as if the wire is now acting like a magnet. Continuous electric currents had been available for 20 years, ever since Alessandro Volta and his battery (**1796**), so it is odd that no one had seen this effect before. It may be that Italian lawyer and amateur physicist Domenico Romagnosi had found something like it back in 1802. The finding was merely written up in his local newspaper and so was easily overlooked; Orsted was a prominent professor, and people took notice.

When the current flowed, the shift in the compass needle indicated that the new magnetic field ran around the wire in a circle. This suggested that the effect could be used to push a current-carrying wire round and round a magnet, as a sort of electric motor. Michael Faraday at London's Royal Institution tried this out in 1821. His motor was not very practical as it needed the wire to drag through a bowl of mercury, but it worked. Unfortunately, Faraday published this work without due acknowledgement to colleagues. The resulting fuss caused him to bypass work on electricity and magnetism for 10 years (**1831**).

Much more useful was the 'electromagnet', made by wrapping many turns of wire into a coil, which greatly multiplied the strength of the field and gave it a north and a south pole like a bar magnet. A compass needle hanging in the middle of such a coil would shift one way or the other in response to even very small currents flowing in the coil. This was the genesis of the 'galvanometer', named after Luigi Galvani (**1786**). In the soon-to-be-invented 'electric telegraph', the flicks of the needle would represent pulses of current grouped according to a code, allowing messages to be sent over hundreds of kilometres.

The French physicist André-Marie Ampère went one stage further. If a current-carrying wire could shift a magnet, would it shift another current-carrying wire, which was like another magnet? Within a week of hearing about Orsted's discovery, he had his answer. Depending on the way the currents were flowing, two current-carrying wires attracted or repelled each other, just like two bar magnets. This observation brought the promise of practical electric motors, and was also the origin of Ampère's continuing familiarity. His name, usually shortened to 'amp', has been given to the unit of electric current.

The Return of Encke's Comet

1822

The best known of all comets is of course Halley's Comet, so called because the English astronomer Edmond Halley predicted in **1705** that it would return in 1759 (which it did). Not for more than a century was another 'periodic comet' identified. Other comets were seen, sometimes spectacular ones, but none seemed to come by twice. Methodical German astronomer Johann Encke combed through years of data to show that appearances by comets in 1786, 1795, 1806 and

1818 were the same comet. He predicted a return in 1822; he had his reward. Comet Encke comes by every three and a bit years. It is not as spectacular as Halley's Comet, which is the only bright comet anyone is likely to see twice in their lifetime.

Later in the century, the Italian astronomer Giovanni Schiaparelli (whose name gained further prominence in **1877** during the close approach of the planet Mars) realised that every passing comet leaves its calling cards, small fragments of rock and dust boiled off by the heat of the Sun. Once or twice a year, as the Earth passes through this trail of debris, some is swept up and pulled earthwards, where it burns up in the atmosphere as a shower of meteors or 'shooting stars'. So each meteor shower, lasting a few days, is a reminder of a particular comet.

Occasionally the showers will contain larger lumps that survive the fiery passage and arrive on the ground as meteorites. Many people found it hard to accept that meteorites could have an origin far beyond the Earth and its atmosphere, especially when hundreds fell at once, as in one spot in France in 1803.

Even more startling was the discovery by the Swedish chemist Jöns Berzelius in 1834 that at least one meteorite contained chemicals that are very similar to those found in living things on Earth. Could there be life elsewhere in the cosmos? Imaginative minds, such as English physicist Lord Kelvin (**1846**), pondered the possibility that life might have been brought to Earth on 'some moss-covered stone broken away from mountains on another world', as he told the Christian Evidence Society in 1889. This vision of 'panspermia' has never lost its appeal.

Sadi Carnot: The Motive Power of Fire

Sadi Carnot had a great pedigree in both engineering and politics. His father Lazare had been Napoleon's Minister for War from 1799 to 1897; he resigned to devote himself to the education of his two sons. Sadi Carnot was

1824

barely out of school before he fought under Napoleon to defend Vincennes (outside Paris) against invaders. His father was exiled following the final fall of Napoleon and Sadi Carnot became an army engineer. Finding the work unsatisfying, he enlisted in the reserves on half pay and devoted his spare time to study and research.

Carnot was a practical man, interested in steam engines—the workhorses of the Industrial Revolution in France as elsewhere. He wanted to learn how to get more useful work from the fuel they consumed. In 1824 this culminated in his book *Reflections on the Motive Power of Fire*. He could see that steam engines were getting steadily more efficient, that more useful work (for example, lifting coal from a mine) was now available for the same amount of fuel burnt. Was there any limit to this advancement? Would we one day extract more work than the energy put in, so achieving perpetual motion?

Carnot started with the concept of heat as a weightless substance called caloric. Although the concept of caloric had been debunked by Count Rumford in **1798**, some scientists had not let it go. Caloric could not be created or destroyed, just transferred from one place to another. In Carnot's 'ideal' steam engine, caloric flowed from a hot place (the boiler) through the cylinder where caloric was given up and useful work done, and on to the cold condenser. The steam going into the condenser was still hot, so it had some remaining caloric that had not done any work. Wastage of energy was therefore inevitable, and we would never get perpetual motion. Making the difference between the starting and finishing temperatures as great as possible would maximise efficiency.

Carnot thought his engine would be 'reversible'; energy could be put in at the lower temperature to reach the higher temperature. This should require exactly the same amount of work as the engine had done when heat was running 'downhill'. These ideas would later take shape as the 'laws of thermodynamics', the science of work and heat. But Carnot died of cholera less than a decade after his book appeared and before he could make further progress. His work was overlooked until **1850**.

Volts, Amps and Ohms

1826

In the new world of electricity, the names of the leading practitioners were soon part of the language. Frenchman André-Marie Ampère (**1820 Orsted**) was immortalised in the unit of electric current (commonly shortened to 'amp'). By definition, a current of one ampere moves one coulomb of charge every second, so also honouring Ampère's eighteenth-century countryman Charles Coulomb, who had found the law of forces between electric charges (**1785**).

We can compare electricity in a wire with water in a pipe. Electric current (as coulombs per second) corresponds to so many litres per second of water. Two things control how much water will flow. One is pressure. As experience shows, more

---

THE LAWS OF ELECTRICAL CURRENTS

Ohm's Law: If V is voltage (potential difference) in volts across a conductor, I is the current in amps through the conductor, and R the resistance of the conductor in ohms, then V = IR.

Joule's Law: If P is the rate of energy loss (through heat) in the conductor in watts (joules per second), then P = IV. It follows that P = I^2R.

So a 2400-watt heater on a 240-volt circuit draws 10 amps and has a resistance of 24 ohms.

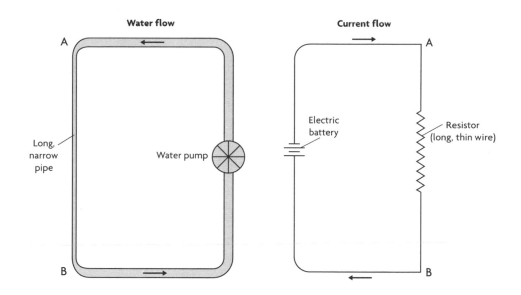

Water flow	Current flow

In the circuit on the left, the water pump creates a difference in pressure between A and B and pushes water through the pipe. On the right, an electric battery does the same; the pressure difference or 'voltage' between A and B drives electric current through the wire.

The rate of water flow (in litres per second) and electrical current flow (in coulombs per second or amperes) depends both on the pressure and on the 'resistance' of the pipe or wire. Long, narrow pipes and long, thin wires have more resistance than short, wide ones.

In each case, flow equals pressure divided by resistance. This is Ohm's Law.

pressure means more water flow (unless the pipe bursts). Electrical pressure is now commonly called 'voltage', though we still hear older words like 'potential difference' or 'tension' (especially 'high tension' or HT). The relevant unit is the 'volt', which recalls Alexandro Volta, who made the first continuous currents with his 'battery' in **1796**.

The other thing affecting water flow is the resistance of the pipe. Generally speaking, the longer and narrower the pipe the less water will get through in a given time, because the pipe resists the flow more. Wires carrying electric current behave the same way. Electrical resistance is measured in ohms. This brings in the name of the Dutchman Georg Ohm. He was a careful and methodical experimenter and in 1826 brought the various units together in his renowned Ohm's Law, based on his observations.

Since more electrical pressure increases the current, and more resistance reduces it, Ohm could write that the current (symbol I) in amperes was equal to the voltage (symbol V) in volts divided by the resistance (symbol R) in ohms. This also defines the ohm: it takes one volt to push one amp through a resistance of one ohm. Long, thin wires have 'more ohms' than short fat ones.

While we use 'v' and 'a' as shorthand for volts and amps, the symbol chosen to represent ohms was the Greek letter omega Ω, presumably because it sounded like the name Ohm. Was this a little electrical joke? Certainly all this complexity was much more than the scientists being pedantic. Precision in measurements and calculations would be vital if electricity was to be used effectively and efficiently, as it later was in so many ways, transforming our lives.

The First Talk about the 'Global Greenhouse'

1827

'Global warming' and the 'greenhouse effect' are hot topics in the twenty-first century, but the debate began nearly 200 years ago. Perhaps the first to express an opinion was the Frenchman Joseph Fourier, who escaped appointments with the guillotine no less than four times during the French Revolution. He did a lot of work on heat, and knew that the surface of the Earth was warmer than it should be, given the amount of energy reaching our planet from the Sun. In 1827 he suggested that perhaps the Earth's atmosphere traps in heat, a bit like the glass of a greenhouse, keeping the Earth warm enough for comfort and indeed for life. So the phrase 'greenhouse effect' first entered the language.

Molecules can soak up heat in various ways; by moving faster, by rotating faster and by vibrating more. The more atoms there are in a molecule, the more ways it can move and so absorb heat. Our air is mostly a mixture of oxygen and nitrogen, whose molecules each have only two atoms; they are almost transparent to heat radiation and let it pass from the Earth's surface back into space. So the likely culprits in the 'greenhouse effect' were water vapour, carbon dioxide and the various oxides of nitrogen, each of which has three atoms to each molecule. These 'greenhouse gases' (as we would call them today) are almost opaque to heat and trap it as it tries to escape, so warming the lower atmosphere.

After Fourier, the debate (among the few who were interested) was over which gas in the atmosphere had the most capacity to trap heat. Irish scientist John Tyndall had done the most study of the heat-trapping abilities of various gases; around 1863 he argued for water vapour. Without its impact, he said, 'the Earth would be in the iron grip of frost'.

The other likely candidate was carbon dioxide. Later, it too had its champion (**1895**).

Robert Brown: A Botanist Finds Atoms

1827

When navigator Matthew Flinders set out in 1801 in his ship *Investigator* to chart the coastline of the continent to which he gave the name Australia, his scientific party included the Scots-born biologist Robert Brown. Brown stayed on in Australia to collect specimens and when he returned to Britain became assistant to Joseph Banks (**1778**) at the Royal Botanical Gardens at Kew. When Banks died in 1820, his great 'herbarium' of pressed plants was donated to the British Museum; Brown became the first Keeper of the Botanical Collection. He was also the first to glimpse the tiny central part of a plant cell that he called the 'nucleus', but that is another story (**1838**).

All this is impressive enough, but it is not the reason Brown's name lives on today. In 1827 Brown was peering through his microscope at some grains of pollen in water. Instead of floating motionless in the water as might be expected, the grains were jiggling back and forth, very slightly but very quickly. Thinking the pollen grains might be agitated because they were 'alive', Brown substituted particles of dust. They also had the shakes. This odd behaviour is still known as 'Brownian motion' and it remained a puzzle for 70 years.

The answer is usually credited to Albert Einstein in **1905**, but others had guessed it and even done the mathematics. They realised that Brown the botanist had provided what the physicists had been seeking for many decades, some real evidence of the existence of atoms and molecules. The 'atomic theory' had of course been around for more than 200 years (**1624**), and had been extremely useful in explaining all sorts of things in physics (**1659**) and chemistry (**1808**). But as with Galileo in the motion of the Earth (**1633**), there was no real evidence that atoms existed. Now here it was. The best explanation for the wobbles in the bits of pollen and dust was that they were being buffeted by the constant impact of minute particles (molecules) of water in motion, countless tiny prods that in total cancelled each other but could be glimpsed as they occurred.

Perhaps we were still in the realm of hypothesis; the particles were jiggling 'as if' they were being battered by atoms. Perhaps there was another explanation, but atoms looked the best bet; here was something you could see, not merely measure.

Friedrich Wohler: The Living and the Non-living

Over the years, the border between the living and non-living, once thought so sharp, became increasingly vague. Experiments had shown, for example, that both a burning candle and a breathing animal consumed oxygen and generated carbon dioxide. Combustion and respiration seemed virtually the same thing. The electricity in lightning seemed basically the same as that released by an electric eel or as shown by Luigi Galvani (**1786**) to make muscles move. Yet some researchers continued to claim that, whatever the physics and chemistry might say, living things still have something special, some 'life force' that was beyond identification and study.

1828

A sensational discovery by Friedrich Wohler struck a big blow against the 'vitalists'. An ambitious young German in search of the best chemical education of his time, Wohler went to Sweden to work with the mighty Jöns Berzelius (**1814**) and remained in close touch with him during nearly 50 years teaching chemistry at

Gottingen back in Germany. In 1828 Wohler was trying to make the chemical compound ammonium cyanate. He mixed silver cyanate with ammonium chloride and ended up with a dish of clear crystals. However, these did not behave like ammonium cyanate; for one thing they did not fizz and release carbon dioxide when some acid was added. In fact, they looked and behaved like crystals of urea, a chemical made by animals, and discarded by them as a waste in their urine, as Herman Boerhaave had found 100 years before (**1724**).

Going a step further, Wohler analysed his compound and found it had the same composition as urea; each molecule had one atom of carbon, one of oxygen, two of nitrogen and four of hydrogen. But so did ammonium cyanate. The two compounds were the same in composition, though very different in behaviour. One came from living things, the other from inanimate chemicals. There seemed to be no place for any 'vital spirit'.

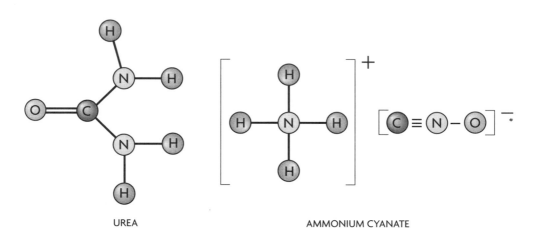

UREA   AMMONIUM CYANATE

Molecules of the two compounds urea and ammonium cyanate contain the same atoms (four of hydrogen, two of nitrogen, one of carbon and one of oxygen), but they are arranged and interconnected differently.

*In urea, all the atoms are connected by 'covalent' bonds (**1928**). Ammonium cyanate is in two parts ('ions') held together by electrical attraction but able to be separated if dissolved or melted. The different organisation gives the two chemicals very different properties.*

Wohler could not contain his excitement. He wrote to his mentor Berzelius. 'I can no longer, so to speak, hold my chemical water. I must tell you I have made urea without needing a kidney, either of man or dog. Ammonium cyanate is urea!' Berzelius had a leaning towards vitalism and may have found his former student's enthusiasm irritating. If he did, he did not show it, perhaps because he already knew the answer to the key question. How can two apparently different compounds have the same composition? The answer was 'structure' (**1848 Berzelius**).

Johann Dobereiner: Ordering the Elements

As the nineteenth century progressed, the number of known chemical elements rose alarmingly. Long gone was the comforting simplicity of the four elements proposed by the ancient Greeks. In fact, even they had

1829

known of nine elements that fitted the definition advanced by 'sceptical chymist' Robert Boyle in **1661**: the seven metals, gold, silver, iron, tin, lead, copper and mercury, together with carbon and sulphur. Arsenic, antimony, bismuth and phosphorus were on the list by 1700; in the eighteenth century the gases oxygen, nitrogen, hydrogen and chlorine were added, together with a swag of new metals including cobalt, platinum, zinc, magnesium, nickel, tungsten and uranium.

In the 1800s many more elements were released from minerals. These included metals like sodium and potassium that Humphrey Davy had liberated from compounds using electricity (**1807**). Later on silicon, bromine and iodine, and still more metals, like aluminium, were discovered. By the 1830s, the list contained 55 substances unable to be broken down into anything simpler. How many more were there to find? A hundred? A thousand? An infinite number? Why were there so many? Was there any order or system among them?

German chemist Johann Dobereiner offered the first insights in 1829. He suggested that chlorine, bromine and iodine formed a sort of family. Bromine lay roughly halfway between the other two in properties. Chlorine was a gas at normal temperatures, bromine a liquid and iodine a solid. Chlorine was the most reactive, iodine the least. And in terms of atomic weight (**1000 Dalton**) bromine was almost exactly in the middle between chlorine and iodine.

This was the first of Dobereiner's 'triads'. He found two more with similar neat gradations of properties: calcium, strontium and barium; and sulphur, selenium and tellurium. But that was where it stopped. None of the rest of the 55 seemed to fit into groups of three. Perhaps the triads Dobereiner had found were just a fluke. Chemists on the whole were not impressed, even though most of them longed for some order among the elements. Rather better evidence arrived in **1864**.

Charles Lyell: It All Takes Time

When Charles Darwin sailed as naturalist on the *Beagle* (**1836**) he carried (and treasured) the first volume of *Principles of Geology* by Englishman Charles Lyell. Volumes two and three, published later, chased him by parcel

1829

post around the globe. He often commented how much he owed to Lyell and the theory of 'gradual change' (uniformitarianism) that he propounded. Alfred Wallace, who also developed a theory of evolution (**1858**), acknowledged the same debt.

Lyell became one of Darwin's strongest supporters (**1871**) and was instrumental in arranging the joint meeting in 1858 at which Darwin's and Wallace's theories were both publicly aired for the first time. He had previously opposed the concept of evolution, disparaged Lamarck's ideas (**1809**) and never accepted Darwin's theory of natural selection (**1859**), though he ultimately became convinced that evolution had taken place.

In the tradition established by James Hutton (**1778**), Lyell believed that rocks accumulate over immense periods of time, and form an enduring record of past times and life forms. Using fossils as markers, as William Smith (**1799**) had done, he divided the young group of rocks said to represent the Tertiary Period into three and named them Eocene, Miocene and Pliocene (for 'dawn of recent', 'less of recent', and 'more of recent'). William Wherwell (**1833**) helped him devise these names.

He studied a major group of rocks that differed from both sedimentary rocks (like sandstone or shale, laid down underwater) and igneous rocks (like basalt and granite, which had once been molten). He called his third group 'metamorphic' rocks, meaning 'changed', and argued they were caused by intense heat and pressure, say from molten rock ('magma') close by. In this way, limestone becomes marble, shale becomes slate.

Lyell's *Principles of Geology* was a most influential book, going through 12 editions over 50 years. Its subtitle explains its purpose: 'Being an Attempt to Explain the Former Changes of the Earth's Surface by Reference to Causes Now in Operation'. The 'gradual change' mindset that imbued it became the accepted view; the rival view, catastrophism (**1778**), disappeared from view, at least until recently.

Lyell trained and worked initially in law, but lectures from William Buckland at Oxford changed the course of his life. With estates in Scotland to support him, he was, like Darwin, a 'gentleman scientist'. He was knighted and made a baron, and, as with the greats of English science, buried near Isaac Newton in Westminster Abbey.

Michael Faraday and Joseph Henry: Magnetism Makes Electricity

1831 Prolific English physicist Michael Faraday (**1813**) returned to studying electricity and magnetism in 1831 after a decade of abstinence following the 'electric motor' affair (**1820 Orsted**). His starting point was again the link that Hans Orsted had found between electricity and magnetism: an electric current in a wire is surrounded by a magnetic field. If electricity can produce magnetism, could magnetism produced electricity?

Anyone can do what he did. He wrapped a wire a number of times around one side of an iron ring, and connected the ends to a battery and switch. He wrapped a second coil around the other side of the ring and connected it to an instrument (a 'galvanometer') to report any current that flowed. He hoped to show that a current flowing in one coil would make the current flow in the other, the connection coming via the magnetic field.

Almost nothing happened. When Faraday turned on the current in the first coil, the needle in the galvanometer flicked quickly one way before settling back to zero. When he turned the current off again, the needle flicked the other way and returned to zero. So the magnetic field had to be changing, growing stronger or weaker, before it would make a current flow. Faraday knew other ways to change the magnetic field through a coil. Simply move a bar magnet in and out. The results fitted. Push the magnet into the coil; a brief current flows. Hold it steady; nothing. Pull it out; another brief current but opposite in direction to the first.

In 10 days of intense work, Faraday had found electromagnetic induction. A changing magnetic field 'induces' a current to flow in the wire. The practical consequences were obvious. A machine with magnets and coils of wire constantly moving relative to each other would produce a sustained electric current (albeit an 'alternating' one, flowing back and forth), The way lay open to building 'generators' or 'dynamos' that would one day make electricity the lifeblood of advanced societies.

An identical discovery was made across the Atlantic in the same year. 'Natural philosopher' Joseph Henry was the next significant figure in the growth of American science, following polymath Benjamin Franklin (**1751**) and adventurer Benjamin Thomson (**1798**). He equalled or surpassed their achievements in research and as a 'statesman of science'. In 1846 he left a post at what was later called Princeton University to become the first secretary of the newly founded Smithsonian Institution in Washington, today one of the world's great museums and laboratories in science and technology. He also helped found the National Academy of Sciences, the North American equivalent of the Royal Society, and was its second president.

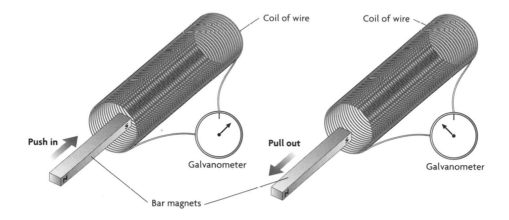

This diagram demonstrates the link between electricity and magnetism. Pushing the bar magnet into the wire coil makes the magnetic field stronger and induces a current flowing one way through the coil (as shown by the galvanometer). Pulling the magnet out weakens the field and sends the current the other way. Current flows only while the magnet is moving (and the magnetic field is changing). The faster the movement, the stronger the current.

These roles took him away from his own research but his reputation was already established. In the late 1820s he built on the discovery of electromagnetism and made powerful 'electromagnets' by wrapping thousands of loops of current-carrying wire around iron bars. Some of these powered the first electric telegraph, which, among other benefits, allowed the rapid reporting of weather information and forecasts; Henry was instrumental in setting up a national system in the USA for the purpose.

In 1831, independently of Faraday, Henry discovered electromagnetic induction. He took the matter further, discovering 'self-induction' before Faraday did. Changing current in a coil of wire makes a changing magnetic field; that field in turn generates another current in the same coil opposing the first. So changing currents (including alternating ones that flow back and forth) are impeded or 'choked' as they try to pass through a coil of wire. The more rapidly the current changes (the higher its 'frequency') the more noticeable is the effect. 'Chokes' remain an important part of electronics today. Henry's contribution to this work is commemorated in the unit of inductance—the henry.

Henry followed up this work (as did Faraday) by using induction in a transformer, a vital part of modern electrical technology. If two coils are wound onto the same piece of iron, the first with a few turns of wire, the second with many, the induced current in the second coil has a much higher 'voltage' (**1826**) than the first. So voltages can be 'stepped up' (or 'stepped down' if the coils are reversed) but only with alternating currents (AC); steady 'direct' currents (DC) are not affected. The current in the second coil is much smaller than the first; otherwise energy would be created (**1847 Joule**). Also known as 'induction coils', these devices were central to early efforts to drive electricity through gases, opening up fascinating new phenomena (**1854 Geissler**). ➤➤**1834**

John Herschel: Like Father, Like Son

1832
It was probably inevitable that John Herschel would become an astronomer. His father, William, was the king's astronomer. He had found the planet Uranus (**1781**) and built the first big telescopes (**1789**). Add to that the influence of John's aunt, Caroline, who shared her brother's passions and talents for both astronomy and music. Young Herschel fought his destiny for a time; after doing brilliantly in mathematics at Cambridge, he turned to law, if only briefly. Even his father had other ideas, hoping John would become a clergyman. But astronomy claimed him in the end.

Herschel's most productive years were spent in South Africa from around 1832. From there, he charted and catalogued many of the southern stars for the first time, using a big reflector telescope he had built himself. He saw Halley's Comet when it came by in 1835, and he argued that some oddities in its behaviour were the result of a gale of hot gases pouring off the Sun. Johannes Kepler had proposed such a 'solar wind' more than 200 years before.

Herschel was prominent in the Royal Society. In 1831 he was almost elected President as the candidate of a strong group of Fellows trying to shake up the venerable 'Royal' and make it more relevant to the times. The traditionalists' candidate won narrowly. It was this disappointment that drove Herschel to South Africa and to work with his close friend Charles Babbage, the inventor of the first computer, to found the British Association for the Advancement of Science (the 'BA') to do what the Royal Society would not.

When nearly 60, Herschel took an odd career move; he stepped out of astronomy to become Master of the Mint, responsible for the reform of the currency. One hundred and fifty years before, Isaac Newton had done the same thing, though the job had probably suited Newton's talents more than it did Herschel's.

Herschel took an early interest in photography, then in its infancy. His discovery of a chemical process to 'fix' the image and stop it from fading was vital. He might have been remembered today as the inventor of photography if he had given the matter proper attention (**1842 Starlight**). He also took note of developments in a new study called 'spectroscopy' (**1814 Fraunhofer**) and was perhaps the first to suggest that analysis of the light from the Sun and the stars could determine what they were made of.

William Wherwell: A Way with Words

You may not have heard of William Wherwell, but you will have used some of his words. Wherwell had a particular talent for crafting terms to cover the many new ideas, concepts and inventions emerging from **1833**

laboratories every day. He drew on a detailed knowledge of Latin and Greek that other researchers lacked. It was Wherwell who advised Michael Faraday (**1834**) in his choice of words like anode, cathode, ion and electrolyte to describe experiments and observations where electricity was passed through liquids. He helped Charles Lyell (**1829**) with words like Eocene, Miocene and Pliocene to name the subdivisions of the rocks of the Tertiary Period in geological history. He also devised the description uniformitarianism for the theory of gradual geological change expounded by Lyell and by James Hutton (**1778**) before him.

Most famously, in 1833 Wherwell coined the term 'scientist' to describe someone engaged in scientific research and investigation. It was not untill the end of the century that the term came into general use, replacing 'man (sic) of science', and before that, 'natural philosopher'.

Wherwell was not a great experimenter but he wrote with authority on almost every aspect of science in his day, including how science should be done. He knew so much that one contemporary wrote, in a jocular tone, 'Science is his forte, omniscience is his foible'. Wherwell was everywhere in the science of his day: Fellow of the Royal Society, President of the Geological Society and Master of Trinity College, Cambridge (Newton's old College). With Charles Babbage and John Herschel (**1832**) he founded the British Association for the Advancement of Science and was an early president.

As a disciple of the popular 'natural theology', which saw the hand of the Creator everywhere in the design of the natural world, Wherwell was a famous opponent of Charles Darwin (**1859**) and his theory of evolution. In 1825, the estate of the Earl of Bridgewater paid for eight 'treatises' to demonstrate the 'power, wisdom and goodness of God as manifested in the works of Creation'. Wherwell was chosen by the Archbishop of Canterbury, the President of the Royal Society and others to write the *Bridgewater Treatise* on astronomy and physics. These volumes represented the comfortable, assured world of early nineteenth-century science. The mood would be very different by the time the century ended.

Michael Faraday:
The New Language of Electrochemistry

1834

Following his discovery of electromagnetic induction (**1831**), the great experimenter Michael Faraday turned back the clock 20 years to the work of his predecessor Humphry Davy (**1807**). Davy had begun the science of electrochemistry, driving chemical changes with electric currents. Faraday took it up again with vigour and his customary care, and soon figured out the basic rules.

Suppose the experiment was causing silver to be 'deposited' from a solution of one of its salts by a flow of the electricity (perhaps to 'silver plate' some metal object). Faraday found that the amount of silver let loose depended on the amount of electric charge consumed (the current flow multiplied by time). If you doubled the charge supplied, the amount of silver doubled.

But the same amount of charge liberated different amounts of different metals. Charge that deposited 108 grams of silver would produce only 23 grams of sodium or 39 grams of potassium. These ratios were constant, whatever the current. And they were the same as the ratios of the atomic weights (**1808 Dalton**) of these elements (sodium 23, potassium 39, silver 108). Did this mean that the current always released the same number of atoms of any element? Faraday confirmed that this was generally so, though sometimes the amount released was cut to exactly one-half or one-third.

Here Faraday was on the edge of a major discovery. Perhaps there was a fundamental unit of electric charge (an 'atom of electricity'); atoms in solution carried one or two or three of these and fell out of solution when they were taken away. That would explain what he had seen. Faraday did not pursue this (he did not believe in 'atoms') but others did (**1884 Arrhenius**).

This new science needed new language. William Wherwell (**1833**) suggested

Faraday call a conducting solution an electrolyte. Electrodes (Greek for 'paths for electricity') carried charge in and out of the electrolyte. The anode ('the way up') was positively charged; the cathode ('the way down') was negatively charged. These terms turned up later in electronics. To explain how the current passed through the liquid, Faraday reluctantly imagined charged particles called ions (Greek meaning 'wanderers'). Cations carried a positive charge—they were attracted to the cathode; anions were negatively charged. Faraday thought ions no more real than atoms but it was a model that worked and would prove very powerful indeed. ➤➤**1845**

Gaspard Coriolis: Why Winds Blow in Circles

Son of a soldier who became an industrialist but died quite young, Gaspard Coriolis gave his name to an 'effect' that is still widely misunderstood. His effect or 'force' is blamed for making the water spin one way or the other as it goes down the plughole or the toilet, though in fact it is too weak to do that. The Coriolis Effect is noticeable only over large distances, tens or hundreds of kilometres, where it turns weather systems and ocean currents in one direction north of the equator and the other way to the south.

1835

Englishman George Hadley first explained in **1735** why trade winds, which ought to blow directly north or south towards the equator, are bent as they blow so they reach the equator coming from the east. The spin of the Earth does it, he said; on a stationary Earth the winds would blow north–south. A century later, Coriolis tried to make the behaviour of the winds fit Galileo's Law of Inertia (**1604**) or Newton's laws of motion (**1686**); these say that any object, such as the molecules of air that make up a wind, continues to move in a straight line at constant speed unless it is acted on by a force. To make the fit, he had to create a force, the Coriolis Force, that is weakest at the poles and strongest along the equator and whose direction depends on which side of the equator you are.

You can argue that this force is 'fictitious'; it is needed only because the rotating Earth is not an 'inertial frame of reference' in which Newton/Galileo must be obeyed. If you could hang in space while the Earth turned beneath you, you would see the winds blowing in straight lines. In the same way, you would see Foucault's pendulum (**1851**) continuing to swing in the one direction. The Coriolis Force has to be invoked to explain why the plane of motion of the pendulum appears to turn when seen from a spot alongside and fixed to the Earth.

All that being said, the Coriolis Force, which he wrote up around 1835, does greatly simplify calculations on how the winds move across the surface of the Earth, and also how the motion of missiles and aircraft is affected by the Earth's rotation. As to the winds, those blowing towards the equator are bent to come from the east; those going towards the poles become westerlies (on both sides of the equator). This makes the large masses of air, the low-pressure cyclones and the high-pressure anti-cyclones, spin to the left or the right depending on the hemisphere.

Much the same thing happens at sea. The masses of water in the major ocean basins rotate as currents within them are deflected (clockwise in the northern hemisphere, anticlockwise in the south). So the 'gyres' in the oceans drive a cold current along the western coast of a continent and a warm current along the eastern coast. This tends to make the western regions of continents drier than the eastern parts at the same latitude.

Despite a rather delicate disposition (he died aged only 51), Coriolis was a very successful teacher, as well as a mathematician and physicist. He popularised the use of words such as 'work' and 'kinetic energy' much as we use them today. To show he could mix business with pleasure, he also wrote a well-known book on the mathematics of billiards.

Charles Darwin: In the Wake of the *Beagle*

1836

The career of Charles Darwin began with a stroke of good fortune. In 1831 he was offered an unpaid post as naturalist on HMS *Beagle* on a two-year round-the-world voyage of exploration. His grandfather, Josiah Wedgwood, offered to pay for the trip when Darwin's own father objected to the enterprise. Darwin had a remarkable, perhaps unique, opportunity to gather the observations that he would later use to support a remarkable theory. Darwin may have been an outstanding scientist without the *Beagle* experience; someone else might have made as much of the opportunity as he did. We will never know.

Darwin had already turned down careers in the Church and medicine, the latter because he was appalled by operations done without anaesthetics. His circumstances were comfortable. He would be among the last of the 'gentleman scientists', free to pursue his researches without concern for income. His father was a prosperous doctor, his mother a member of the famous Wedgwood pottery-making family. His two grandfathers (Erasmus Darwin and Josiah Wedgwood) had become acquainted through the Lunar Society in Birmingham (**1765**). Darwin married his cousin Emma Wedgwood.

The *Beagle* returned to England in October 1836, the planned two-year voyage ultimately taking five years. Darwin had amassed a great collection of observations and specimens, mostly from South America but also from New Zealand and Australia. The ship's captain, Robert Fitzroy, and others, had also collected assiduously. As a voyage of exploration and discovery, it had been a great success.

The *Beagle*'s visit to the Galapagos Islands off the South American coast is now legendary. The impact on Darwin was summed up in his observations of 'Darwin's finches'. The birds on each of the many islands had distinctive characteristics, such as the shape of their beaks, which suited their lifestyle, including available food. The famous Galapagos tortoises also showed subtle variations from island to island.

> *Ignorance more frequently begets confidence than does knowledge: It is those who know little, and not those who know much, who so positively assert that this or that problem will never be solved by science.*
>
> CHARLES DARWIN

Many naturalists had observed such adaptations in many environments. Some took them as evidence of the wisdom of the Creator, as indeed Darwin did early in life. He also saw hints of another possible explanation. The islands were too widespread for birds to fly from one to the other, so the populations had been isolated for countless generations.

Darwin's account of his discoveries, part of the official voyage report, was published separately as *The Voyage of the Beagle* and became a bestseller. He also wrote up his thoughts about the many coral atolls, he had visited. He proposed that since coral grows only in shallow water, the great depth of coral limestone found in many places indicated the land beneath the atolls had been steadily sinking. At the same time he was pondering the broader implications of his observations in the Galapagos and elsewhere. These ideas were to grow and develop over 20 years to finally burst upon the world in **1859** with the publication of *The Origin of Species*.

Darwin carried another legacy from his time away. While in South America he had been bitten many times by insects, so contracting a tropical disease. This was not fatal but it continued to weaken him from time to time throughout life. He retired with Emma and the first of their 10 children to Downe, a village outside London, where he lived quietly but never stopped working. It was here that *The Origin of Species* took shape. ➤➤**1859**

Jean Charpentier and Jens Esmark: Landscapes of Ice

Throughout the nineteenth century, the steady recovery of bones and other remains of long-extinct animals suggested that conditions on Earth had changed greatly over the ages. Why else would these animals (and plants) have died out? Fragments of ancient plants found in beds of coal required a climate much warmer and wetter than it is now, as did fossils of tropical palm trees found in far-from-tropical England.

1836

During the same period suspicion grew that the climate may also have been much colder at times in the past. Folktales had been a stimulus. The Swiss naturalist Jean Charpentier was reputedly told by a woodcutter that the glaciers in the Alps had once been much larger, not that he knew from personal experience. But Charpentier was intrigued and started to gather more convincing evidence.

Jens Esmark in Scandinavia did the same. By 1824 he had concluded that glaciers must once have covered much of Norway and the adjacent sea. Huge boulders sat in the middle of nowhere ('erratics'), as if they had been dragged or pushed there and dumped. Moraines, the long lines of stones usually found around or within glaciers, crossed the landscape far from any current glacier. Great areas of barren rock had been polished smooth or scarred with long grooves. This was the work of vast, moving sheets of ice, Esmark argued, when the world was much colder. Glaciers had also cut valleys deep into the rock along the coast, later flooded by rising seas to form fiords. It seemed a compelling case.

Charpentier in Switzerland found the same sorts of clues. In 1836 he convinced his eminent countryman Louis Agassiz (**1851**). In his *Study of the Glaciers*, published in 1840, Agassiz gave a name to these long-ago periods of intense cold: 'ice ages'. Even with Agassiz' energetic support, the 'glacial theory' had a cool reception. Many experts opposed such a radical idea. They could not conceive of much of Europe under ice hundreds of metres thick; the vast ice sheets of Greenland were still unexplored, those in Antarctica unknown.

In the end, the evidence became overwhelming. By 1900, few still opposed it. Major questions remained, such as why the ice would have waxed and waned so remorselessly, not just once but perhaps many times. A possible answer was elaborated in **1920 (Croll and Milankovich)**.

'All Cells Come from Cells'

1838

There is a story that in 1838 German biologists Theodore Swann and Matthias Schleiden, who had been students together, were chatting over coffee and talking about their work. Schleiden had been following up a report by the English biologist Robert Brown (**1827**) that all plant cells had a tiny opaque spot that Brown had called a nucleus.

Swann was looking at animal tissues at the time, especially from tadpole embryos where the spinal cord would later develop. Tiny 'opaque spots' were plentiful. The two reportedly went to Swann's laboratory to look at his specimens. Certainly the 'nuclei', if that is what they were, looked just like those in Schleiden's plant cells. The spots were regularly arranged; looking hard through the microscope, Swann could sometimes see a faint wall or membrane around each of them.

The penny dropped. Swann went away and wrote up the discovery, which he published in 1839, sadly, without reference to Schleiden or anyone else. All living things are built up of immense numbers of simple, almost identical units. Cells are to living things what atoms and molecules are to matter. All cells have a wall or membrane and a nucleus. The cells in a muscle might look different from those in the brain, but all brain cells look the same and likewise all muscle cells.

The 'cell theory' soon won everyone over. All of the available evidence supported

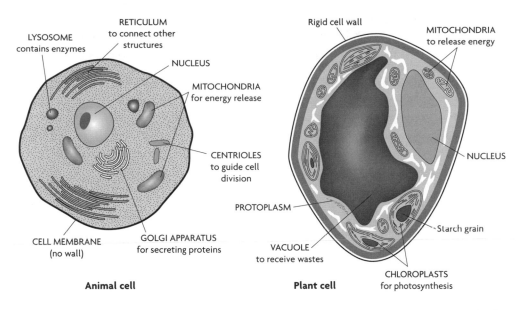

LYSOSOME	RETICULUM		Rigid cell wall		MITOCHONDRIA
contains enzymes	to connect other				to release energy
	structures				

Animal cell

Labels (animal cell): LYSOSOME contains enzymes — RETICULUM to connect other structures — NUCLEUS — MITOCHONDRIA for energy release — CENTRIOLES to guide cell division — GOLGI APPARATUS for secreting proteins — CELL MEMBRANE (no wall)

Plant cell

Labels (plant cell): Rigid cell wall — MITOCHONDRIA to release energy — NUCLEUS — Starch grain — CHLOROPLASTS for photosynthesis — VACUOLE to receive wastes — PROTOPLASM

All living things are made of cells too small to be seen with the unaided eye. The two types of cells (plant and animal) have some common features: a cell membrane, a nucleus and mitochondria. Plant cells also have chloroplasts, vacuoles and a rigid cell wall made of cellulose; animals cells have a golgi apparatus to secrete proteins. All cells reproduce by dividing in two.

it; in fact any number of people, such as Caspar Wolff (**1759**), could legitimately claim to have thought of it first. One big question remained; where did the cells come from? Swann's answer was soon rejected. He said they just grew out of chemicals sloshing around, like crystals did. This was very much like the doctrine of 'spontaneous generation', then popular but soon to be silenced by Louis Pasteur (**1862**) and others.

Evidence was soon against this theory. Whatever new cells could be seen to appear, they were formed by the division of existing cells. By 1858 German pathologist Rudolph Virchow was ready to state the right answer unequivocally; in Latin '*omnis cellula e cellula*'—'all cells come from existing cells'. Set alongside Louis Pasteur's later pronouncement against spontaneous generation—'all life comes from existing life'—this implied that all life is associated with cells. That idea would suffer a challenge before the century was out (**1898 Beijerinck**).

Friedrich Bessel: How Far to the Stars?

Friedrich Bessel was never trained in astronomy. Yet he achieved what astronomers had sought to do for 300 years; he directly measured the distance to another star. That distance had previously only been estimated but was known to be immense.

1838

From the age of 14 Bessel worked for an import–export firm. The company's ships needed to know where they were at sea; that led Bessel to astronomy as a way of finding latitude and longitude. He got a job as an observatory assistant and did some notable work on finding precise positions of stars. That secured him a post in charge of the observatory at Königsburg (now in Russia), where he pinpointed 50 000 stars over 20 years.

If Nicolaus Copernicus (**1543**) was right that the Earth went round the Sun once a year, we should be able to see 'stellar parallax'. This was a slight wobble in the apparent position of a nearby star against stars more distant, caused by the changing point of view of astronomers on a moving Earth. Even as telescopes improved, parallax remained undetected. This showed that even the nearest stars had to lie at immense distances. Only the most precise and careful observers could expect to detect parallax and perhaps not even they could detect it. Perhaps parallax was beyond detection.

Bessel was the man of the hour. In 1838 he 'won' the race. Using years of observations he found that the star 61 Cygni moved backwards and forwards in the sky over 12 months by a minute amount—0.3 of a second of arc (there are nearly 2000 seconds of arc in the width of your little fingernail held at arm's length). This made the star three 'parsecs' away or, on a more familiar measure, a bit over 10 'light-years'. At 300 000 kilometres per second, light had taken 10 years to cross the void from 61 Cygni to Earth.

Hearing of the discovery, the great English astronomer John Herschel described it as 'the greatest, most glorious triumph which practical astronomy has ever witnessed'. Bessel's margin of victory was narrow. Later that year, the parallaxes of the stars Vega and Alpha Centauri were measured by others. Alpha Centauri was found to be just over four light-years away and the closest star to the Sun. With the orbit of the most distant planet in our solar system only a few light-hours across, we can clearly see how great are the gulfs between the stars. If our Sun were the size of a cricket ball, then the next star would be 600 metres away.

James Ross: In Search of the Magnetic Poles

1839

In the mid-nineteenth century, the issue of magnetism again became a spur to discovery. The magnetic compass remained the key to navigation at sea, but steam propulsion and increasing use of iron had raised troublesome issues. It became imperative to know as much as possible about the Earth's magnetic field, its strength, declination and dip at any location at sea, and how those might change over time. Edmond Halley had drawn the first charts back in **1702** (**The Complex Behaviour of the Everyday Compass**); now they needed updating.

One target was to determine the exact location of the two 'magnetic poles', towards which the hands of a compass point. The Flemish mapmaker Gerard

Mercator had been first to suggest that such poles existed, and Englishman William Gilbert (**1600**) postulated that the Earth as a whole was like a bar magnet with two ends. The geographical poles had nothing like the same significance.

English navigator James Ross had found the North Magnetic Pole in 1831, then located high in the Canadian Arctic. In 1839 he went south in the small, wooden-sided but well-equipped ships *Erebus* and *Terror*. These were the first to be specifically strengthened for working in ice. For this reason his voyage was more successful than others at much the same time by the French under Dumont d'Urville and the Americans under Charles Wilkes.

With favourable weather, Ross passed through the floating ice that had barred the way to James Cook (**1769**) and penetrated, largely by chance, deep into a major embayment in the coast south of New Zealand, now called the Ross Sea. On Ross Island, he found two volcanoes, one active, one extinct, which he named after his ships.

He reached a point close to the position of the South Magnetic Pole predicted by the German physicist Carl Gauss, but the location was inland, high on a plateau among steep icy mountains. Ross was not able to reach it. More than 60 years would pass before an expedition under Australian Douglas Mawson in 1909 would find the spot where the compass needle points straight into the ground.

Unlike the geographical poles, the magnetic poles are not fixed. They wander, reflecting some long-term trends in the forces within the Earth that generate its magnetic field. Since 1909 the South Magnetic Pole has moved hundreds of kilometres to the north, and now lies out to sea.

Justus von Liebig: Teaching Chemistry to the World

German chemist Justus von Liebig, who became one of the most eminent researchers and teachers of his time, was judged 'hopelessly useless' by his teachers. He preferred to spend his time in the laboratory beneath his **1840** father's shop, which sold painters' supplies and common chemicals. There he could experiment with an exciting substance called silver fulminate, which he could make explode with a satisfying bang. One story says his school career ended when he blew a window out of a classroom with his homemade explosives.

His brilliance asserted itself later. After studying under the eminent Frenchman Joseph Gay-Lussac (**1808**), he was made Assistant Professor at the University of Giessen when only 21. Because of his youth and his reputation as a student radical, his initial reception by the other staff was hostile. By the end of his life, he was acknowledged as the greatest living teacher of chemistry, having pioneered the methodical experimental approach that became the accepted model. He took students from all over the world, several of whom later won Nobel Prizes in chemistry and biology. His American students went home to set up research stations and agricultural universities based on his ideas.

Liebig was very practical; he was concerned with artificial fertilisers such as ammonium nitrate (he published a book about these in 1840) and with issues of human nutrition. He was perhaps the first applied chemist, and helped revolutionise food production. But he was sometimes quite wrong. For example, he argued that physical exercise consumed only protein for energy rather than sugars or fats. Even when no protein was being consumed, exercise generated urine containing nitrogen, which could only have come from the breakdown of muscle tissue. So athletes needed to eat lots of protein to repair the damage. His authority was such that an enterprising entrepreneur got him to endorse 'Liebig's Meat Extract', perhaps the first protein supplement.

Even Liebig was not beyond challenge. A couple of physiologists did a simple experiment. They measured the amount of nitrogen in the urine from someone climbing a mountain. This showed that the energy the climber had used could not have come just from the breakdown of proteins. Clearly at least some energy came from fat and sugars as well. Brilliant as he was, Liebig was prone to be dogmatic and dismissive of the work of others when it challenged his own. His reputation suffered but much of his achievement remains, particularly his inspiration to others. He was a giant in his time.

Getting the Most from Starlight

1842 The growth of photography from around 1840 was both a consequence of scientific discovery and a great stimulus to it. Discoveries made through microscopes or telescopes could now be permanently recorded.

Photography means 'making pictures with light'. In the eighteenth century certain compounds containing silver were found to go black when exposed to light. Johann Ritter used this effect to discover ultraviolet light (**1802**). Photography grew from there. To be useful, any image had to be both 'developed' (that is, made visible) and 'fixed' (prevented from fading). Chemical processes were invented for those purposes, with the astronomer John Herschel making a major contribution (**1832**).

Astronomers were among the first to exploit this new capacity to capture and store images. The first picture of the Sun was taken in 1840, of the Moon in 1842 and of a cluster of stars in 1851. Astronomers no longer had to labour over drawings of, say, the surface of the Moon. More observations could be made in a night or when the 'seeing' was good. As photography improved over the rest of the century and the various plates and films became both more sensitive and easy to use, astronomical photography grew in power and sophistication. It surpassed the power of the human eye to gather light; 'plates' could be exposed to one part of the sky for many minutes or even hours and exceedingly faint objects became visible.

The number of known stars skyrocketed. In the region of the Pleiades, where Galileo in **1610** had triumphantly found 40 previously unseen stars, a nineteenth-century photograph revealed 2000. Another photograph showed that the faint smear of light in the constellation of Andromeda, known as a 'nebula', had a spiral shape—a significant discovery in the light of later events (**1924 Hubble**). By 1900, 120 000 such nebulas could be detected, spread right across the sky.

The use of photography grew alongside another technology of light: spectroscopy (**1814**). John Herschel proposed analysing the light from the heavenly bodies to find out what they were made of. When the light of the stars was analysed, the 'spectra' proved familiar—they were continuous spreads of colour crossed by some fine dark lines, just as with sunlight. So stars were indeed, as had been suspected, other suns made tiny by immense distance. Slight differences between the spectra showed that not all stars were the same. For a permanent record, the spectrum of a star could be photographed.

Nebulas in general were a puzzle. The leading practitioner in this area was English astronomer William Huggins. Led by the influential Englishman William Herschel, astronomers thought that these dim, misty patches were nothing more than collections of glowing gases. So Huggins expected to see the bright, sharp lines in the spectra that a hot gas would show in the laboratory. Some nebulas did show this, but many more showed the continuous spectra typical of stars. So, many nebulas contained stars too faint or too close together to see. The mystery deepened (**1924 Hubble**).

Christian Doppler Has an Effect

Poor health prevented Christian Doppler from following the family tradition and becoming a stonemason. If he had, this Austrian would be unknown today. As it is, his name is quite familiar. Though he became a

1842

teacher and researcher in various areas of physics and mathematics, he is remembered primarily for the Doppler Effect, which he first wrote about in 1842. His fundamental idea now seems quite obvious: any sort of wave motion, in air, water or even in space, is affected by a movement of the source of the waves or of the receiver.

Consider sound waves in air. If the source of the sound, say an ambulance siren, is moving towards you, the sound waves tend to pile up ahead of the ambulance, shortening the length of the waves (the distance crest to crest). This increases their frequency and so raises the pitch of the siren. Likewise if the ambulance is moving away, the waves are stretched out behind, resulting in longer wavelength and lower pitch. Hence the 'ambulance siren effect'; the pitch of the siren drops suddenly as the ambulance races past. According to one story, Doppler tested 'his' effect with a carriage full of musicians that passed by as he stood beside a railroad track. The pitch of the music slid down at the moment of passing by the precise amount he predicted.

We now have many practical uses of the Doppler Effect: ultrasound scanners that can measure the speed of flow of blood in arteries and veins, radar sets that can clock the speed of winds and waves and catch cars that are going too fast.

As Doppler himself believed, the effect also applies to light, showing up as a change of colour. If the source is coming closer, the colour becomes bluer (that is, the frequency of the radiation increases); movement away causes a 'red shift'. This has been a powerful tool in astronomy. It has proved that the universe is expanding (**1929**) and that planets exist around stars other than the Sun (**1995**).

Marceline Bertollet: Untangling the Chemistry of Life

1845

The discovery in **1828** by Friedrich Wohler that the life-related compound urea was chemically identical to the purely mineral ammonium cyanate did much more than strike a blow at 'vitalism', the old philosophy that living things have some inherent quality beyond scientific scrutiny. It began the methodical study of the vast range of chemical substances that are associated with life, so-called organic chemicals, leading to efforts to create them in the laboratory.

Among the first to be made (in 1845) from its elements (carbon, oxygen and hydrogen) was acetic acid, previously only available from the fermentation of wine by yeasts (hence its popular name, vinegar). The champion at this game was the French chemist Marceline Bertollet, whose father had worked with Lavoisier. He made, from scratch, various alcohols (fermentation again), the gas methane (released from coal in mines as 'fire damp') and the liquid benzene (first found by Michael Faraday in whale oil). The dividing line between 'organic' and 'inorganic' chemicals was crossed so often it was hard to see where it lay.

Most of the organic compounds were much more complex than these. Already a rough threefold division had been made into proteins, fats and sugars. All of these contained carbon, oxygen and hydrogen; proteins had nitrogen as well. Sugars contained hydrogen and oxygen atoms in a ratio of two to one, like water, and so were named more generally 'carbohydrates'.

It was soon found that heating these big compounds with some weak acid or alkali could break them down into much smaller, simpler compounds. Already in 1812 a Russian chemist had heated the carbohydrate starch with acid and ended up with a very simple sugar, later called glucose. Eight years later a French chemist did the same with the protein gelatine (made by boiling up bones) and isolated a simple substance he called glycine, one of the first of the amino acids. About the same time, asparagine was extracted from asparagus and cysteine from a kidney stone. Even earlier, another French chemist had done the same with fats, and produced fatty acids. All fats seemed to consist of three fatty acids coupled with a compound called glycerol.

Bertollet made the startling step forward. He began to put some of these small compounds together in ways that did not happen naturally. For example, he combined glycerol with compounds that behaved like fatty acids but had not been extracted from fats. He ended up with 'synthetic fats', compounds clearly in the same family as naturally occurring fats, but which nature had never made. The chemist could do what nature had not, and without any help from 'vital spirits'. 'Vitalism' was now dead and buried; in its place was the new science of 'organic chemistry', the methodical study of compounds made by living things, all of which contained carbon. It seemed inevitable that in time all their secrets would be revealed.

Michael Faraday:
Linking Electromagnetism with Light

English physicist Michael Faraday (**1813**) was a devout member of a small church group called the Sandemans. The faith encouraged modesty and self-effacement; Faraday declined personal honours like a knighthood and the presidency of the Royal Society.

1845

His religion inspired a belief in the unity of nature that drove much of his work. In **1831** he completed the unification of electricity and magnetism by discovering electromagnetic induction; by **1834** chemistry and electricity were firmly linked through his research. A still greater achievement, bringing together electro-magnetism and light, had to await his recovery from a nervous breakdown, a consequence of his furious pace of work.

When Faraday returned to serious work at the Royal Institution in 1845, a letter arrived from Lord Kelvin (**1846**) suggesting that light rays are driven by interacting electric and magnetic fields. These vibrate back and forth at right angles to the direction a ray moves. When light is 'polarised', the electric field has a fixed direction in space (the 'plane of polarisation'), with the magnetic field at right angles to it. So a strong external magnetic field should twist the light ray's internal one and so rotate the associated plane of polarisation. This shift could be measured directly (**1808 Malus**). Faraday was the greatest experimental physicist of his day, Kelvin thought; no one was more likely to find Kelvin's effect than Faraday was.

Faraday had looked for this effect before; the electricity–magnetism–light linkage was not really a new idea. Perhaps now it could be proved. Spurred on by Kelvin, Faraday tried again; this time he found it. A strong magnetic field sent through a piece of glass parallel to a ray of light did turn the plane of polarisation, an effect now known as 'Faraday Rotation'. The stronger the field, the more the plane shifted. So light really was bound up with electromagnetism.

This finding, along with Faraday's many other discoveries, inspired the Scots maths wizard James Clerk Maxwell (**1871**). He not only proved mathematically that light was an electromagnetic wave, but predicted that other forms of such waves existed. In **1888** Heinrich Hertz would find these waves, later called radio waves. This was how far the ripples ran from Faraday's laboratory.

Faraday Rotation was one of his last major discoveries. From the mid-1850s, Faraday's physical and mental powers began to decline; he died in 1867 aged 76. His association with the Royal Institution had lasted more than half a century.

The Search for Neptune: England Versus France

1846

The discovery of the planet Uranus by William Herschel in **1781** had doubled the size of the solar system and increased the number of known planets from six to seven. No one had expected a new planet then; Herschel had found it by accident. No one knew if any more planets existed, but a close watch was kept on Uranus as it trudged along its 84-year orbit around the Sun.

It was soon apparent that something else had to be out there in the half-dark. Johannes Kepler (**1609**) had argued convincingly that the planets followed a precise path, known as an ellipse. Isaac Newton had proved that it had to be so, since the gravity of the Sun was the only significant force at work. Uranus seemed to have other ideas. It began to deviate significantly from its expected path, and the only reasonable explanation was that yet another planet, further still from the Sun, was pulling it off course. Such was the sophistication of 'celestial mechanics' that two mathematicians, John Couch Adams in Britain and Urbain Leverrier in France, were able to calculate where the new planet had to be. It seemed a simple thing to point a telescope in its direction and find it.

Adams, who completed his calculations first, tried to get Astronomer Royal George Airey to take a look, but Airey thought the Royal Observatory at Greenwich should stick to timekeeping and navigation. He changed his mind when Leverrier's calculation came out a few years later, confirming Adams' prediction. He quickly asked Cambridge University observatory to make a search, but it was too late. By now, Leverrier had the observatory in Berlin on the case and on the evening of 27 September 1846, Johann Galle made history by locating the eighth planet. Neptune, as it was soon called, proved to be far more distant than Uranus and took nearly twice as long (165 years) to complete its 'year'.

Galle found the planet on the very first night he looked. Cambridge had in fact seen the planet, looking like a dim star, four times during the search but had failed to recognise it. Nonetheless there was something of a national scandal when it was recognised that the French and Germans had beaten the British to the prize and that the Astronomer Royal was mostly to blame.

Within a few months, Neptune was proved to be not alone. A companion moon was found for it and dubbed Triton for the great horn that the God on the Sea was reputed to blow. The British missed out there too. Triton was discovered by a Frenchman.

Lord Kelvin: The Great Man

Irish-born William Thomson, later Lord Kelvin, was an icon of nineteenth-century science. He gained every available accolade: knighted at the age of 42, a baron at 70, among the first to receive the Order of Merit, President of the Royal Society, awarded dozens of medals and honorary degrees, ultimately buried in Westminster Abbey near Isaac Newton. For 53 years from 1846 he held the Chair of Natural Philosophy at the University of Glasgow. The jubilee of his appointment (he had been only 22 at the time) was marked in 1896 by an unprecedented assembly of notables.

This recognition flowed inevitably from a career in which he contributed to every area of the physics of his day. Others built their careers and reputations on ideas that flew from Kelvin like 'sparks from his anvil', as one commentator put it. His work on the laws of thermodynamics (**1865**) and on defining absolute zero (**1848**) would

> *I often say that when you can measure what you're speaking about, and express it in numbers, you know something about it; but when you cannot measure it, when you cannot express it in numbers, your knowledge is of a meagre and unsatisfactory kind.*
>
> LORD KELVIN

probably have won him the Nobel Prize, had it existed at the time. He knew the vital importance of research at a time when the only well-equipped laboratories in Britain were in the Royal Institution in London; in Glasgow he started experimenting in the basement of a pub.

Kelvin was vastly influential in technology as well. His understanding of the physics and mathematics of electrical currents in very long wires made the first transatlantic telegraphic cable possible in 1869; he invented vital pieces of equipment, such as the mirror galvanometer and the siphon recorder, for the enterprise. Kelvin developed machines to measure, analyse and simulate the complex patterns of the ocean tides. These were the first 'analog computers', built long before anyone thought of the phrase.

Perhaps most importantly, Kelvin was preoccupied throughout his career with precise measurements and standards, and that spirit spread to his colleagues. He once said that 'only when you can measure something do you know anything really useful about it'. He said lots of other things too, some of which he might have wished he hadn't (**1900**). ➤➤**1848**

Louis Pasteur: The 'Handedness' of Nature

1847   Louis Pasteur was another great scientist who showed little promise in his youth. Teachers described this son of a poor French tanner as 'mediocre'. Yet he did well enough at university to secure a doctorate and employment as a chemist.

He first made a name for himself in 1847 (aged 27) by studying tartaric acid, found in the 'lees' of wine (an appropriate subject for a Frenchman and one he returned to several times). He made crystals from the acid and examined it in several ways, including with 'polarised light' (**1808 Malus**). Briefly, the electric field that drives a beam of light oscillates from side to side and generally at all angles. Light can be 'polarised' in passing it through a special filter, which lets the electric field move back and forth in one direction only, called the plane of polarisation.

Pasteur passed a ray of light through a solution of his tartaric acid, and found that the plane of polarisation now pointed in a slightly different direction (it had been 'rotated' to the right). At much the same time, Michael Faraday in England was finding that magnetic fields had the same effect (**1845**). Like most chemicals associated with living things (**1828**), tartaric acid could also be made chemically. Pasteur was puzzled; the chemically made acid did not affect the polarised light. He examined a heap of crystals under the microscope. To his surprise he saw that there were two sorts of crystals, identical in shape but reversed, as if one type was the mirror image of the other.

He laboriously sorted these into two piles, and made solutions of each. One form of the acid turned the plane of polarisation to the left, the other to the right. Chemically made tartaric acid was a mixture of the two types and so ended up doing nothing at all. Since the shapes of the crystals reflect the shape or arrangement of the molecules within them (**1784**), those molecules had to be similar but different, like a left-handed and a right-handed glove.

Pasteur had grasped a profound truth about nature: it is 'one-handed'. Since Pasteur, 150 years of research have shown that most chemicals extracted from nature have the same quirk. Left-handed versions of some organic chemicals can be made easily in the laboratory, but not by living things; they only make right-handed versions. For other organic chemicals, the opposite it true; those made naturally are all left-handed. Why is it so? The answer must lie far back in the history of life on Earth. ➤➤**1856**

Ignaz Semmelweis: 'Doctors Should Wash Their Hands'

'Childbed fever' was a nasty disease in nineteenth-century maternity hospitals, killing as many as one in 10 women within a few days of childbirth. Most doctors thought nothing could be done about it, as they

1847

had no idea as to the cause. Hungarian-born obstetrician Ignaz Semmelweis did not agree. In 1847, while working in the big maternity hospital in Vienna, he noticed that the mortality rate varied between two wards, from around 2 per cent to more than 13 per cent. The wards had similar patients and used similar techniques. Semmelweis decided to investigate the differences, though he did so against the wishes of superiors, who thought the exercise pointless.

Semmelweis soon found a possible explanation. The ward with the high rate trained medical students who had to do post-mortems on mothers who had died from the fever. The other ward trained midwives. Perhaps doctors, going straight from the autopsy room to the maternity ward, carried on their hands some 'infectious particles' that caused the lethal fever in the new mothers. When the students in his ward washed their hands in a solution of chlorinated lime, the death rate fell sharply to only 2 per cent. Washing surgical instruments before use reduced it to less than 1 per cent.

His success was not well received by colleagues. Many thought that disease came from 'miasmas' (**1854 Snow**) or an 'imbalance of humours', and hand-washing could have little effect on those. Besides, washing hands between patients was thought to take too much time for busy doctors. Semmelweis was not a persuasive advocate for his ideas, and was reluctant to defend them in public. Perhaps many doctors would not accept his views because it required recognising that previously accepted practice may have killed many people.

Rejected in Vienna, and perhaps short of money, Semmelweis returned to Hungary. He established a good practice and reputation there and demonstrated how effective his methods were in reducing childbed deaths, but he was overtaken by mental illness and died aged only 47. The legend grew that he had died of childbed fever, contracted through a cut on his hand. By then, the work of Louis Pasteur (**1862**) and others was putting a solid foundation under the 'germ theory' of disease and finding the sort of 'infective particles' that had spread the disease. In time, the medical establishment came to accept Semmelweis' findings, ending much needless loss of life among young mothers.

The Earl of Rosse: The Leviathan of Parsonstown

1847

Since Galileo and others had first used the telescope to sweep the night skies in **1610**, it had grown greatly in size, power and sophistication. Size was always important: bigger telescopes with wider mirrors and lenses were able to capture more light, see further, image fainter objects and map finer detail. They were also more expensive, and often more difficult to use, but the 'drive for aperture' continued. This had culminated in **1789** with William Herschel's monster built in Slough, west of London, with a tube 12 metres long and more than a metre in diameter. Even Herschel often found that telescope to be more trouble than it was worth.

Herschel was bettered only by a combination of astronomical passion and vast wealth. The Irish landlord William Parsons, Earl of Rosse, wanted the biggest telescope ever built. By 1847, despite delays caused by the Potato Famine, he had it. 'The Leviathan of Parsonstown' boasted a 20-metre tube and a polished metal mirror 2 metres in diameter (mirrors made of silvered glass were not yet available). The huge instrument could not be pointed to all parts of the sky; instead it hung from massive masonry walls and wooden scaffolding so that stars and planets could be observed for two hours as they passed overhead.

The telescope performed magnificently. It revealed stars 10 000 times fainter than were visible with the unaided eye. Rosse could see tiny craters and rills on the surface of the Moon, and new and delicate details on the surfaces of Mars, Jupiter and Saturn. In **1877**, the Leviathan confirmed the existence of the newly discovered moons of Mars.

Its biggest impact was out among the stars. Some of the faintly glowing nebulas (**1750**) were turned from featureless blobs to stunning spirals of light. Soon hundreds of such 'spiral' nebulas were known. Here was some evidence to support the assertion of Immanuel Kant 100 years before (**1750**) that among the nebulas were 'island universes', whole congregations of stars lying beyond the one in which we live. That remained a radical thought, unaccepted by most astronomers, for another 80 years (**1924**).

The Leviathan had a relatively short life. Maintenance costs and other factors saw it pass out of service around 1878, though nothing surpassed it for size for another seven decades. It still stands in the middle of Ireland and has recently been reactivated for tourists.

James Joule: Work Is Heat and Heat Is Work

Anyone who has been on a diet has heard of James Joule, spare-time Manchester brewer and full-time physicist. His name (multiplied a thousandfold to make 'kilojoules') provides the unit for measuring the energy content of food. Energy in all forms is now measured in joules.

1847

Joule had studied science with John Dalton (**1801**), inventor of the 'atomic theory' and Manchester's most famous citizen. An electric motor in his brewery sparked Joule's interest in 'energy' (a new term in this time) and its close associate 'work', and in trying to link both of them with 'heat'. Joule was a devout Christian, like many scientists in his day. He may have pursued the connection through a desire to find some unity among the diverse phenomena of nature.

Joule knew that raising a weight against the pull of gravity required work, and so a falling weight would make energy available. He also knew that heat 'energy' made water hot. So he built a clever apparatus in which falling weights made paddles stir water in an insulated vessel and warm it up. He compared the energy liberated by the falling weight with the heat represented by the rising temperature of water. Performed many times through the 1840s, the experiment showed the ratio between the two to be always the same. So much work produced so much heat, according to a formula. Joule had established a fixed 'exchange rate' between work and heat, which he called 'The Mechanical Equivalent of Heat', and others called Joule's Equivalent.

More profoundly, work/energy and heat were two sides of the same coin. 'Heat is work and work is heat' now says the First Law of Thermodynamics. Work can become heat, or heat can become work, say through a steam engine, but (and this is crucial) nothing is ever lost in translation. Energy cannot be destroyed or created, but simply converted from one form to another. Any gain or loss is only apparent.

Joule found a similar exchange worked with electricity. Here it was the motion of

JOULE

The joule is the unit of energy or work: 1 joule of work is done when an object is moved with a force of 1 newton through 1 metre; 1 joule of energy is released every second via an electrical current of 1 ampere flowing through a resistance of 1 ohm.

WATT

The watt (named after the eighteenth-century pioneer of the steam engine) is the unit of power, that is, the rate at which energy is taken in or given out: 1 watt equals 1 joule per second; 1 horsepower equals 745 watts.

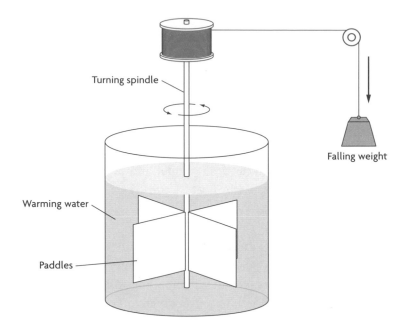

The diagram shows how James Joule demonstrated the conservation of energy. The weight (falling at a steady speed) loses potential (stored) energy as it gets nearer the centre of the Earth. Through the cord driving the spindle and attached paddles, this energy is transferred to the water and converted into heat that warms the water. Under ideal conditions no energy is lost.

Turning spindle

Falling weight

Warming water

Paddles

electric charges that could be turned into heat, through a sort of electrical 'friction' called resistance (**1826**). The rate at which heat was released (the 'power', measured in watts), was equal to the voltage multiplied by the current. Written in mathematical symbols, this is the Joule Equation.

Joule's 'Principle of Conservation of Energy' first appeared in a public lecture in 1847 after years of experiment. He faced scepticism from those who doubted his experimental accuracy or clung to the 'caloric' theory of heat that he was undermining. But by the end of the 1840s, the struggle was over. Thanks to Michael Faraday, the Royal Society heard of Joule's work in 1849, and he was soon a Fellow. He had also become close to Lord Kelvin and that association would prove fruitful. ➤➤**1852**

Jöns Berzelius: A Man of Many Words

1848 It is hard to overestimate the impact of Jöns Berzelius on the growth of chemistry. The Swede, who died in 1848, was in the same league as Antoine Lavoisier and John Dalton. He contributed powerful new ideas to explain the rising tide of observations and summed these up in new words, since no words existed that fitted the new phenomena.

There was a downside to this. His influence was so great (like Newton's) that he controlled which theories were acceptable. As a consequence, ideas he did not care for were consigned to obscurity, though many were later proved valid. Avogadro's Hypothesis was one such (**1811**).

But his overwhelming influence was progressive; until very late in life he was open to innovation. In **1814** he had introduced a whole new way to summarise chemical reactions by symbols written in equations. Friedrich Wohler's discovery in **1828** that urea and ammonium cyanate had the same chemical composition but very different properties led him to coin the word 'isomers' for compounds with the same combination of atoms put together in different ways. He knew that structure and organisation are least as important as composition.

In 1835 he pulled together evidence to declare that chemical reactions, both in life and in the test tube, can be speeded up by the presence of some chemical that is not itself consumed. Most of these reactions broke complex substances down into simple ones, so he called the phenomenon 'catalysis' (Greek for 'breaking down') and the chemicals involved were 'catalysts'. In 1838 he created the word 'protein' as an omnibus term for the many compounds found in plant and animal tissues and which all contained nitrogen. They seemed so fundamental that he incorporated the Greek word 'protos' for 'first' (also used in protoplasm, protozoa and, later, proton).

In 1840 his new word was 'allotrope', for the different forms an element like carbon can take, each with distinctive properties. Soft and greasy graphite (sometimes called 'lead' as in lead pencils), powdery soot and hardest of all minerals, diamond, are all allotropes of pure carbon. Englishman Humphry Davy had proved that was so for diamond by setting fire to one; he ended up with just carbon dioxide. As with isomers, the way the atoms are arranged makes all the difference.

Edouard Roche: Why Saturn Has Rings

Edouard Roche of France had his name added to the dictionary of astronomy by defining the Roche Limit, the closest distance at which a
satellite can orbit a planet before being torn apart by the planet's gravity.

The force of gravity depends very much on the distances involved. If the satellite is inside the Roche Limit, the difference in the pull of the planet's gravity on the different parts of the satellite is greater than the forces that hold the satellite together. So it will be shredded.

This helps us understand the rings of Saturn, first glimpsed (and misunderstood) by Galileo in **1610** and correctly described by Jean Cassini (**1669**). These lie inside the Roche Limit for Saturn and therefore cannot be solid; rather, they are made out of an immense number of fragments orbiting like tiny moons.

It is not that there was once a moon that was later pulled apart, but rather that no moon has been allowed to form; any efforts at 'moon building' within the limit will be thwarted. All the planets, even the closest, Mercury, lie outside the Roche Limit for the Sun and were allowed to build up from fragments when the solar system was very young. Likewise our Moon would have to be only 18 000 kilometres away rather than the present 400 000 kilometres before it was in any danger.

But visits by space probes in the late twentieth century found rings around all the other big planets (Jupiter, Neptune, Uranus), though nowhere near as spectacular as Saturn's. And we have seen examples of comets coming to grief by passing too close to one of the big planets.

William Thomson (Lord Kelvin): The Lowest Possible Temperature

If you ask what the temperature is, you will usually get the answer in degrees Centigrade (Celsius), say 20°C (**1742**). An older person, or someone in the United States, might use degrees Fahrenheit (**1701**). No one will say 293 degrees Kelvin. Yet the Kelvin scale gives the 'absolute temperature'. You can quote a temperature of so many degrees below zero in the other scales but not on the absolute scale: 0 degrees Kelvin is as cold as you can go, 'absolute zero'.

English physicist Lord Kelvin proposed the new temperature scale in 1848 (it was named in honour of him 50 years later), but crucial evidence came from careful gas experiments by the Frenchmen Jacques Charles (**1783**) and Joseph Gay-Lussac (**1808**). The idea of a lowest possible temperature goes back 100 years more to Frenchman Guilliame Amontons (**1702**). Experiment found that for every degree the temperature falls, the volume shrinks by one part in 273 (of its volume at 0°C). So a gas cooled to 273°C below zero will presumably shrink into nothing.

Does this prove that −273°C is the lowest achievable temperature? Not of itself, perhaps. The next step was made by the German Ludwig Boltzmann. He knew that the pressure of a gas is linked to the energy of its particles, and he linked that energy to the temperature on Kelvin's suggested new scale. So at −273°C, the pressure of the gas and the energy of its particles would be zero, like the volume. Particles would stop moving. No cooling machine or process could still work and so reach a lower temperature. We would be at absolute zero. Of course, no gas actually goes to zero volume and pressure. Well above absolute zero, gases become liquids and even solids and start to behave very differently. But the reasoning holds.

Some physicists were so terrified of everything stopping at absolute zero that they rejected the very link between temperature and the energy of particles. Papers expounding that idea were rejected by learned bodies like the Royal Society. Some careers were frozen as a result.

Two postscripts. By international agreement, temperatures of the Kelvin scale are marked, say, 200K. And Kelvin's name in temperature measurement lives on in the brand name of an American refrigerator: the Kelvinator. ➤**1852**

Leon Foucault and Hippolyte Fizeau: The Speed of Light Yet Again

1849

French physicists Leon Foucault (he of the Pendulum, **1851**) and Hippolyte Fizeau were in a sense amateurs, 'gentlemen scientists' like their contemporary Charles Darwin, men of means who held few official positions and for whom science was a consuming hobby. But they were no dabblers. Collectively and individually they focussed (literally) on light, exploring its properties, developing the new technology of photography for use in astronomy and devising ways to measure the intensity of sunlight.

Their major achievement was to measure the speed of light without leaving the surface of the Earth. Previous measurements by Ole Roemer in **1676** and James Bradley in **1729** had relied on objects and events in space. The two Frenchmen had a 'terrestrial' approach. They cut a beam of light up into short pulses with very high speed toothed wheels or revolving mirrors, and sent the fragments on a 20-kilometre journey to a mirror on a nearby mountain. They could measure how long the journey out and back had taken (less than one-thousandth of a second), and so compute the speed of light in air; their figure of around 300 000 kilometres per second is very close to the value accepted today.

Foucault and Fizeau used similar methods to show that light moved more slowly in water than in air. This helped to settle (for the moment) the long-running debate between the wave and particle theories of light in favour of the former. The particle theory may have had the support of Isaac Newton (**1672**) but it required light to go faster in water.

It also explained the law of refraction of light found more than two centuries earlier by Willebrord Snell (**1621**): the more slowly light travels in something transparent like water or glass, the more the light will bend as it crosses a boundary between that medium and air or a vacuum; to be more technical, that is, the higher will be its 'refractive index'.

Rudolf Clausius: The Measure of Disorder

1850

The word entropy is an odd one, and the concept takes some understanding. But it is worth the effort, given its importance. The great Austrian physicist Ludwig Boltzmann had the equation for entropy carved on his tombstone. But it was the German Rudolf Clausius who first grasped the idea around 1850 and devised the word. 'Entropy' is made out of 'en-' for 'energy' and the Greek word tropos, meaning 'change' or 'turning'; it is fundamentally involved with what happens when energy is changed from one form to another.

Clausius had recently come across the **1824** book by Frenchman Sadi Carnot on the efficiency of steam engines. Carnot found that no engine could be 100 per cent efficient. Some energy was always wasted, since the steam coming out of the engine still had heat in it that had not been used.

Pondering this, Clausius glimpsed some more fundamental principle. The energy transformed by the steam engine had come originally from the complex chemical compounds in coal or wood, reduced by combustion to simple gases such as water vapour and carbon dioxide and disbursed into the atmosphere. Order had been replaced by disorder. The confined molecules of steam in the boiler had been reduced to water and let out into the environment. Again disorder had increased: precise information about those molecules was now lost and any energy they still contained could not be recovered.

Clausius invented the term entropy to describe this disorder or 'loss of information'. He argued that entropy increases whenever energy is released. And entropy increases wherever 'a system' is left totally to itself. A wall of neatly ordered bricks can become over time a disorderly pile of rubble. A discrete drop of dye released into water spreads until all the water is faintly coloured, and the whereabouts of the original drop is lost. Increasing entropy and the loss of order seem to be everywhere.

We can of course use the pile of bricks to rebuild the wall, so apparently increasing order and decreasing entropy. But that takes energy, and so increases entropy elsewhere. If the various actions are all taken into account, overall entropy will increase. Indeed, the entropy of the whole universe steadily increases.

This powerful principle is commonly called the Second Law of Thermodynamics (the Conservation of Energy being the First Law). It explains much: why perpetual motion is impossible (it would decrease entropy), and why the 'heat death of the universe' is coming (**1854**). But there was still much to learn about the nature of entropy, for example from Ludwig Boltzmann (**1877**).

Trains, Telephones and Electricity

NINETEENTH-CENTURY TECHNOLOGY

In the nineteenth century new technology impacted more and more on everyday human activities. In transport, steam power began to supplement, and then replace, sail in shipping; in 1833 a ship crossed the Atlantic under steam power alone for the first time. From the 1840s iron railway tracks began to connect cities and towns across most advanced nations; by the 1860s they were bridging continents.

Towards the end of the century, the first electric trams came into use in cities, and soon after, the first 'horseless carriages'; they would transform personal transport in the century to come. These inventions were all supported by new knowledge: of thermodynamics to make steam engines more efficient; of the electromagnetism that drives electric motors; of distillation of oil to provide liquid fuels for 'automobiles'. Science, long supported by the 'technology of knowledge', was beginning to repay the debt.

In communications, advances in the understanding of electricity and magnetism made possible the telegraph from the 1840s (with a cable under the Atlantic in the 1860s) and the telephone from the late 1870s. The promise of 'wireless' emerged tentatively in the 1890s. Late in the century came the first recorded sound, on cylinders, disks and tapes, and the first 'moving pictures', though those would not make much impact until the next century.

In medicine, the practice of surgery was totally transformed from the middle of the century by the introduction of anaesthetics, and a little later by the new 'antiseptic' methods pioneered by Joseph Lister among others, and made imperative by the 'germ theory' of disease. Operations could now be undertaken without suffering to the patient, so surgeons could take more time and care; surgical wounds were much less likely to become infected. Emboldened, surgeons began to work within, and not merely on, the body.

The appalling state of public health (early in the century the average age of death in London was only 27 years) was affected by the same realisation—that 'dirt brings disease'—leading, for example, to the first sewers to keep human waste away from drinking water. But air and water quality plummeted in the growing cities, with smoke from countless chimneys and effluent from factories.

Doctors got their first tools to find out what was happening inside the body early in the century when the stethoscope, and then the clinical thermometer, were invented. Late in the century came the promise of X-rays to help diagnose injury and disease, and radioactivity to cure some of it, but there would be perils as well. Louis Pasteur and

Robert Koch developed ways to provide protection against many serious diseases through vaccination.

Alfred Nobel's dynamite and other new explosives provided major tools for mining and engineering, but also greater destruction in war. Carnage on the battlefield multiplied with the invention in the 1840s of breech-loading rifles that were much quicker to use, and the first machine guns arrived in time for the American Civil War. Samuel Col's famous 'revolver' came earlier.

At the beginning of the century, houses and factories were lit by candles and oil lamps; streets were mostly dark at night. Gas lighting arrived in the 1810s, and later Humphry Davy's 'electric arc' would light public spaces. The revolution came in the 1870s with Thomas Edison's incandescent 'light bulb' (though Englishman Joseph Swan had invented it first). This provided the motivation to make and distribute electric power on a large scale, to power industry and transportation and in the next century to transform the home and office.

Chemical science and optics lay behind the invention of photography, which began to have an impact on science, especially astronomy, around 1840, and on everyday life later in the century, with George Eastman's 'film' and box cameras.

Agriculture, which still employed most people, felt the impact of new technology: steel ploughs, horse-drawn reaping machines, artificial fertilisers and intensive plant breeding all raised productivity. Industry was shaped not only by steam (for example, James Naismith's 'steam hammer' to shape metal), but by new materials such as the dyes liberated from coal tar.

The shape of cities began to change. Buildings went up and up thanks to the invention of steel frames (and helped by better steels from Henry Bessemer's converter), Joseph Monier's reinforced concrete and the passenger elevator protected by Elisha Otis' safety brake.

The list of life-shaping (and changing) inventions of the nineteenth century is impressive and much too long to enumerate in full, but includes: Howe's sewing machine, Pitman's shorthand, Braille's writing for the blind, Sax's saxophone, Scholes' typewriter, Higginson's hypodermic syringe, McIntosh's rubberised cloth, Westinghouse's air brake for trains, Goodyear's vulcanised rubber and Walker's safety matches.

1851–1900

The World Stage

DURING THIS PERIOD THERE WERE two great movements for national unity. A series of conflicts in the Italian peninsula between 1859 and 1870 led to the formation of the modern Italian state; Rome became the national capital. The political unification of German-speaking states (excluding Austria), led by Prussia and its Chancellor Otto von Bismarck, created the German Empire, with its capital in Berlin. Along the way there were wars with Denmark (1864), Austria (1866) and lastly with France (Franco-Prussian War 1870–71) to secure territory and borders.

Russia continued to push south, seeking access to the Black Sea and the Mediterranean. This precipitated the Crimean War (1854–56) with France and Britain supporting Turkey (the Ottoman Empire) in the conflict. The declining Ottoman Empire was now the 'sick man of Europe', and its weakness encouraged struggles for independence by various Balkan and eastern Mediterranean peoples.

While Europe was uniting, the United States was in danger of coming apart. The American Civil War (1862–1965) between the northern ('Union') and southern ('Confederate') states was largely over the issue of slavery. The more industrialised Union states were victorious and the slaves emancipated by President Lincoln. The post-war years saw a major westward expansion of settlement and growing conflict with Native Americans. Colonial tensions precipitated war between Spain and the United States in 1898.

In the midst of conflict, there were signs of greater international cooperation, with the establishment of the International Red Cross (1864), the Universal Postal Union (1874), and an agreed system of international time zones (1883).

ARTS AND IDEAS

France remained pre-eminent in painting, with the rise of impressionism and the works of Manet, Monet, Degas, Renoir, Cezanne, Millet and Saurat. Elsewhere we saw the religious tones of Holman Hunt in England, the early Picasso in Spain, the revolutionary Van Gogh in Holland, Munch's *The Scream*, and Whistler's mother. Leading the sculptors was France's Rodin. Skylines changed with the early skyscrapers in America, the Eiffel Tower in Paris, the Statue of Liberty in New York and London's Crystal Palace.

France was active in music too, with Berlioz, Saint-Saens, Gounod, Franck, Debussy and Fauré. High romantic music was strong in Germany and Austria: Schumann, Brahms, Bruckner, Mahler and Richard Strauss. Orchestras grew in size, compositions in length and complexity. Elsewhere music drew on nationalism and folk traditions: Glinka, Tchaikovsky, Rimsky Korsakov and Borodin in Russia; Liszt in Hungary; Smetana and Dvorak in Bohemia; Grieg in Norway; and Sibelius in Finland. Opera flourished; Wagner, Verdi, Puccini, Massenet, Bizet, Offenbach, Gilbert and Sullivan. The Strauss family made the world waltz. America had its own classical music through Ives and McDowell, along with Foster's popular songs and Sousa's marches.

Theatres staged new plays by Ibsen, Strindberg, Shaw, Chekhov, Barrie and Feydeau, spanning high drama to farce. Leading poets included Britain's Tennyson, Swinburne, Hopkins and Morris; Rimbaud and Baudelaire of France; and America's Whitman and Dickinson. Readers of novels had abundant choice: Twain, Hawthorne, Melville, James and Alcott from America; Kipling, Stevenson, Carroll, Eliot, Hardy, Dickens, Wells, Trollope and Conrad from Britain; Tolstoy, Turgenev and Dostoyevsky from Russia, Wilde from Ireland, Flaubert, Verne, Hugo, Zola, Maupassant and Dumas from France:

Among new ideas stirring were the philosophies of Nietzsche, Stuart Mill and Compte, the religious writings of Newman, the political theories of Marx and the Zionist Herzl, and the psychoanalysis of Freud. The Salvation Army and the Christian Science movement were founded.

SEEKING UNITY

I n one stunning half-decade of discovery from 1895, researchers across Europe discovered X-rays and radioactivity, first used radio waves to transmit a message, proved the existence of the electron as the fundamental unit of electric charge, isolated radium, found the first viruses and postulated the 'quantum' of energy. These discoveries were almost all unexpected, though some built on decades of work. It was a fitting end to a half-century of burgeoning understanding about the natural world.

Yet each decade of this half-century opened our eyes. In the 1850s we knew for certain that the Earth turns, began experiments with cathode rays, found that disease could be carried by contaminated water, and heard two scientists, working independently, expound the key theory of natural selection to explain the pattern of life on Earth.

Through the 1860s the venerable theory of 'spontaneous generation', of life coming from non-life, was finally dispatched; the Periodic Table of Elements took final shape; the laws of biological inheritance were set down for the first time; a new chemical called nucleic acid was found in cells; and an element unknown on Earth was found in the Sun.

In the decade from 1871, we began to appreciate the place of the human race in the evolution of life forms on our planet, speculated about the possible existence of an unknown form of invisible radiation similar to light, started to make sense of the new word entropy, had a close encounter with the red planet, Mars, and found that malaria is caused by something other than bad air.

In the 1880s, we failed to detect the aether which had been accepted for so long. But the triumphs were more than compensation. Great publicity attended the finding of a vaccine against rabies and the first 'radio waves' were created in the laboratory. Scientists began to unravel the complexities of the chemicals from which living things are made: proteins, fats and carbohydrates. We saw the dance of chromosomes in dividing plant and animal cells and first heard of the puzzling 'Edison effect' which would in time deliver the technology of electronics.

Leon Foucault: The Earth Does Turn!

In 1851 the French physicist Leon Foucault found indisputably what no one had found before: firm proof that Earth turns. A spinning Earth had been an article of faith among astronomers and physicists since the time of Copernicus (**1543**), as had an Earth that went round the Sun. Galileo had faced the Inquisition in **1633** believing firmly in both, though he had no real evidence. Both Edmond Halley's explanation of the trade winds (**1676**) and Isaac Newton's figuring of the shape of our planet (**1736**) depended on an Earth that rotated once a day, but there may have been other explanations.

Foucault put the matter beyond doubt in 1851 when he hung a heavy weight by a long wire from the dome of the Pantheon in Paris. Newton's laws of motion (**1686**) decreed that once set in motion, the pendulum would swing back and forth in the same plane. But it did not, or seemed not to. Over hours and days, the direction of the pendulum's swing slowly changed, pointing in succession to various parts of the chamber in which it hung.

If Newton was right, only one explanation was possible. The chamber itself was moving, turning beneath the swinging pendulum as the Earth turned. 'Yet it does move', Galileo had reputedly muttered under his breath when forced to pubicly deny it. Two hundred years later, Foucault's Pendulum vindicated him.

All this raises profound questions. How can it be that the pendulum ignores the Earth in whose close proximity it hangs and instead holds its course in line with the vastly more distant stars or even the universe as a whole? We are challenged to ask about the origin of inertia. Does this apparently fundamental property of any bit of matter depend on the existence of the rest of the universe? Ernst Mach was one of those who wrestled with that conundrum (**1877**).

Louis Agassiz: From Heidelberg to Harvard

1851 Swiss-born biologist Louis Agassiz was Charles Darwin's best known and most vocal opponent. He remained unconverted to the concept of biological evolution until the day he died. A devout man, he saw the hand of God everywhere in nature, and could not accept a view like Darwin's that left no room for divine design. A species of plant or animal was, he said, 'a thought of God'. Yet Darwin and others used the evidence from Agassiz' own extensive research to support the theory that Agassiz so strenuously opposed. That is often how science progresses.

Agassiz was an expert, the expert, on fossil fish, found with increasing richness in the old rocks from what was now called the Devonian period. His 1843 book about them was so detailed and comprehensive that it increased tenfold the total number of vertebrates (animals with backbones) described by science. His knowledge of the pattern of ancient life was immense and he distilled it into a system of understanding quite different from Darwin's, and which he first published in 1851.

Agassiz saw a pattern of growth and development from the oldest, simplest organisms to the most recent and most complex, and not only in rocks. It was evident in the growth of the human embryo, which went through stages of looking like the young of a succession of simpler organisms, and in the way plants and animals were spread across the planet from the poles to the equator. The same pattern everywhere, he proclaimed. Surely evidence of divine design.

Agassiz was born in Germany but ended his life in the United States as a professor

> *Every scientific truth goes through three states: first, people say it conflicts with the Bible; next, they say it has been discovered before; lastly, they say they always believed it.*
>
> LOUIS AGASSIZ

at Harvard for a quarter-century. There he was mentor to dozens of American naturalists and became immensely popular and well known through public lectures.

Agassiz' own mentor was French biologist Georges Cuvier (**1812**), arch-apostle of the theory of catastrophism (**1778**). Cuvier, Agassiz and others saw Earth history as a series of immense violent events, including even the Biblical flood. Nothing happened gradually or slowly as the 'uniformitarians' believed. For Agassiz, the concept of ice ages (**1836**), which he did much to promote and justify through research (and which he once called 'God's great plough'), simply continued the pattern of recurring catastrophes.

Claude Bernard:
The Importance of the 'Milieu Interieur'

Frenchman Claude Bernard wanted to be a successful playwright but lacked the talent, despite spending every spare moment outside his boring job as an apothecary's apprentice attending the theatre. He showed his epic verse drama *Arthur de Bretagne* to a Paris critic and was bluntly told he should try something else.

1851

Bernard chose medicine, and despite being both poor and shy, and having trouble passing exams, he secured a post as a research assistant to a leading doctor. He married a rich physician's daughter to bolster his meagre pay, but the marriage was stormy. Nonetheless, his career blossomed. He was a skilful and dedicated researcher and important discoveries in human physiology began to flow.

His first triumph was with the pancreas, whose function was unknown at the time. Bernard found it makes a chemical (an enzyme) that helps break down fats when secreted into the gut. He also discovered that most digestion takes place not in the stomach but in the small intestine. The liver then attracted his attention. He found that it manufactures a carbohydrate he called 'glycogen', made up of small units of blood sugar (glucose) joined together. The liver stores this energy-rich substance and releases it back into the blood as needed.

His most powerful insight, made around 1851, came from studying the blood supply to the various organs and the skin. This appeared to be affected by the external temperature, with the blood vessels widening or narrowing to increase or decrease the blood flow. In very cold weather, oxygen-rich blood is withdrawn from the skin (which is why we go white or even blue with cold) and concentrated into the vital organs to keep them going. When it is hot, the skin is flooded with blood (we 'flush') so that excess heat can be released to the air.

In this way, we keep our internal body temperature almost constant, at the level where our systems work best. Shivering while cold and perspiring while hot are part of this temperature control mechanism. If we stray too far from this preferred temperature, we can die from heatstroke or hypothermia. Bernard realised that the body has a series of such controls, affecting the various body systems and keeping our internal environment (the *milieu interieur*) stable. He called this mechanism 'homeostasis'.

To establish this and other discoveries (such as that the protein haemoglobin in red blood cells carries oxygen which can be replaced by carbon monoxide with lethal consequences), Bernard used both observation and careful experiment to test hypotheses, believing that the same laws of physics and chemistry apply to living things as to inanimate matter. He was perhaps the first medical scientist in the modern sense and his textbook on his methods was the 'Bible' on the subject for many years.

Despite his peasant origins, Bernard became one of the most famous and prolific scientists of his day. When he died in 1865 from a kidney infection, the French Government paid for his funeral—he was the first French scientist so honoured.

Edward Frankland: Valency Is the Key to Chemistry

1852

It was in the nineteenth century that people first realised the importance of clean water. Water free from contamination is not only pleasanter to drink and wash in; it does not spread disease or cause illness. John Snow's experiment with the water pump in Broad Street, London, during an outbreak of cholera (**1854**) showed that water fouled by sewage or industrial waste is dangerous. This naturally led to an interest in measuring biological or chemical contamination in water, and in ways of purifying it, say by filtration.

English chemist Edward Frankland was a key figure in all this. As a member of a Royal Commission into the pollution of rivers, he spent six years at the laboratory bench documenting in detail the sorts of contamination commonly found. He was already providing monthly reports on the quality of London's water; he continued to do this till the end of his life and was able to document the steady improvement in

> *I am convinced that the future progress of chemistry as an exact science depends very much on the alliance with mathematics.*
>
> EDWARD FRANKLAND

quality as government regulations imposed ever-tighter controls on what could be dumped into rivers and streams.

Frankland's reputation was already well established. As a chemist concerned with practical matters (like Humphry Davy, he had begun his career as an apothecary's apprentice) Frankland had few equals, but his contribution went further. In 1852 he defined a fundamental feature of chemicals when they react and combine with each other. The idea came from his intensive study of the 'organo-metallic compounds' made by reacting metals like zinc, mercury, arsenic, antimony and tin with an 'organic' (carbon-containing) compound like those made by living things. Many of these have proved quite nasty; methyl mercury in drinking water can cause severe deformities in newborn babies.

Frankland found very regular patterns in the composition of these compounds; the ratios between numbers of atoms of nitrogen, phosphorus, arsenic and antimony and the numbers of atoms of other elements always seemed to involve three or five. Such patterns turned up elsewhere, so much so that Franklin was able to define an atom's 'combining power', in terms of the number of atoms of other elements it would combine with. These numbers appeared to be fixed. For nitrogen and phosphorus and their compounds the ratio commonly included three or five, for oxygen and sulphur it included two or perhaps six, for carbon and silicon four, for active 'alkali' metals like sodium it was one, as it was for the halogens (chlorine, bromine and iodine).

We now call this property 'valency' and use as a reference the valency of hydrogen, taken to be one. So the formulas HCl, H_2O, NH_3, CH_4 show that chlorine, oxygen, nitrogen and carbon have valencies one, two, three and four respectively. This insight enabled chemists to make sense of the compositions of hundreds of different compounds and even to predict the composition of compounds not yet discovered. Valency has proved a major unifying principle in chemistry.

Frankland admitted that he could not explain valency and he did not try. It was enough that it was so. The principle turned up again in the Periodic Table of Dimitri Mendeleyev (**1869**). The elements in a column of this table, also known as a group (such as the quartet of nitrogen, phosphorus, arsenic and antinomy) all have the same valency. As for an explanation, that had to wait until the twentieth century and the genius of Niels Bohr (**1926**).

James Joule and Lord Kelvin: Gases Expanding and Cooling

Try this little experiment. Blow on the back of your hand—the air will feel cool. Blow with an open mouth—the air will feel warm. Pump up a bicycle tyre; the barrel of the pump will get hot. Let the air out of the tyre (or a balloon)—it will feel cold. These are simple, everyday observations, but not till 1852

1852

did English scientists Lord Kelvin (**1846**) and James Joule (**1847**) take real notice. So the tendency for a gas to get colder as it expands (increases in volume) and hotter when it's compressed (forced into a smaller space) is called the Joule–Thomson Effect (Kelvin was baptised William Thomson).

The effect explains a lot of things, such as why it is colder at the top of a mountain than at the bottom, even though at the top you are closer to the Sun. Heated air rising from the ground expands (because pressure gets less) and so it cools. The effect became a great way to turn gases into liquids. A gas is first compressed (and allowed to cool down again, since it will have got hot). It is then expanded through a nozzle and can chill so much that it drops its boiling point, condensing into a liquid. By 1900 air had been liquefied, as had nitrogen and oxygen separately.

The Joule–Thomson Effect should not happen at all, not if the gas particles behave like tiny ping-pong balls as scientists had thought until that time. Such particles would bounce off each other when they collide but otherwise not interact at all. To explain the Joule–Thomson Effect, something had to be added to the 'model': the particles of a real gas must have a slight attraction for each other, though not enough to make them stick together (that happens in liquids). You have to expend energy to lift a stone up against the pull of gravity. Likewise, when a gas expands and the particles move further apart, energy is needed to fight the mutual attraction. This has to come from the energy of motion of the particles, so they slow down, and the gas becomes colder.

The Joule–Thompson explanation also handles the 'champagne cork' effect. Pop the cork from some sparkling wine and you often see a patch of mist inside the neck of the bottle. The sudden release of pressure cools the air enough to turn the water vapour in it into tiny liquid drops and so make a small cloud. ➤➤**1862**

Heinrich Geissler: Electricity through the Air

1854

As the 'electric century' unrolled (**1800**), some physicists were keen to make electricity pass through air in the laboratory, just as happened in nature in a stroke of lightning. It was not easy, and needed a current at very high voltage (**1826**). They soon found that the current travelled more easily if some of the air was first pumped out. A partial vacuum was more willing to carry the current than air at normal pressure.

In 1854 the German physicist Heinrich Geissler showed the way with his 'discharge tubes'. He sealed two metal plates (electrodes) into a tube closed at both ends and connected the plates to an induction coil, a sort of transformer (**1831**) to give lots of voltage. One electrode was positively charged (the anode), the other negatively charged (the cathode).

The tube was also connected to an air pump. These had come a long way since Otto von Guericke and Robert Boyle 200 years before (**1657, 1659**). Back then they were just pistons and produced only a mediocre vacuum. Now Geissler had designed a

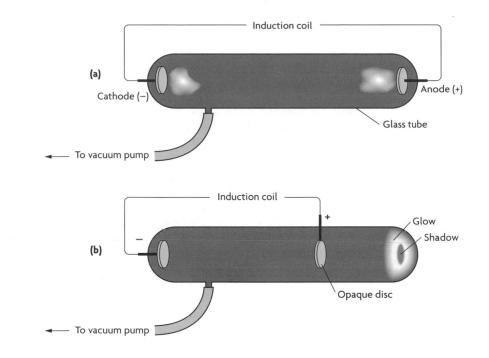

(a) The simplest 'discharge tube' is a sealed glass container with two embedded metal plates (cathode and anode) connected to a high-voltage source. A vacuum pump extracts air from the tube. As the pressure falls, patches of light appear with dark spaces between.

(b) At very low pressures, the tube is dark but the glass at the end opposite the cathode begins to glow. The anode casts a shadow in this glow. Electric and magnetic fields move the shadow about, showing that the invisible 'cathode rays' that cause the glow are charged particles.

vacuum pump that used mercury vapour to collect and sweep away any remaining air particles. Later models would leave behind only one in every 100 000 air particles.

As he pumped the gas out, Geissler saw the gas begin to glow; the electric discharge produced different-coloured glows in different gases; this was a foretaste of later neon signs. As more gas was pumped and the pressure fell further, the glow was concentrated in certain places with darkness in between. With virtually all the air gone, he saw something really fascinating. The discharge tube itself had become dark, but the walls at one end of the tube glowed with a mysterious green light.

Other experimenters, such as Julius Plucker in Germany and William Crookes in England (he was later also active in psychic research), worked hard to understand what made the glow. An early guess was that some invisible form of energy was coming from the cathode, since the walls up that end of the tube did not glow. Soon the talk was of 'cathode rays'.

But what sort of rays were these? The evidence was not clear. Right behind the anode the tube walls did not glow. Some tubes were made with an anode in the shape of a Maltese cross to highlight this shadow effect. The Germans claimed that to cast a sharp shadow the rays had to travel in straight lines, as light did. William Crookes made a tube which let the rays turn a tiny paddle wheel placed in their path. This meant that they had momentum and could exert a force, as particles could.

So were cathode rays like light waves or like particles? Further research (**1895 Cathode Rays**) would answer that question and also lead us to a host of discoveries: X-rays, electronics, television and the ultimate structure of matter. Quite a list.

Hermann Helmholz: The End of the Universe

1854   German physicist Hermann Helmholz, well known for his studies of musical notes and the operation of the human eye, was first to predict the 'heat death of the universe'. In his dead universe, there would be plenty of energy, but we would be unable to use it and so everything would stop. In the biggest picture, this death appeared inevitable.

In his time, a new understanding of heat linked it with motion; to be specific, the motion of atoms and molecules in matter. Gases make up most of the universe. According to the kinetic theory of gases, one sample of a gas is hotter than another because its particles are moving faster, at least on average. Mix two samples of gas at different temperatures and the particles will start to collide, sharing their energy of motion. Fast-moving particles from the hotter gas will give up energy to slower particles in the colder gas. Given enough time, particles throughout the mixture will have the same average speed; any temperature differences will vanish, leaving 'thermal equilibrium'.

Helmholz believed this smoothing-out process went on all the time in the wider universe. Over the eons, initial temperature differences must disappear, leaving bland uniformity. In other words, entropy (meaning disorder) must reach a maximum (**1865**). Why does it matter? The particles of gas are still rattling around, so there is plenty of energy out there. The Law of Conservation of Energy forbids any to be lost.

But that misses the point. A gas turbine operates because the gases going in are hotter than those coming out. Winds blow towards a region which is hotter than the place they leave behind. Differences in temperature are crucial in engines and in nature. Without them, no useful work can be done. Such a 'heat death', said Helmholz, awaits the universe at large.

But sometimes the biggest picture is too big. Another German physicist, Ludwig Boltzmann, had better news. In 1897 he saw no reason why large chunks of the universe, as big as galaxies or more, should not avoid 'thermal equilibrium' and stay full of diversity, life and action for long periods of time. As long as there is energy to spend on maintaining temperature differences, we can fight off the 'heat death'. Our corner of the universe is currently doing just that.

John Snow: Cholera in the Water

In 1854 cholera broke out again in the Soho area of London. Epidemics of cholera, typhus and typhoid cost the crowded cities of Europe dearly, with hundreds of thousands dying in the worst outbreaks. Many doctors agreed that disease and filth were somehow associated. The primitive sanitation of the time saw human excrement and urine, mixed with other rotting garbage, pile up in stinking cesspools in the streets. Contagion often seemed to follow. The loss of life was most severe among the poor in crowded, dirty and damp suburbs without proper drainage.

But how was the disease actually generated and spread from the decomposing wastes to humans? The 'miasma' theory was the most popular. The rotting wastes sent out some sort of malignant vapours that oozed into houses and brought disease. Supporters of the miasma theory included leading health authorities and the pioneering nurse Florence Nightingale. Officials in charge of sanitation responded by building sewers to trap the noxious airs (and the wastes that produced them) in underground pipes and carry them out of harm's way, such as into the nearest river.

London doctor John Snow had a different theory. He thought that such disease was carried by water rather than by air, perhaps because the water contained tiny living things like the 'animalcules' which Dutchman Anthony van Leeuwenhoek had seen with his microscope (**1673**). In the eighteenth century, Austrian doctor Markus Plenciz had claimed that 'every disease has its organism' (**1762**). But no one had linked any particular organism to a disease, so the 'germ theory' struggled for credibility.

In the 1854 outbreak Snow used a map to plot where the victims had lived. The houses of nearly all of the dead stood very close to a certain well and pump in Broad Street. People near other pumps did not get the disease unless they had taken their water from the Broad Street pump or sent their children to school near it. Snow arranged to have the handle of the pump removed so no one could drink from it. The cholera outbreak died out almost at once.

Snow had not only confirmed that cholera passed through water rather than air and so saved many lives; he had initiated a new approach to public health known later as epidemiology. Careful study of the distribution of a disease often helps track down the cause. Snow did not know what it was in the water that caused the disease. Though neither he nor anyone else in London knew it, the culprit was a tiny living organism called *Vibrum cholerae*, identified that same year.

Wonders from Coal Tar

Coal tar does not look very promising; it's a thick, black, smelly, oily mass left behind when coal is strongly heated ('destructively distilled)' in a closed container to make coke (for smelting iron ore) and coal gas (for lighting and heating). Chemists soon guessed that coal tar is not a single substance but a mixture of compounds, some of which could be liberated by gentler ('non-destructive') distillation. If the heating was done carefully and at the right temperature, the various compounds could be driven off and condensed one by one.

An early discovery was the compound aniline and its close relative phenol. The German Friedlieb Runge found those in 1834. Phenol would later be called 'carbolic acid' and valued for its ability to kill the micro-organisms that could spread disease. It played a major role in the early growth of antiseptic surgery.

Guessing a bit how the colourless, oily aniline was put together, English chemists thought it might be similar to quinine, the only known defence against malaria, then a scourge of the British Empire. In 1845, the leading German chemist Wilhelm Hoffman was imported to run the newly established Royal School of Chemistry in London, with the major task of somehow converting aniline into quinine.

In 1856 William Perkin, a laboratory assistant still in his teens, reacted aniline with potassium dichromate and washed the result in alcohol. He found not quinine, but a beautiful purple dye, which he called Perkin's mauve or mauveine. It was the first synthetic coloured dye, and the source of Perkin's personal fortune. He dropped out of the college to set up a dye-making business with his father's money. Mauveine became immensely fashionable and popular for women's clothing. Perkin prospered. Biology benefited too. Perkin's mauve proved very useful in staining plant and animal specimens and revealing hidden details (**1869 Miescher and Flemming**).

The synthetic or coal-tar dye industry was born. It grew rapidly, especially in Germany once Hoffman returned to his homeland. Synthetic dyes in many other colours were soon developed; together with drugs and other useful products made from coal tar, they would be the well-spring of German industrial and economic strength in the twentieth century. There was a downside; many compounds distilled from coal tar would prove to be cancer-causing.

Finding Our Ancestors

1856 In 1856 some puzzling fragments of bone turned up in a limestone quarry in the Neander Valley (Neanderthal in German), reputedly named after composer Joachim Neander, who lived nearby. The workers thought the bones, which included pieces of a skull, arms, legs and ribs, might be from a bear, but they looked more human, if not quite. As more fragments of 'Neanderthal Man' accumulated, it appeared that these creatures, apparently now extinct, could plausibly be our ancestors. They were short, stocky and powerfully muscled, with large noses, rugged skulls, sloping foreheads and receding jaws, and they had larger brains than we have. They seemed adapted to living in cold climates.

Over 400 specimens of these creatures are now known, spread throughout Asia and Europe. They seemed to have flourished over several hundred thousand years, when the regions where they lived were often covered in ice. The name now given to them, *Homo sapiens neanderthalensis*, shows that they were of our species. Dressed in furs, they would have been hard to distinguish from a modern Eskimo.

Another group of 'pre-humans' was first found in France in 1868. 'Cro-Magnons' apparently coexisted with the Neanderthalers but were almost identical to us in appearance. We now know that with Cro-Magnons came the first evidence of culture, such as delicate jewellery and the great cave paintings of France and Spain. It seems that, in time, the Neanderthalers died out and the Cro-Magnons became us, but the details of what went on while they shared parts of the planet are still being argued.

Today we have no problem in accepting that these creatures are mostly likely our ancestors, but it was a controversial idea at the time they were first discovered, When Charles Darwin proposed in *The Descent of Man* in **1871** that we humans shared an ancestry with other 'primates', notably the great apes such as chimpanzees and gorillas, many people were outraged. The notion was offensive, and even heretical, contrary to the word of Scripture that humans were a special and separate race created by God. But in science a theory has to stand or fall on how well it fits evidence, and evidence was growing steadily.

In 1891 Dutch doctor René Dubois, spurred on by ideas from Darwin's contemporary, Alfred Wallace (**1858**) went hunting for evidence in the East Indies (now Indonesia). In East Java he found bones (a skull cap and a femur) of a creature he first thought was an ape but later decided was an 'ape man' (*Pithecanthropus*), dating back perhaps a million years.

Dubois' assertions about his 'Java Man', such as that he walked upright, were greeted with disbelief; they went against both the religious orthodoxy and the scientific understanding of the time. To damp down the controversy, Dubois re-interred the bones under his own house, where they remained for 30 years.

But similar bones were uncovered in China in 1927 ('Peking Man') and others have been found in Africa, including an almost complete skeleton in Kenya ('Turkana Boy') in 1975. The creature is now called *Homo erectus* ('upright man'), making it close to modern humans (that is, a 'hominid') but not of our species. *Homo erectus* was tall and thin, suited for life in hot climates.

So a believable story of human ancestry, based on the bones, was beginning to emerge; with *Homo erectus* descending from the same forebears as the great apes, evolving into the Neanderthalers and Cro-Magnons and then into us. But many gaps remained (**1964 Leakey**).

Louis Pasteur: Unravelling Fermentation

Louis Pasteur was a rising star in French science when in 1856 he was appointed Professor of Chemistry at the University of Lille, an industrial town. Pasteur had a strong interest in practical matters and was quick to offer problem-solving help to people running factories and distilleries. The father of

1856

> *There does not exist the category of science which one can give the name applied science. There is science and the applications of science, bound together as the fruit is to the tree that bears it.*
>
> LOUIS PASTEUR

one of his students called him in; he was making alcohol by fermenting beetroot, but sometimes got lactic acid instead. What was going on?

At the time, no one was really sure what fermentation was, only that it went on. It clearly involved some sort of chemical change which turned sugar into alcohol and released carbon dioxide, but what caused it? Some argued that yeasts found floating in the vats were involved, but the majority opinion said yeasts were either irrelevant or a result of fermentation. To say they were the cause was too much like 'vitalism', the old and increasingly discredited philosophy that gave living things properties beyond scientific study.

Pasteur soon had some evidence on the matter. His microscope revealed only yeasts in the vats making alcohol. In the vats where the fermentation was off track and making lactic acid, the yeasts were mixed with other tiny rod-shaped creatures. He also found that some of the chemicals being made had molecules twisted only one way, that is, they were asymmetric. Based on his earlier work (**1847**), Pasteur believed that the presence of a chemical with only one 'handedness' (not both) was an indication that living things were at work.

Pasteur reached two conclusions. The alcohol and the acid were actually being made by the micro-organisms ('microbes'). Otherwise they would not have been asymmetric. Also, while the yeasts made alcohol, the other microbes made acid. Keep the offending microbes out of the mix and all would be well. Pasteur's findings, followed up over the years, transformed quality control in industries which used fermentation. Manufacturers could be confident about their product if the microbe mix was right.

Soon any sort of fermentation, such as the souring of milk or the rotting of human wastes, was being blamed on microbes, and many of them were being found under the microscope. Pasteur found that microbes could be killed by heat, even moderate heat for a short time. This was the beginning of 'pasteurisation', now universally used to ensure the safety of foods and prolong shelf life by killing off the bacteria that cause them to go bad. ➤➤**1862**

Alfred Wallace: In the Shadow of Darwin

1858

The relative obscurity in which English biologist Alfred Wallace languishes is an accident of history. The theory of evolution, one of the crowning achievements in biology, was developed by Wallace quite independently of Charles Darwin (**1859**). Yet Darwin is the name on most lips today.

The matter runs deeper. Darwin had been reluctant to promote his decades-old ideas because he feared a hostile response. Only when Wallace sent him an essay with the typically Victorian title *The Tendency of Varieties to Depart Indefinitely from the Original Type* did Darwin decide to 'go public' with his very similar concepts. That decision sparked a joint presentation of separate papers describing the findings of each man to the Linnean Society in London in July 1858 (at which neither author was present), and drove Darwin to pull all his abundance of evidence together and go to print with *The Origin of Species* the next year (**1859**). Wallace did not bring out such a book until 1870.

The ideas presented were strikingly similar. Both Darwin and Wallace followed up the gloomy thoughts of the English economist Thomas Malthus concerning the never-ending struggle for existence. From that flowed the insight that the 'fittest survive' by taking advantage of small variations in their bodies or behaviour. The two men drew on a similar background of exploring and collecting. Darwin's three years on the *Beagle* (**1836**) were matched by Wallace's four years in the Amazon rainforest and eight years in the East Indies.

Wallace's career mixed tragedy with triumph. A fire and shipwreck destroyed all of his Brazilian specimens and most of his notes. Yet his studies in the Malay Archipelago stimulated his thoughts on biological evolution and generated the theory of the Wallace Line, running between Borneo and Sulawesi and south between Bali and Lombok. The flora and fauna on either side of the line are very different, telling much about the history of life and land movements in the region. Wallace was the founder of 'biogeography'.

Wallace lived to be nearly 91. He had his share of honours, including the Order of Merit and a string of medals from learned societies. He was certainly well regarded in his time. Yet today, Darwin has the popular reputation as the founder of evolution, perhaps because of *The Origin of Species*, perhaps because of what came later (**1871**).

Friedrich Kekulé: Hydrocarbons and Snakes

It seems that German scientist Friedrich Kekulé was not much of a practical chemist. Nor could he be described as an inspiring teacher. Yet he left an indelible mark on chemistry which has held to this day. The way he made his big find is an intriguing tale.

1858

Kekulé wanted to explain the variety and growing number of compounds collectively called hydrocarbons. Once American chemist Benjamin Silliman started distilling crude oil in 1854, Silliman identified a dozen or more pure substances, mostly liquids, which seemed to be made only of carbon and hydrogen. They had different properties. Some of these hydrocarbons were light, others heavy (more dense). Some boiled easily, others needed to be heated to a high temperature.

Kekulé worked on the assumption that carbon had a 'valency' of four (**1852**), meaning that a carbon atom could combine with four atoms of hydrogen. That explained the formation of methane with a formula CH_4. What about all the others? Kekulé imagined that each carbon atom had four 'hooks' to connect to other atoms. Suppose one hook connected to a second carbon atom, with the three other hooks on each carbon atom snaring hydrogen atoms. We would have a compound with two

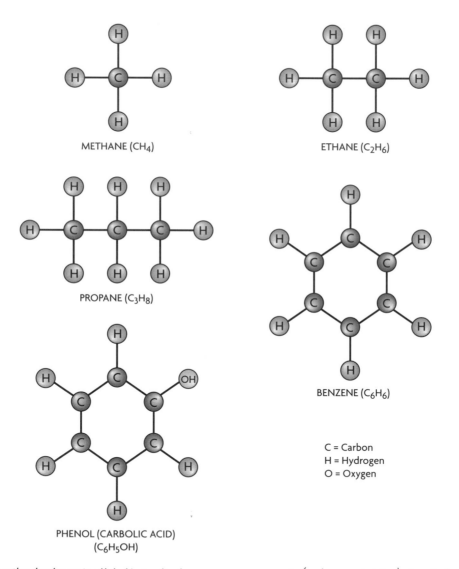

*How the daydreaming Kekulé visualised some common organic (carbon-containing) chemical compounds, including the three lightest hydrocarbon gases, and benzene and phenol. The atoms in these compounds are held together by 'covalent bonds' (**1928**), marked by the bars. Phenol was one of the first compounds extracted from coal tar (**1856**) and was later used to kill bacteria.*

carbon and six hydrogen atoms and a formula C_2H_6, formally called ethane. Three atoms of carbon connected in a line and joining with hydrogen atoms would produce C_3H_8 or propane; C_6H_{14} would be hexane; C_8H_{18} would be octane. The chains could probably be of any length. Each additional carbon atom would make the liquid thicker and harder to boil.

Kekulé once said the idea of linking carbon atoms in chains came to him during a daydream on a London bus. A better known variation of the story deals with a hydrocarbon called benzene, which Michael Faraday had discovered (**1813**) which would not obey his rule. Analysis of benzene showed its formula should be C_6H_6. No linking in long chains would give that. It seems that as Kekulé dozed on that bus (or another) he had a vision of snakes, one of which suddenly took its tail in its mouth. He awoke to proclaim that the six carbon atoms in benzene were linked in a ring (more precisely a hexagon). Each clutched a hydrogen atom and linked with either one or two hooks with the carbon atoms on either side. The 'benzene ring' is the most important structure in the vast range of compounds that make up organic chemistry.

In another version of the story, Kekulé dreamt of six monkeys in a ring, each holding another's tail. Different animals, same outcome. The details do not really matter.

Charles Darwin: *The Origin of Species*

The Origin of Species was published in 1859. Advance publicity by Darwin's friends, who were influential in science and society, combined with the book's accessibility to non-scientific audiences to ensure a lot of public interest. Reviews appeared in prominent newspapers. The first print run of 1000 copies sold out in a single day. *The Origin of Species* remains the most discussed scientific book ever published.

1859

Darwin had been working on his 'big book of species' for 20 years, since he returned from his voyage on the *Beagle* (**1836**). Anticipating the controversy his ideas would generate, he had been reluctant to go into print, much like Copernicus (**1543**). But he had little option once fellow English biologist Alfred Wallace put forward his very similar ideas in **1858**.

The full title of the book sums up Darwin's vision: *On the Origin of Species by Means of Natural Selection or the Preservation of Favoured Races in the Struggle for Life.* That new species of plants or animals could appear, that the pattern of life on Earth was subject to change, was not now controversial. Most people saw evidence for change, gradual or violent. Darwin opted for gradual change; he was in the camp of the 'uniformitarians', like Charles Lyell (**1829**).

He rejected (because the evidence did not fit) the style of evolution expounded by Jean Lamarck (**1809**), where characteristics acquired by animals and plants during

> *I see no good reason why the views given in this volume*
> *should shock the religious feelings of anyone.*
>
> CHARLES DARWIN (IN *THE ORIGIN OF SPECIES*)

life were passed on to their offspring. Yet he saw, as Lamarck did, environment as the crucial dynamic of change, especially through the struggle for existence. 'Favoured races' would be better fitted by small variations in body or behaviour to secure food, mates and living space and so perpetuate their kind. Over time, favourable adaptations in an isolated population would accumulate to the point where the plants or animals could no longer interbreed with other populations. They would have become a 'new species'.

Darwin saw this process of 'natural selection' as comparable to the work of plant and animal breeders who consistently chose for future propagation those that best suited their needs and so were best 'fitted to survive'. *The Origin of Species* begins with an exposition on the breeding of pigeons.

This new view had a far-reaching implication. If new species continually form from existing ones by natural selection, then all life has a common origin. Evolution has produced the most complex forms of life from the simplest. ➤➤**1871**

John Phillips: The Three Layers of Life

1860

William 'Strata' Smith (**1799**), the self-taught and hard-pressed English geologist, left a powerful legacy; the first use of fossils to identify, map and order the many known layers of rock, the first detailed geological maps made anywhere. The legacy also includes his nephew John Phillips, who came under his care when the young man's parents died. Phillips accompanied Smith on his many trips across England to chart rocks, and later wrote Smith's first biography.

Inspired by Smith, Phillips went on to establish a career and reputation of his own. He gained a professorship at King's College in London, as well as the Geological Society's Wollaston Medal, which his uncle had been the first to receive. Most notably, he continued to name and classify the various large-scale groupings of rock strata; increasingly these were being seen as containing the enduring record of ancient events, including the development of life.

In 1860, with an eye to the big picture, he divided the rock layers that carried fossils into three big groups. The oldest he called Palaeozoic for 'old life'; in those rocks the most prominent and plentiful fossils were of fish. Next in terms of years were the rocks of the Mesozoic or 'middle life' age. Here, remnants of reptiles, including the newly named dinosaurs (**1811**), were the distinctive feature. Lastly (and uppermost) came the Cenozoic or 'new life' rocks, where mammals first became

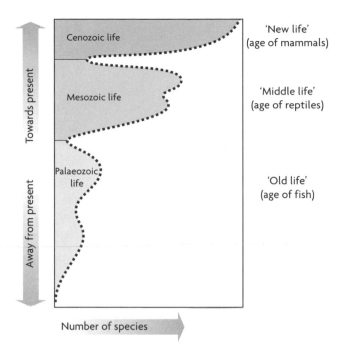

John Phillips mapped out the changing diversity of life through three ages represented by successive layers of rocks. Note the significant fall-off in diversity between the three major eras. This was the first real recognition of 'extinction events'.

abundant. In terms of Giovanni Arduino's earlier work (**1759**), Palaeozoic and Mesozoic rocks were 'secondary' and Cenozoic rocks were 'tertiary'.

Philips noted something more. In terms of the numbers of different fossils, each successive era had a greater diversity of life, especially in the ocean. There were more different forms of life in the Mesozoic Era than in the Palaeozoic Era, and more again in the Cenozoic Era.

More sobering was the fact that between these eras the diversity of life was decimated. Many (or even most) life forms simply died out. For reasons not yet known, the river of life on Earth had ebbed and flowed over the eons of time, sometimes gently, sometimes catastrophically. William Smith had already noted the mass extinction, the greatest known, between the Permian period that ended the Palaeozoic Era and the Triassic period that began the Mesozoic Era. To that extent, Philips carried on where his uncle had left off.

Lord Kelvin and the Age of the Earth

In 1862 the ever-confident Lord Kelvin laid down a challenge to Charles Darwin, whose book *The Origin of Species* had appeared only three years before (**1859**). Kelvin the physicist argued that Darwin the biologist was wrong to state that the many and varied species of plants and animals around today were a result of a very slow process of biological evolution. There had not been anything like enough time for this to occur, Kelvin proclaimed.

1862

The basis of this assertion was simple: if the Earth had begun as a molten globe of rock, physicists could calculate the time taken for it to cool to its present temperature, if not with precision, at least to an order of magnitude. The sums indicated a range of 20 million to at most 400 million years. If true, this was a headache not only to the biologists, who needed much more time than that for the many steps in biological evolution, but also to the geologists, who saw the Earth-shaping processes of erosion and mountain-building as equally slow.

Kelvin had support from astronomers like the American Simon Newcombe. If the Sun had formed from a vast nebula of gas and dust (**1796**), it must have been shining for at most 100 million years. Any longer and it would have gone cold. This assumed that all the Sun's heat came from the contraction of the nebula under gravity; the true source of the Sun's energy, which can keep it shining for billions of years, was unknown until **1938 (Bethe and von Weiszacher)**.

Kelvin's confidence in his prediction should have been undermined by an event in his lifetime—the discovery of radioactivity by Henri Becquerel in **1896**. Radioactive elements like radium (**1898**) give off heat: French physicist Pierre Curie found in 1903 that radium could melt its own weight of ice in an hour. It was soon clear that radioactivity was widely spread throughout the Earth and could therefore have provided enough continuous heat to keep the Earth warm for billions of years, enough time for both biological evolution and slow geological change to have their impact. One of the first to acknowledge this was Charles Darwin's son George, then an astronomer at Cambridge.

Kelvin, being Kelvin, found it hard to admit publicly that he had been wrong and went to his grave in 1907 supporting a 'young Earth'. But he was prepared, so it seems, to say privately that his calculations had been based on 'incomplete knowledge'. **➡1900**

Louis Pasteur: 'Life Comes From Life'

1862

Over the centuries, the notion of 'spontaneous generation'—that living things could pop into existence from non-living matter—had remained popular. Rats appeared around decomposing garbage, eels in stagnant water, maggots in rotting meat. Even when it was argued that animals hatched from eggs laid by their parents, there were still issues. The crust of growth that covered the skin of a rotting orange seemed to come from nowhere. And now there were the microbes, tiny organisms found in water and elsewhere but visible only under the microscope. Where did they come from?

Spontaneous generation had been challenged often enough. Francesco Redi (**1670 Swammerdam**), Lazzaro Spallanzani (**1768**) and others had found that meat covered to keep the flies away did not generate maggots, and that boiled meat broth sealed from the air did not go bad. The knockout blow was delivered by Louis Pasteur, already famous for his work on fermentation (**1856**) in wine, beer and milk. He knew the microbes caused those, so maybe other forms of decay involved them as well.

No one one could seriously argue with his 'swan-necked' flask experiment. A meat broth was boiled and placed in a flask whose top was drawn out into a long tube, dipping down then up, like the neck of a swan. If the end of the tube was sealed to exclude air, the broth did not 'go off'. If the very end was then broken off, so letting in air, the broth still stayed fresh, because (as Pasteur argued) the dust particles carrying the microbes causing the fermentation were trapped in the curved tube and could not reach the broth.

Pasteur backed this up with any number of other demonstrations. Arguing that the fermentation of wine was due to yeasts that collected on the skin of the grapes, he showed that grapes grown under a fine gauze that kept off the dust (and the yeast) would not make wine. Grape juice drawn from under the skin would not ferment.

After all this, no one could really dispute Pasteur when he said that spontaneous generation was a myth. Microbes arose from other microbes like themselves: 'Life comes from life'. That left open the question about the origin of the very first life, but no one was ready to tackle that yet. ➤**1886**

Ordering the Elements Part 2

Thirty-five years after the German Johann Dobereiner tried in **1829** to find some order among the swelling ranks of the chemical elements (and failed to convince most of his colleagues), English chemist John Newlands took

1864

up the challenge. By this time, the idea that atoms of the various elements had a distinctive 'atomic weight' was well established, and the values of these were much better known. Newlands arranged the known elements, by then about 60 in all, into a list according to atomic weight. At the top was the lightest atom, hydrogen; at the bottom the heaviest then known, uranium.

Playing around with this list, Newlands found that arranging them into columns of seven produced an interesting pattern. Each row held elements with similar properties: potassium sat next to sodium, sulphur next to selenium, calcium alongside magnesium. And each of Dobereiner's triads had a place. Making an analogy with the seven notes of the musical scale, Newlands called his layout The Law of Octaves. But there were dissonances among the harmonies. Some of the rows contained elements with wildly differing properties. Again the majority was not convinced. As with Dobereiner's triads, Newland's octaves were dismissed as a coincidence.

Next to try was the German chemist Julius Meyer. He calculated how much space the atoms of each element took up (the 'atomic volume'), and arranged those figures according to atomic number. The numbers rose and fell like a wave, peaking with the 'alkali metals' like potassium and sodium, which had larger atoms than other elements near them in the list. A run of elements from one peak to another defined a 'period'. The first two periods had seven elements in each and corresponded to Newland's octaves. But the third and later periods were much larger. Some order was becoming apparent but the code had still not been cracked. The breakthrough was to come in **1869**.

Gregor Mendel: The Rules of Inheritance

1865   Austrian monk Gregor Mendel wanted to teach biology, but spent most of his time in his garden. He pondered one key question. What determines how the various characteristics of a plant are passed from one generation to another? In a now famous series of experiments begun in 1857, Mendel grew varieties of peas, concentrating on a few obvious characteristics such as the height of the plants, whether the seeds were smooth or wrinkled, the colour of the flowers and so on.

He started with 'true-breeding' lines, where every plant in every generation was the same. For example all plants in the lines were either tall or short. He then 'cross-bred' the various strains, transferring pollen from one strain to another to produce hybrids, while preventing the plants from self-pollinating. In the first generation all the offspring were the same, say, tall. Mendel argued that tallness was dominant over shortness. Smooth seeds were dominant over wrinkled seeds and so on. If he cross-bred the progeny to make a second generation, the results were different. Three-quarters of the plants had the dominant characteristics but a quarter showed the alternative, say short, which Mendel said was 'recessive'.

To explain his findings, Mendel set up 'laws of inheritance'. A plant received a piece of genetic information, or 'factor', from each parent. Separate pieces of information controlled each characteristic. Cross-breeding true lines of tall and short plants gave each of the offspring a factor for tallness and a factor for shortness. Since tall was 'dominant', the plants in the first generation had to be tall. But in the second generation, 25 per cent of the plants would (on average) have two tall factors, 25 per cent two short factors and 50 per cent one of each. So 75 per cent of the plants would be tall and 25 per cent short.

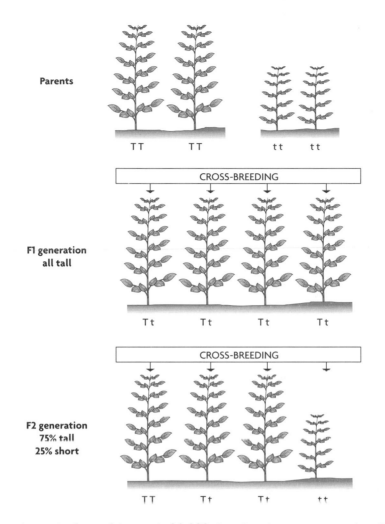

Parents

TT   TT   tt   tt

CROSS-BREEDING

F1 generation
all tall

Tt   Tt   Tt   Tt

CROSS-BREEDING

F2 generation
75% tall
25% short

TT   Tt   Tt   tt

This shows the result of one of the nearly 30 000 plant-breeding experiments that Gregor Mendel did over 10 years in his monastery garden. True-breeding tall plants are cross-pollinated with true-breeding short plants. All the first-generation plants are tall, but a quarter of the second-generation are short.

The pairs of letters labelling the inherited 'factors' (genes) explain why. The TT and Tt 'genotypes' are both tall, since T is dominant over t. But 25 per cent of the second-generation have the genotype tt and so show the recessive 'phenotype' of shortness.

Mendel was ahead of his time, though a century earlier Joseph Koelreuter (**1761**) had glimpsed some of the same patterns of inheritance. But Mendel's findings, published in 1865, were ignored until rediscovered in **1900**. Sceptics still wonder about the precision of the numbers he quoted. Did he 'polish the data' to make his conclusions stand out more clearly? It was fortunate that he chose characteristics that were independently inherited. If two of them had been linked, so that inheriting one meant that you also inherited the other, his data would have been very messy. Still, he found what he found, and his pioneering research launched the science of genetics with all its implications. ➤➤**1900**

The Twin Laws of Thermodynamics

1865

The idea, first propounded by Antoine Lavoisier (**1789**) that heat was a weightless substance called 'caloric', may have proved ultimately inadequate and unable to fit the facts, but it helped in one significant way. As a substance, heat could not be created or destroyed. It would have to be 'conserved', as matter is.

By the 1840s the evidence was supporting this conviction, though heat itself was now being thought of as mostly a form of motion. The substance being conserved was now called 'energy', which could exist in a variety of interchangeable forms. Englishman James Joule (**1847**) had shown that the energy carried by an electric current could be converted into heat in a predictable manner and, more famously, that energy released when something fell under the influence of gravity could also be converted into heat (so, the water at the bottom of a waterfall is just a little warmer than the water at the top).

In 1847 the German Herman Helmholtz was perhaps first to set down 'conservation of energy' as a general principle. Energy in any form, from motion, heat, light, electricity, chemical reactions, could be changed into any other form at a fixed 'rate of exchange', but none could be created or destroyed. This was later proclaimed the First Law of Thermodynamics (the term comes from Greek words for 'heat' and 'motion').

At the same time, a second law was being firmed up. Sadi Carnot knew that not all the heat in a gas was available to do work (**1824**). English physicist Lord Kelvin claimed to see a general tendency for useful mechanical energy to be turned into waste heat, say through friction. A hot gas gets colder as it expands (**1852**), suggesting more waste. Helmholtz saw much the same thing and predicted the 'the death of the universe' (**1854**) as a result. Another German, Rudolf Clausius, noted

---

THE LAWS OF THERMODYNAMICS

First law: Energy cannot be created or destroyed but it can be converted from one form to another or stored for later release.

Second law: Heat cannot of itself pass from one body to a hotter body. This is equivalent to saying that energy becomes steadily less available, that is, that entropy increases.

that hot objects naturally get colder and that energy has to be supplied if they are to get warmer—'Heat cannot of itself pass from one body to a hotter body'.

Clausius later wrapped these various statements up in his notion of entropy, a measure of disorder in the natural world (**1850**).

It was Clausius who provided the most succinct statement of these two vital laws of thermodynamics. In 1865 he declared: (1) The energy of the universe is constant; (2) The entropy of the universe tends to a maximum. Rarely was so much of importance said in so few words.

Norman Lockyer and Jules Janssen: The Element from the Sun

In 1869 we discovered something unexpected about the Sun. Some of the stuff in it does not (apparently) exist on Earth. In ancient times, of course, the Sun and the Earth were believed to be made of very different sorts of matter; but that idea seemed long buried. The same laws of physics and chemistry were now thought to hold throughout the solar system and probably the universe.

1868

There is quite a background to this puzzle. It arose through spectroscopy, the study of colours in light given off by various objects such as the Sun. This had come a long way since the eminent German researcher Joseph Fraunhofer had started taking it seriously in **1814**. Passing the light from different objects through prisms and later diffraction gratings to separate the colours, researchers knew that spectra from glowing solids had all the colours, but those from incandescent gases had only a few bright lines in very pure colours against a black background. The Sun seemed to be a special case: across its spread of colours were many thin, dark lines, as if some of the colours were missing.

Around 1860 two Germans, Robert Bunsen (think 'bunsen burner') and Gustav Kirchoff put all this together. They looked at the spectrum of a flame into which they had sprinkled some bits of, say, sodium. This 'emission spectrum' had just the expected bright, coloured lines. But passing the light from a very hot lamp through the same flame gave a continuous band of colours crossed by black lines superimposed where some colours had been absorbed.

Suddenly the penny dropped. The black lines and the bright lines were in the same places in the spectrum. The same colours—distinctive of particular elements— were involved. So the dark lines in the solar spectrum identified elements in the cooler outer parts of the Sun that had absorbed some of the light from the hot surface.

This meant that by analysing sunlight, we could find out what the Sun was made of. Soon all the dark lines had been tagged with the names of elements known on Earth; all except one, first seen in 1868 by astronomers Norman Lockyer of England and Jules Janssen of France. It had no known earthly counterpart, and so was named 'helium', from the Greek word for 'sun'.

In time, helium was found on Earth. Thirty years later, its spectrum turned up associated with a mineral containing uranium. But that is another story, one with a very significant ending (**1898**).

Why Gases Behave as They Do

1868

Explaining the behaviour of gases like everyday air has been a long-running issue for science, especially once we understood that gases were a distinct category of substances and the third state of matter (**1727**). Irishman Robert Boyle and Swiss Daniel Bernoulli were first to have a go. Boyle thought that the particles in a gas were connected together by springs which could be squeezed or allowed to expand. This explained Boyle's Law (**1659**) but not much else.

Bernoulli (**1738**) had a better idea. He argued that the particles in a gas are all free to move and that they go faster when the gas is hotter. They repeatedly bounce off the walls of the container and those blows collectively constitute the pressure of the gas. Doing the mathematics gives us not only Boyle's Law but also Charles' Law (**1783**) which connects the volume of a gas with its temperature. So it was a very successful 'model', even if real gases do not behave exactly like that (**1852**).

Great minds worked on this challenge, steadily refining the mathematics. They linked the pressure of the gas directly with the energy of motion of the particles. Having an idea what the particles weighed, they could estimate how fast they travelled. It was surprisingly fast, around 500 metres a second, about the speed of a rifle bullet. But particles in ordinary air do not go far in one direction (much less than a millimetre) before colliding with another particle (about six billion times each second) and bouncing off in another direction. So the movement of a mass of gas, say when wind is blowing, is slow compared with the speeds of individual particles.

The Austrian physicist Ludwig Boltzmann took it a stage further, and linked the

---

BOLTZMAN'S CONSTANT (k)

This number connects the energy of motion of an atom or molecule in a solid, liquid or gas with the temperature of the substance, measured in degrees absolute or kelvin. The number is exceedingly small, a consequence of Avogadro's number (**1811**) being so large. Each degree rise in temperature increases the energy of a typical particle by less than a million million million millionth of a joule.

The energy of an Avogadro number of molecules (as many as there are in 2 grams of hydrogen or 12 grams of carbon, in other words, in one mole) increases by 8 joules for each degree rise in temperature.

---

energy of the particles to the temperature of the gas (the absolute temperature, that is, as defined by Lord Kelvin in **1848**). To quote one equation:

$$\text{average energy of each particle} = 3/2kT$$

where T is the temperature in degrees kelvin and k is Boltzman's constant (so many joules of energy for each degree of temperature).

Around 1868 Boltzman realised that a gas particle can move in three dimensions: up and down, front to back and side to side. He called these 'degrees of freedom' and said that each would have an equal share of the available energy; that is, $1/2kT$. This was his famous rule of the Equipartition of Energy; we will meet it again, including in the saga of the 'ultraviolet catastrophe' (**1900 Planck**).

Francis Galton: The Measurer of All Things

Perhaps being the cousin of Charles Darwin has not been good for the reputation of Francis Galton. Despite an extraordinarily diverse range of achievements, he is all but forgotten today, while everybody knows his cousin's name. Those who do remember him are most likely to recall that he was an early disciple of 'eugenics', the selective breeding of humans to 'improve the stock'. That philosophy has gained bad press over the years.

1869

Charles Darwin and Francis Galton shared a grandfather (Erasmus Darwin) and much the same comfortable middle class life, able to pursue their research interests without any real need to earn money. Unlike Darwin, who concentrated on biology, Galton had a hand (and sometimes an arm and leg) in many things. He could be listed as an explorer of tropical Africa, anthropologist, geographer, psychologist and statistician. He strongly promoted, if he did not actually invent, the technique of fingerprinting to establish who was present at the scene of a crime.

He was, for example, the first to suggest that drawing maps of the distribution of air pressure could be very useful in understanding and even predicting the weather. He gave the name 'anticyclones' to the large masses of high-pressure air lying on either side of the equator, the name contrasting with the term 'cyclone' already in use for smaller, more intense regions of low pressure. In a further fascinating connection, the first such maps to be publicly available were drawn on behalf of the Admiralty by Robert Fitzroy, who had been Darwin's captain on the Beagle (**1836**).

Perhaps his greatest contribution was in human measurement. He measured everything—heights, weights, head and limb sizes—for thousands of people and plotted the results to see how these factors were distributed. He found that they tended to be spread throughout the population in a predictable manner; predictability became more pronounced the more measurements he took. He developed the technique of regression to see how likely it was that a person with one characteristic would also have another. For example, could the length of the forearm be used to predict a person's height?

Most controversial were his thoughts on the relationship between heredity and environment in determining things like intelligence. Unlike John Locke (**1689**), who thought that everything in our minds comes from experience and that a baby's mind is a 'blank sheet', Galton was convinced (based on his studies of identical twins) that many abilities (including mental ones) are inherited. His own ancestry seemed to show that. He proclaimed a 'law of inheritance': each parent provides a quarter of a child's abilities, with grandparents providing the rest.

The 'nature versus nurture' argument continues to this day. It was Galton's conviction that nature is far more important than nurture, expounded in his 1869 book *Hereditary Genius*, that led him to eugenics, and put his reputation into an eclipse from which it never really recovered.

Dimitri Mendeleyev: Ordering the Elements...the Finale

1869

Just about anyone who has done science at school has heard of the Periodic Table of Elements. You may recall having to learn the early parts of it by heart. You may also remember the name of its inventor, Dimitri

THE PERIODIC TABLE OF ELEMENTS				
Group No.	**One**	**Two**	**Three**	**Four**
	Alkali metals	*'Alkaline earth' metals*		
Common valency	One	Two	Three	Four
Period 1	1 Hydrogen			
Period 2	3 Lithium	4 Beryllium	5 Boron	6 Carbon
Period 3	11 Sodium	12 Magnesium	13 Aluminium	14 Silicon
Period 4	19 Potassium	20 Calcium	31 Gallium	32 Germanium
Period 5	37 Rubidium	38 Strontium	40 Indium	50 Tin
Period 6	55 Caesium	56 Barium	81 Thallium	81 Lead
Period 7	87 Francium	88 Radium		

In this cut-down version of the Periodic Table, elements are listed by atomic number (the number of electrons they contain). In a more complete table, the elements not listed here (such as silver, gold, iron, copper and uranium) fit between groups two and three from period four onward. Elements in a group generally have similar properties; those in a period show a steady

Mendeleyev, one of the few Russians to appear in the story of science so far.

At the time of his great discovery, Mendeleyev was a professor at the University of St Petersburg. Born in Siberia and orphaned at an early age, he was reputedly an outstanding teacher. His 1869 chemistry textbook, which introduced his Periodic Table, was a bestseller for decades. Mendeleyev had broad interests: in later life he ran the national weights and measures organisation in Russia and played a major role in the hunt for oil and gas.

Before Mendeleyev came on the scene, several people had tried to arrange the known elements, which were steadily growing in number, into a useful pattern. Among them were Johann Dobereiner of Germany (**1829**), John Newlands of England and Julius Meyer of Germany (**1864**). They had arranged the elements by atomic weight and looked for some regular pattern in chemical properties or the size of the atoms. Mendeleyev did the same, but for him the key property was 'valency', or chemical combining power (**1852**).

On the valency scale, the elements could be ranked from one to four. Plotting valency against atomic weight, Mendeleyev found that it rose and fell in waves, a bit like Newlands' 'octaves' or Meyer's 'periods'. Seven steps, taken in order of atomic weight, covered the elements from lithium (valency 1) to carbon (valency 4) and on to fluorine (valency 1). Another seven steps ran from sodium (1) to silicon (4) and on to chlorine (1). The next period, and the one after that, were much longer, but they rose and fell in the same way as the first two periods at each end, with a lot of elements of the same valency in the middle.

Mendeleyev published his Periodic Table in 1869. He believed that some gaps in his table represented elements still to be found. When these were in fact found, the first in **1875**, Mendeleyev's triumph was complete.

THE PERIODIC TABLE OF ELEMENTS			
Five	**Six**	**Seven**	**Eight**
		Halogens ('salt-forming')	*Inert, rare or noble gases*
Three or five	Two	One	Zero
			2 Helium
7 Nitrogen	8 Oxygen	9 Fluorine	10 Neon
15 Phosphorus	16 Sulphur	17 Chlorine	18 Argon
33 Arsenic	34 Selenium	35 Bromine	36 Krypton
51 Antimony	52 Tellurium	31 Iodine	54 Xenon
83 Bismuth	84 Polonium	85 Astatine	86 Radon

change in properties from metals on the left to non-metals on the right. All elements above 82 Bismuth are radioactive. The table contains many elements unknown to Mendeleyev but which neatly fitted into his scheme once they were found. These include 31 Gallium and 32 Germanium, both of which he predicted, and the rare gases, which he did not.

Johann Miescher and Walter Flemming: What Goes On inside the Nucleus of a Cell?

1869

It was the Scottish botanist Robert Brown (**1827**) who had first seen the tiny opaque spot inside each plant cell (of orchids to begin with) which he called the nucleus, meaning a 'little nut'. A decade or so later, the 'cell theory' was well established (**1838**). All plants and animals were composed of cells, and all cells had a nucleus, which suggested it played a significant role in what the cell did.

But what was in the nucleus? What was it made from? In an early step, cells were soaked in the newly discovered chemical pepsin, which is secreted by the stomach lining of animals and speeds up digestion of proteins found in foods. The pepsin ate away the membrane around the cell, most of the contents (the protoplasm) and even the membrane around the nucleus, but not its contents. So whatever it was that filled up a nucleus, it was not protein.

Johann Miescher now enters the story. He would have followed in the family tradition and become a doctor, but an attack of typhus had left him partially deaf. He thought this a disadvantage in a physician, so he took to chemistry. He was to make his mark in another way. In 1869, while still a graduate student, Miescher broke open the nuclei of some white blood cells (these came from the bandages of hospital patients, where pus containing white cells was plentiful). Finding that the material inside the nucleus would dissolve in an alkali solution but not in an acid one, he called it 'nucleic acid'.

We should fast forward to 1882, and Miescher's countryman Walter Flemming. He, too, was keen to follow up what Robert Brown had found but it was hard to see

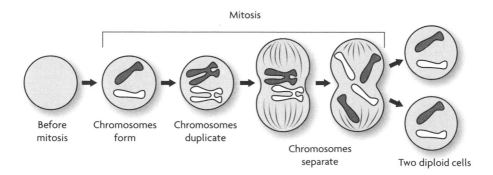

Mitosis

Before mitosis

Chromosomes form

Chromosomes duplicate

Chromosomes separate

Two diploid cells

Before mitosis begins, the chromosomes are not visible. In this diagram (which shows only the nucleus of a cell) a pair of chromosomes condenses from the chromatin and is copied. One copy of each chromosome is drawn to each end of the nucleus, which then divides to form two identical (diploid) nuclei. The rest of the cell divides at the same time. In meiosis, the chromosomes are not first copied, so the resulting (haploid) cells have only one chromosome of each pair.

much inside the nucleus. Seeking a clearer view, he soaked some plant cells in the purple dye that English chemist William Perkin had extracted from coal tar (**1856**). The effect was brilliant. The nucleus took up the dye; the rest of the cell did not. So the nucleus was clearly visible, and so highly coloured that Flemming called the chemical inside it 'chromatin', from the Greek for 'colour'.

Through his microscope, Flemming saw many cells in the process of dividing in two to make a new cell. It was now well established that 'all cells came from cells' (**1838**). Something extraordinary happened during division. The featureless chromatin came to life, forming up into short thread-like bunches he called chromosomes (meaning 'coloured bodies'). Just before division, the bunches doubled in number, dividing end to end. As the cell divided, the bunches of chromosomes also split, with one half of the collection going into each new cell. The chromosomes then faded away, apparently dissolving back into the chromatin.

Flemming called the whole process 'mitosis', from the Greek for 'thread', referring to the appearance of the chromosomes. It happened every time the cells divided, producing two apparently identical cells with the same collection of chromosomes.

So in little more than a decade, our understanding had leapt forward. The nucleus of a cell contains nucleic acid, and the material (the chromatin) in the nucleus can form up into chromosomes. It was reasonable to argue that there was nucleic acid in the chromosomes, and that a lot of nucleic acid was made quickly when a cell divided.

But what was it all for? (**1884 Hedwig and Weismann**)

Charles Darwin: *The Descent of Man*

The publication of *The Origin of Species* in **1859**, with its compelling evidence and imposing presentation, was enough to convince most biologists that Charles Darwin and his fellow travellers were on the right

1871

track. Biological evolution had occurred, slowly and over immense periods of time. New species had been brought to life by the struggle for survival. The assent was not universal. Louis Agassiz (**1851**) for one remained a fierce and vocal critic, and there were others. Argument continued over how natural selection operated and it was not settled until the discovery of mutations (**1901**). At the same time a strong and growing band of Darwin's friends and disciples, such as Charles Lyell and Thomas Huxley (aka 'Darwin's bulldog') defended him in public, which ill health and a

> *We must, however, acknowledge, as it seems to me,*
> *that man with all his noble qualities...still bears in his bodily*
> *frame the indelible stamp of his lowly origin.*
>
> CHARLES DARWIN

certain reticence prevented him from doing himself.

Some people objected on religious grounds. Darwin's biological universe seemed to run purely on chance and randomness, leaving no place for a Creator with a divine plan. Many scientists had struggled to reconcile the often conflicting evidence from the 'book of nature' and the 'book of revelation'—that is, the Bible— over things like the age of the Earth and the reality of the Great Flood, before concluding, as Darwin did, that the best explanation about how nature works is to be found in nature itself. Darwin, an orthodox believer in his youth and destined at one stage for the Church, ended his life an agnostic.

The conflict with the established Church was sometimes lively, especially when Darwin specifically included the human race in the evolutionary picture in *The Descent of Man*, published in 1871. This saw no clear division between humans and the rest of nature; man was closely linked in both body and behaviour with the other 'primates', especially the great apes like the chimpanzee and the gorilla. In one exchange, Bishop Wilberforce quizzed Huxley as to whether Darwin claimed descent from an ape through his grandfather or his grandmother. 'Darwin's bulldog' famously retorted that he would rather be an ape than a bishop who 'used his gifts of eloquence and culture in the service of prejudice and falsehood'.

But the Church had lost any real power to influence the growth of science. Giordano Bruno and Michael Servetus had been burnt alive, Copernicus was censored, Galileo was subject to house arrest, Buffon had his books burnt. Now the separate roles of science and religion were more clearly defined. Conflict was likely only when one party claimed too much.

THE ELECROMAGNETIC SPECTRUM									
		1 km			1 m			1 mm		
Wave length (m)	10^4	10^3	10^2	10	1	10^{-1}	10^{-2}	10^{-3}	10^{-4}	10^{-5}
Type of radiation	Radio waves				Microwaves			Infra-red		
How produced	Alternating electric currents				Accelerated electric charges			By vibrating atoms or electron transition within atoms		
Effects on matter	Induce electric currents							Cause heating		
Effect on human body	None proven							Shorter wavelengths detected as heat on the skin		
Discovery	Heinrich Hertz (1888)							William Herschel (1800)		

Many types of EM radiation (from microwaves to X-rays) can be created when energetic electrons tangle with magnetic fields ('synchrotron radiation').

James Clerk Maxwell: Understanding Light, Predicting Radio

The first day that James Clerk Maxwell arrived at school in Edinburgh, he was wearing homemade shoes. This quickly gained him the nickname 'Dafty', confirmed by his shy, dull manner. Yet Maxwell became one of the greatest minds of his time, his insight into the true nature of light perhaps the outstanding scientific achievement of the century.

In 1871 the University of Cambridge set up the Cavendish Laboratory, destined to become world famous under Lord Rayleigh, JJ Thomson and Ernest Rutherford. Confirming his eminence, Maxwell was appointed the first professor and director. The Cavendish family (the Dukes of Devonshire), who sponsored the laboratory, had included the eccentric eighteenth-century genius Henry Cavendish (**1766**). By editing and annotating papers which had lain unappreciated for nearly 200 years, Maxwell was able to demonstrate for the first time how much Cavendish had done.

Building on the work of others, Michael Faraday at London's Royal Institution and Joseph Henry in the USA (**1831**) had shown that electricity and magnetism were closely linked. Not only could an electric current behave like a magnet (**1820**), but a moving magnet could also make a current flow. To explain this, Faraday turned to the idea of a 'field', a zone of influence around a magnet or an electric charge in which another magnet or charge felt a force. A changing magnetic field generated an electric field (and so made a current flow in a wire); a changing electric field produced a magnetic field.

THE ELECROMAGNETIC SPECTRUM									
1 μm			1 nm				1 pm		
10^{-6}	10^{-7}	10^{-8}	10^{-9}	10^{-10}	10^{-11}	10^{-12}	10^{-13}	10^{-14}	
Light	**Ultraviolet**		**X-rays**		**Gamma rays**				
Electron transmission within atoms	Electron transition within atoms		High-speed electrons hitting solid target Very hot gases		Energy shifts within atomic nuclei (radioactive atoms) Astrophysical events				
Powers photosynthesis	Ionisation Cause some chemical reactions Break down compounds		Ionisation Expose photographic plates		Induce radioactivity				
Detected by eyes	Sunburn Skin cancer Vitamin D production Cataracts		Tissue damage Mutations Cancer Cancer		Mutations				
Antiquity	Johann Ritter (1802)		Wilhelm Roentgen (1895)		Paul Villard (1900)				

Key: μm = micrometre; nm = nanometre; pm = picometre

> *The only laws of matter are those which our minds must fabricate,*
> *and the only laws of mind are fabricated for it by matter.*
>
> JAMES CLERK MAXWELL

Maxwell was a brilliant mathematician. Writing down the relationships between electric and magnetic fields in the form of equations, he noted a remarkable consequence. Under certain circumstances, this interplay of shifting electric and magnetic fields could generate a sort of wave, able to carry energy across empty space. Maxwell showed that the speed of his predicted wave depended on some numbers that physicists knew well, numbers which represented the ease with which electric and magnetic fields can spread through space. Slotting those numbers into his calculations gave a speed of 300 000 kilometres per second, familiar as the well-established speed of light (**1676, 1729, 1849**). The implication was irresistible. Light (infra-red and ultraviolet as well as visible) was an 'electromagnetic wave', spreading its influence through interlinked electric and magnetic fields, oscillating in synchrony.

So at last we could be sure how light worked. And there was more. Publishing his theory in 1871, Maxwell predicted that there might be other forms of such 'electromagnetic radiation', generated perhaps by rapidly moving electric charges, such as in sparks. The discovery of those waves less than 20 years later (**1888**) showed how stupendous Maxwell's insight has been.

Filling the Gaps in the Periodic Table

1875

Dimitri Mendeleyev's Periodic Table of Elements, first published in **1869**, was a tantalising document. Not only did it catalogue the known chemical elements, then numbering around 60, more successfully than anyone had done to date, it also predicted that some other elements as yet unknown must exist. The evidence came from gaps in the table where an element should sit but none was known to fit. From the properties of the elements on all sides of the gap, Mendeleyev even predicted how the missing elements would behave.

No one went off immediately to hunt for these elements; they were found mostly by chance. In 1875 a French chemist found what appeared to be a new element in some zinc ores. He suspected it was new because its spectrum, as examined with a spectroscope (**1814 Fraunhofer**), was not like any other known. The discoverer called it gallium after his homeland, Gaul being the old name for France. Mendeleyev was quick to point out that gallium behaved just like the element he said should sit under aluminium in his table, and which he had called eka-aluminium.

Over the next few years two more gaps in the table were filled, one with scandium (found by a Swedish chemist), the other with germanium (its discoverer was German). These had the properties that Mendeleyev had predicted for his eka-boron and eka-silicon. Intriguingly, all three of the new elements were given geographical names. As elements, these were more than curiosities. For example, germanium and gallium have great value nowadays for making advanced computer chips.

With such a close fit between prediction and reality, few now doubted that Mendeleyev had got it right. His table was able to swallow up the 'rare gases' when these were found by the English pair Lord Rayleigh and William Ramsay (**1898**). That upstart family of elements would give the periodic table eight columns (or 'groups') instead of seven.

Robert Koch: Microbe Hunter

By the time German doctor Robert Koch arrived on the scene around 1870, his illustrious French contemporary Louis Pasteur (**1862**) had won the early battles on behalf of the 'germ theory' of disease. Evidence was growing that many, if not most, diseases were caused by tiny living things, now known as microbes or bacteria. This confirmed what the Austrian Markus Plenciz had argued in **1762** and the Italian Girolamo Fracastoro had speculated more than 200 years earlier still (**1546**).

1876

Koch, at the time a busy medical officer in a rural area, still found time to do research on the nasty disease anthrax, suffered by animals and capable of spreading to humans. Inspired by Pasteur, others had found a rod-shaped micro-organism (a 'bacillus') living in the blood of infected animals. Using a microscope his wife had given him, Koch set out to prove it was the culprit. Samples of blood from animals which had died from anthrax were placed in cuts in the skins of mice. This blood contained the suspect microbe. The mice quickly died of anthrax. Mice inoculated with blood from healthy animals (blood free from the anthrax bacillus) did not die.

Koch was not quite convinced; the disease could have been spread by something else in the blood. So he grew a 'culture' of the bacillus through several generations as free as possible from other contamination. Even though the micro-organisms had never directly contacted a diseased animal, they could still spread the disease, as tests with mice showed. The case appeared proven, but it was not until Pasteur found an anthrax vaccine in **1886** that Koch's findings were really accepted.

Koch added to his achievements by identifying beyond doubt the microbes that caused other diseases, including tuberculosis (TB), cholera and septicaemia. TB, known as 'consumption', killed one in every seven people in Europe at the time. Koch's work was judged so important that he won the 1905 Nobel Prize for Medicine.

Koch developed ingenious and powerful methods to separate and culture various sorts of bacteria. He also formulated Koch's Postulates, the evidence needed to prove that an organism 'caused' a disease. The organism had to be found in all cases of the disease examined; it had to be able to cause the disease even after several generations in culture. Armed with his methods and his postulates, Koch's students identified the germs that cause diphtheria, typhoid, bubonic plague, tetanus, syphilis and pneumonia. So his influence outlived his death in 1910.

A Visitation by Mars

1877

Mars, the 'red planet', normally comes close to Earth every two years. The approach of 1877 was closer than average, making it an exciting time for Earth-bound watchers of Mars. The best telescopes and keenest eyes were prepared for whatever the encounter might reveal.

Two major findings were announced. In Italy, Giovanni Schiaparelli fancied that he saw, when viewing was steady enough, thin dark lines on the surface of Mars. He called them 'canali' in Italian, which means 'channels'. The matter might have rested there had not the American astronomer Percival Lowell seized on the word (and the observations), claiming not only that he too could see the 'canals' but that they were artificial. Mars, he said, was inhabited by an intelligent but dying race, which had built the vast network of canals to bring water from the Martian poles to the arid deserts. No one else could see any such thing, but that seemed to make Lowell more convinced that he was right.

Later investigation showed that the canals were a figment of Lowell's imagination. His reputation suffered, though he was a well-regarded astronomer and predicted the existence of the ninth planet, Pluto, which was found in 1930. There was a literary footnote to all this: the talk of 'Martians' at this time was a major stimulus to the young English novelist H. G. Wells, who wrote *The War of the Worlds* a few decades later.

The other reported finding proved more enduring. American astronomer Aspeth Hall found two tiny moons orbiting Mars; these were named Deimos and Phobos after the two dogs that according to ancient legend accompanied the God of War in battle. These moons have since been seen up close by spacecraft; there is no doubt they exist. The moons are very small compared with Mars; they are most likely asteroids (**1801 Piazza**) that strayed too close and were captured by Mars' gravity.

There is a literary connection here too. Long before Hall found the moons, the English writer Jonathan Swift, in his novel *Gulliver's Travels,* reported that Mars has two companions. It was presumably just a guess. The Earth has one moon, Jupiter four (at least, four were known when Swift was writing). So if Mars had moons, two would be an appropriate number, lying between one and four. Swift did predict something else more striking: that one of the moons went round Mars faster than the planet rotated so it would appear to rise in the west and set in the east. Remarkably, this is actually the case.

Ernst Mach: His Number, His Principle

The speed of an aircraft or projectile is often given a 'Mach number'; this compares its speed with the speed of sound in the same conditions. Mach 2 means twice the speed of sound. The number is one of the ways we recall one of the most influential scientists of the nineteenth century, the Czech-born Ernst Mach. Mach was versatile; his work on projectiles led him around 1877 to postulate and then to detect the cone-shaped shock wave generated when a projectile exceeds the speed of sound. The noise it causes has been dubbed the 'sonic boom', the action 'breaking the sound barrier'.

1877

As well known as this is, it is far from all we can say about Ernst Mach. Around the same time he tackled one of the most profound mysteries in science: what gives objects the quality known as inertia, the resistance to a change in their state of motion, first defined by Galileo (**1604**)? Mach gave the answer as 'the universe'. Everything in the universe pulls on everything else, even at vast distances, and the network of forces generates a stability we call inertia. Later, Albert Einstein was greatly influenced by this idea; it contributed to his notion in the General Theory of Relativity (**1915**) that matter bends space, and so the presence of matter somewhere affects the motion of matter elsewhere. It was Einstein who first called this concept the 'Mach Principle'.

The challenge of inertia had been summed up by the early nineteenth century English philosopher/Bishop George Berkley, in the conundrum known as 'Bishop Berkley's Bucket'. Hang a bucket of water from a twisted rope and set it spinning. After a time, the water starts to spin too and forms a concave surface. Stop the bucket, and the water continues to spin for a while, still retaining its curved surface. How does the water surface 'know' it is spinning and should stay curved? Not by reference to the bucket, since it is curved if it is spinning relative to the bucket or not. Berkley's answer was the same as Mach's. The water measures this motion against the distant universe of stars (to which we would now add galaxies and quasars).

For Mach, this idea about inertia satisfied a fundamental criterion. Scientific theories need to be as simple as possible, stripped of all unnecessary trappings and based purely on observation. What we call 'laws of nature' are simply summaries of observations, developed so we can hold the observations in our minds. The 'laws' say more about us than they do about the universe. Since ideas are valuable only if they can be verified by experience, Mach rejected notions like absolute space and time, to

> *Every statement in physics has to state relations*
> *between observable quantities.*
>
> ERNST MACH

which Newton had been attracted (**1716**), and even the existence of atoms. From his ideas flowed the tough-minded empirical outlook called 'logical positivism'.

Mach's rejection of absolutes also influenced Einstein, opening the way for his Special Theory of Relativity, which says that measurements of time, space and mass can be altered by relative movement between the observer and what is being measured (**1904**). Nothing is absolute.

Ludwig Boltzmann: Understanding Entropy

1877 Entropy was a big thing in the life of Austrian physicist Ludwig Boltzmann. He arranged to have the equation for entropy engraved on his tombstone. And though he did not invent the idea or the term (the German Rudolf Clausius had done that in **1850**), he did give us a new understanding of it.

Entropy is a measure of disorder, or of a lack of information. Consider Leonardo da Vinci's famous painting *The Last Supper*, painted on a monastery refectory wall in Milan, Italy. When first completed, this fresco was packed with information, every square centimetre of colour arranged in precise order. Sadly, the painting has decayed badly over the years: many details are now lost, the colours have faded. Information and order have drained away, entropy has increased. In time the painting may fade away entirely, leaving a blank wall; no information, no order, entropy at a maximum.

Suppose the painting was originally done on a million detachable tiles. Initially there would have been only one way to arrange the tiles so that the painting looked right, since every tile would then have been different. Order would have been high, entropy low. As they faded, many of the tiles would begin to look like other tiles and could have been interchanged without changing the look of the painting. More 'states' would now be available, entropy would be increasing and would reach a maximum when the wall was blank and all the tiles were therefore identical. Then the tiles could therefore be arranged in an almost infinite number of ways, leaving the wall looking the same. In Boltzman's language, there are a vast number of 'available states' and the high entropy of the 'system' of tiles reflects this.

What we have imagined happened to *The Last Supper* happens in the natural world. Decay of order, loss of information goes on all the time. Increase of entropy seems universal, at least in the biggest picture, as physical systems tend towards a uniformity of temperature (thermal equilibrium). Differences disappear. Embedded information degrades. Entropy increases inevitably. Hence the threat of the 'heat death of the universe' (**1854**).

Powerful figures in the German scientific establishment ridiculed these ideas. Depressed by the attacks and in poor health, Boltzmann hanged himself in 1906 while on holiday. His tombstone in Vienna's Central Cemetery carries the entropy equation, put in place once his own entropy began to increase.

Alphonse Laveran and Ronald Ross: Mosquitoes and Malaria

French physician Alponse Laveran had doctoring in his blood. Both his father and his grandfather had been medical men. When war broke out between France and Prussia in 1871, Laveran became an ambulance officer in the army and later spent time as a military doctor in Algeria, where he first came face to face with malaria.

1880

In 1880 he made an important discovery. Autopsies of many malaria victims revealed that their blood contained a single-celled living organism like the 'animalcules' first seen by Anthony van Leeuwenhoek 200 years earlier (**1673**). By now they were being called protozoa (for 'first animals'), and plasmodia, the name they carry today. The blood of victims of other diseases did not carry this particular parasite, so the connection appeared clear.

Laveran's argument that the parasite caused malaria was rejected at first; 'miasmas' of infected air were still widely regarded as the cause of such diseases (malaria means 'bad air'). But other evidence soon supported him and within a decade he had proven his point. Laveran's work on malaria and on other diseases carried by protozoa, such as 'sleeping sickness', earned him the Nobel Prize for Medicine in 1907. He gave half the prize money to establish a tropical medicine unit at the Pasteur Institute in Paris, which had been funded by a public grateful for Louis Pasteur's triumph over rabies (**1886**).

If the plasmodia caused the malaria, how did they get into the victim's blood? English Army doctor Ronald Ross had spent time in India, where malaria was rife. He had even suffered it himself. Early on, he guessed that mosquitoes were involved; they were plentiful in malaria-ridden areas. Learning of Laveran's discovery, he went looking for the parasites inside mosquitoes. On what he referred to as 'Mosquito Day', 20 August 1897, he found the parasites in the stomachs of mosquitoes which had been allowed to suck blood from a man with malaria (the man was paid for his discomfort). Later still they were found in the salivary glands of the mosquitoes, which explained how they were injected into the humans they bit.

Like Laveran, Ross won the Nobel Prize for Medicine (in 1911) for his discovery. The control of malaria has benefited greatly from both of these discoveries, yet today, more than 100 years later, malaria remains a serious health problem in many parts of the world.

Thomas Edison: A Mysterious 'Effect'

1882 American inventor Thomas Edison is often eulogised as the 'man who created the future'. Certainly the list of machines and devices he created is immense, including the phonograph and the electric light bulb. He pioneered the mass generation and distribution of electricity and experimented with 'moving pictures' (the 'Vitascope'). During his lifetime he took out more than 1000 patents. Perhaps his greatest legacy was the modern industrial research laboratory, where teams of researchers and engineers could create and improve useful (and profitable) inventions in a methodical manner.

Edison has his place in the history of science through the 'Edison Effect', an accidental discovery made around 1882 and based on his light bulb. At the time, the challenge was to stop carbon particles coming off the glowing filaments (made of carbonised cardboard at the time) and blackening the inside of the glass. Edison's lamps used direct current; one end of the filament was therefore positively charged, the other negatively. Edison noted that the carbon particles seemed to come mostly from the negative end. He wondered if there was a way to repel them before they hit the glass. He had light bulbs made with an extra metal plate inside. Charging the plate negatively might send the carbon particles back where they came from. When the plate was positively charged, he found a tiny but steady current flowing out of the plate, apparently coming from the filament through the near vacuum inside the bulb.

Edison was busy with other matters at the time, including setting up the first commercial power station, in Lower Manhattan. Though he patented his effect and discussed it with colleagues, he did nothing much about it. In reality, what he was seeing was akin to the 'cathode rays' which were so intriguing scientists in Germany and England (**1854 Geissler**). The current was due to a flow of what would soon be called electrons (**1897**), driven out of the filament by heat. The phenomenon was dubbed 'thermionic emission'.

As for the effect itself, it was not until the new century that the English engineer Ambrose Fleming refined the light globe with an extra plate into the first 'valve', which let a current pass only one way. A couple of years later, American Lee de Forrest added a wire mesh ('a grid') between the other plates to make a triode valve to amplify currents. From this came the increasingly complex technology of 'electronics' which would drive communications in the century to come, that is, until the arrival of semiconductors and the transistor.

Ilya Metchnikoff: White Cells for the Defence

One of the more striking theories about the workings of the human body came from an odd source—baby starfish in the hands of a brooding, politically radical Russian, wandering the Mediterranean shores near

Messina in Sicily. Ilya Metchnikoff was often suicidal and even more frequently in trouble, partly because of his arrogant conviction that he was always right. Fortunately, his wife's money enabled him to escape from Odessa in the Crimea, where he was teaching, and spend time under the Mediterranean sun.

Studying the transparent starfish larvae under his microscope, Metchnikoff (a zoologist by training) noted tiny cells inside the starfish, cells that seemed able to move anywhere throughout the creature. They looked very much like the white cells found among the red ones in human blood, but whose purpose was unknown. Metchnikoff wondered if the cells inside the starfish had anything to do with digestion. He squirted red dye in and saw the mobile cells surround the droplets of dye and eat them up.

This provoked a radical thought. Suppose the purpose of the white cells in human blood is similar? Perhaps they surround and swallow up foreign objects, such as invading disease-causing bacteria. To test this, Metchnikoff pushed rose thorns into the skins of the starfish larvae. Within a few hours hundreds of the white cells had gathered around the thorns. Better proof came when he infected the larvae with bacteria. As he watched under the microscope, the cells surrounded the bacteria and devoured them.

The phenomenon needed a name. Metchnikoff coined the term 'phagocytosis', from Greek words meaning 'eating cells'. Here then was one way the body could defend itself against disease. It also explained the origin and purpose of pus, the white fluid, brimful of white cells, that accumulates around wounds. The white cells (or 'phagocytes') congregate to prevent bacteria entering the body through the break in the skin.

The theory was understandably controversial, and it had a rival. Evidence was growing that another form of defence was circulating in the blood, some sort of 'humour', to use an old word (**1891**). But it was not a case of either/or. In time it was shown that we are protected from disease by both forms of defence. Metchnikoff shared the 1908 Nobel Prize for Medicine with Paul Erlich.

Metchnikoff has another memorial, familiar to many today. He argued that the presence of certain bacteria in the human gut, especially those that release lactic acid, is essential for good health. These bugs aid digestion and prevent food from fermenting and so releasing toxins. Metchnikoff thought they would in fact prolong life, and from words meaning 'in favour of life' he created the name 'probiotics'. Yogurt is full of them.

Svante Arrhenius: 'Ions' Are Real

1884

When 25-year-old Swedish student Svante Arrhenius submitted his 150-page thesis for his doctorate in 1884, the whole thing went right over the heads of the professors at the University of Uppsala. They so misunderstood what he was saying that he received the lowest possible mark short of an outright failure. How wrong they were was apparent in 1903, when Arrhenius was awarded the Nobel Prize for Chemistry for essentially the same theory his examiners had dismissed 20 years before.

Arrhenius put forward an explanation for why water mixed with a little salt or acid will allow an electric current through, while very pure water blocks it. His idea was not really new. Forty years before, Michael Faraday had coined the word 'ions' for hypothetical particles with an electric charge that wandered through a conducting liquid (an electrolyte), effectively forming an electric current (**1834**). But Faraday did not really think these particles existed (he was sceptical about atoms too) and if they did, it was the act of trying to pass the current that formed them.

Arrhenius thought otherwise. He argued that if ordinary table salt (sodium chloride) was dissolved in water, each molecule of it broke apart into a positively charged part (a sodium ion) and a negatively charged part (a chloride ion). When not in solution (or molten) the molecules of salt were held together by the attraction of two oppositely charged ions. Again, this was not a new idea: Jöns Berzelius (**1814**), and before him Humphry Davy (**1807**), had argued that molecules had positive and negative bits, but the idea had been totally forgotten—hence the incomprehension of Arrhenius' professors.

As proof, Arrhenius pointed out that a salt solution boiled at a higher temperature than pure water did, and the freezing point was lower. This is just what would be expected as a result of the presence of the extra particles produced when the molecules broke apart ('disassociated'). This showed that the charged particles were real, and that they were there all the time.

This view is now called the 'ionic theory'. For example, all acids in dilute solution release positively charged hydrogen ions, with the remainder of the molecule forming a negative ion. These hydrogen ions give acids their distinctive properties (sour taste, turning litmus red, releasing hydrogen when reacted with metals). In a strong acid (like sulphuric acid) nearly all the molecules dissociate into ions; in a weak acid (like citric acid in lemon juice) only a few do. Since water molecules do not break into ions, they must be held together in some other way (**1928 Pauling**).

Fortunately for Arrhenius, researchers in other countries understood him rather better and he was soon offered jobs and chances to travel. His fellow Swedes were slow to accept his theories, but his career never looked back. Arrhenius had a very active mind: once his ionic theory was accepted he turned to pondering the origin of ice ages (**1895 Global Greenhouse**) and wondering if life was spread from planet to planet by hardy spores travelling through space (the notion of panspermia [**1822 Encke's Comet**]). He was very good at talking about science to non-scientists.

Emil Fischer: Sugars, Proteins and Enzymes

Emil Fischer's father ran a successful timber business in Germany and thought his clever son would join the family firm. Young Fischer was not so keen and wanted to study science, but he yielded to his father's wishes. For whatever reason, it did not work out. Fisher senior reportedly claimed that his son was too stupid to be a businessman but bright enough to be a chemist.

1884

The choice was obviously right: Fischer had a brilliant career over 40 years, culminating in a Nobel Prize in 1902. His real legacy lies in insights as to how the various complex chemicals that make up, and are made by, living things are put together. Experiments (**1845 Bertollet**) had already shown that the major categories of proteins, carbohydrates and fats could be broken down into smaller units—amino acids, simple sugars and fatty acids respectively—and presumably could be reassembled to make the original big molecules. Nature must be doing that all the time.

In 1884 the 30-something Fischer began working on sugars. He was soon able to ring the changes between the various closely related simple sugars such as glucose, mannose and fructose. Of these, glucose seemed the simplest, and some of the others could be created by joining several glucose units together. Perhaps that was true of all the carbohydrates, including even starch and cellulose, the basic building material in plants.

Fischer also discovered that the amino acid molecules in proteins link together through 'peptide' bonds. A number of amino acids so joined make up a polypeptide; a number of these suitably arranged and linked make a complete protein like the haemoglobin in red blood cells or casein in milk. The name 'peptide' goes back to pepsin, the chemical in digestive juices able to break these bonds and separate the amino acids (**1869 Miescher and Flemming**).

Pepsin was one of many proteins proving to be catalysts, as Jöns Berzelius (**1848**) defined them: compounds that sped up chemical reactions but were not themselves used up. For a while pepsin and similar chemicals were called 'soluble ferments', but they later acquired the name 'enzymes' from the Greek meaning 'in yeast'. One of them, called 'zymase', was extracted from the bakers' yeast or leaven that helps the sugar and starch in dough break down a little, releasing carbon dioxide to make the bread rise.

It was Fischer who discovered how these enzymes work, around 1894. Enzyme molecules have a special shape that can fit into the various chemicals involved in the reaction 'like a key in a lock' and get them interacting. So every reaction needed its own enzyme. Enzymes speed up reactions sometimes hundreds or thousands of times, making them run fast enough to keep organisms alive.

While Fischer was happy and successful in his work, particularly at Wutzburg, where he loved to walk in the Black Forest, his private life was not peaceful. His wife died after seven years of marriage, one son was killed in World War I, another killed himself from the stress of compulsory military training. Fischer died of cancer in 1919.

Oscar Hedwig and August Weismann: Almost Home in Genetics

1884 It is hard to put a date to the first time someone guessed how one generation of animals or plants tells the next generation what it will look like and how it will behave. The understanding grew over many years and involved many people; by the end of the nineteenth century we had a good idea of the broad picture, if not the details.

Perhaps Karl von Nageli in 1842 began it, when he first observed the splitting of plant and animal cells, with two apparently identical cells taking the place of one. (It was also von Nageli who told Gregor Mendel that his voluminous observations on plant breeding 'needed more work'.) Cell theory had just been established thanks to Matthias Schleiden, Theodore Swann and Rudolph Virchow (**1838**): everything is made up of cells, all cells come from other cells, all cells have a nucleus that is copied when the cells divide. Most people began to think that the nucleus held the key.

A little later, work was proceeding on three fronts, independently at first. While Mendel (**1865**) was defining the laws of genetics and postulating some 'unit' of hereditary information, Johann Miescher (**1869**) was picking the nucleus apart and finding in it a special chemical called nucleic acid, and a little later Walter Flemming was seeing the intricacies of chromosomes forming and dividing within the nucleus as cells divided.

By the 1880s things were coming together, with the Germans in the lead. Oscar Hedwig and August Weismann were among the biggest names. In 1882 Weismann stated that chromosomes were where the action was as far as inheritance; in 1884 Hedwig announced that nucleic acid is the substance that transmits the genetic information.

Under the microscope, Hedwig had seen the union of an egg cell (ovum) and a sperm; he claimed this was the moment of fertilisation. The whole purpose of sex, he declared, was to get those two together.

Meanwhile, Edward van Beneden, a Belgian cytologist (a new title for the new study of cells), found that all the cells in a species of animal or plant have the same number of chromosomes: 46 in a human, 22 in a giant sequoia tree, eight in a fruit fly. The exception to that rule was his other great discovery. Cells involved in reproduction, such as sperm, pollen or eggs before fertilisation, had only half the normal number of chromosomes. So when fertilisation occurred between, say, sperm and egg, each contributed half of the chromosomes to the fertilised cell.

It would have been easy now to make a connection to the passing of inherited characteristics from one generation to the next. But sadly the vital work of Gregor Mendel on that matter (**1865**) had been overlooked or ignored, and had to await rediscovery (**1900**).

Louis Pasteur: The Triumph Over Rabies

Eminent French biologist Louis Pasteur was one of the first to react in **1876** when German Robert Koch identified the micro-organism (the 'bacillus') that caused anthrax, a serious animal disease. He quickly corroborated Koch but also noted that not all animals catch anthrax. Chickens are immune. Was this because their body temperature is around 43°C rather than the 37°C in other animals? Perhaps heat weakened or even killed the anthrax germs. Pasteur chilled the body of a living chicken to 37°C—it lost its immunity.

1886

In a trial, publicly staged to answer his critics, Pasteur heated anthrax bacilli and injected them into 25 sheep. Nothing obvious happened. He then injected the same sheep and 25 others with fresh anthrax germs. The first lot lived, the second lot died. The heated bacteria had been unable to cause the disease, but somehow made the animals immune. Soon millions of sheep and cattle were being protected against anthrax. Someone estimated that the savings were equivalent to the reparations France had to pay after losing the war against Prussia.

Pasteur repeated the methods successfully with chicken cholera, another costly animal disease. He called the treatment 'vaccination', so honouring Edward Jenner, who 90 years before had pioneered something similar, using cowpox to protect people against the more virulent smallpox (**1796**). He called the cultures of weakened bacteria 'vaccines'.

Pasteur's greatest triumph was over rabies (or 'hydrophobia'), which could kill humans bitten by an infected dog. He made a vaccine from the spinal cords of diseased rabbits, though he had not yet clearly identified the bacteria. He could soon stop the disease from spreading from dog to dog but would not try the treatment on humans for fear of something going wrong. He took the step only in 1886, when

> *In the field of observation, chance only favours those*
> *minds which have been prepared.*
>
> LOUIS PASTEUR

nine-year-old Joseph Miester was brought to his laboratory after being mauled by a rabid dog.

The boy seemed certain to die, so Pasteur treated him (though he was not licensed to practise medicine). Miester made a complete recovery, as did a young shepherd similarly bitten a few months later. Publicity was intense; Pasteur became perhaps the most famous scientist in the world. Public gratitude supported the founding of the Pasteur Institute, where Joseph Miester was to serve as gatekeeper for many years. Pasteur died from the complications of a stroke a decade later, aged 75.

Albert Michelson and Arthur Morley: Searching for the Aether

1887

For centuries many scientists had imagined that apparently empty space was filled with an extremely tenuous stuff known as the 'aether'. The French philosopher René Descartes had called it 'subtle matter' and claimed that the planets were driven in their orbits by vast whirlpools or vortices in it (**1644**). Others thought that light was simply a chain of ripples in the 'luminiferous' (light-bearing) aether. Others still thought the aether might be part of the 'absolute space', against which, according to Isaac Newton (**1716**), the real movement of objects could be measured.

With so many potential uses, no wonder the concept was popular, but from the start, it was a tricky idea. Physicists struggled to reconcile the various properties it

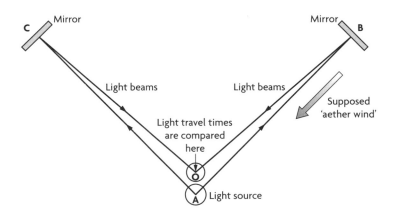

In their famous experiment, Michelson and Morley raced beams of light over equal distances from A to mirrors at B and C and then back to O, where their times of arrival were compared.

If an 'aether wind' (produced by the motion of the Earth through the supposed aether) was blowing from B towards O and A, the round–trip ABO should have taken minutely longer than the ACO journey across the aether wind. But despite many trials, no difference in the times of travel was ever found. The aether wind could not be detected.

needed to have. To carry light at its vast speed, the aether had to be very rigid. At the same time, it had to be very flexible so as not to slow down the planets as they ploughed through it.

A crucial test came in 1887. The American physicists Albert Michelson and Arthur Morley tried to detect the aether by very carefully measuring the speed of light in different directions. They argued that the aether should be streaming by the Earth as it moved around the Sun, like a current in a river. Light should be like a boat in that river, travelling at different speeds against, with or across the current.

In essence, Michelson and Morley raced beams of light out and back in directions parallel or at right angles to the movement of the Earth. They compared the returning light beams in an 'interferometer', looking for the very small speed changes that would reveal the presence of the aether. Despite a great many such attempts, they found no such changes.

It seemed that the aether did not exist, or at least that it could not be detected, which is perhaps the same thing. Put another way, any attempt to measure the speed of light, say in a vacuum, would always give the same result. The speed of light was constant, no matter how the source of the light or the detector moved. On this profound discovery, Albert Einstein would build his Special Theory of Relativity (**1904**).

Heinrich Hertz: Discovering Radio Waves

Today we know the name of German physicist Heinrich Hertz as the unit of frequency, the number of times a second something happens. The clock in a one gigahertz (gHz) processor in a computer ticks a thousand million times a second. In his day, Hertz gained fame by discovering 'Hertzian waves', though he personally doubted they would ever be of any value. Finding what we call radio waves confirmed a prediction made by the Scots genius James Clerk Maxwell (**1871**), who thought that electric sparks might produce radiation akin to light and travelling at the same speed.

1888

To test Maxwell's prediction, Hertz set up a narrow gap between two metal rods and attached the rods to an induction coil, a form of transformer (**1831**). The voltage from the coil was high enough to break down the resistance of the air. He argued that sparks would jump back and forth across the gap (the 'transmitter') and perhaps make Maxwell's predicted waves. To trap those waves, if there were any, Hertz set up a loop of wire with a small gap (the 'receiver'). Sure enough, when sparks were arcing in the transmitter, smaller sparks simultaneously jumped the gap in the receiver. Clearly something was carrying energy across the room.

Hertz then bounced his new energy from a metal sheet on the far wall and moved the receiver back and forth between the mirror and the transmitter. He found locations within the room where the receiver found lots of sparks, others where it recorded none. These were like the light and dark fringes that Thomas Young had seen in his 'double slit' experiment in **1801**, the result of interference between two rays of light, and clear evidence that light was a wave. They also let Young measure how far apart the crests of the waves were (the 'wavelength').

Hertz did the same for his new waves. He knew the frequency of the waves from the properties of the induction coil. Frequency multiplied by wavelength gives speed. When he did the sum, the discovery was complete. The new waves travelled at the speed of light, and were indeed what Maxwell had predicted. Later experiments showed that they could be reflected and refracted just like light.

From such beginnings grew the technology of radio, beginning with the work of the Italian Guglielmo Marconi (**1895**). For generations afterwards, long after better ways had been found to make the waves, the radio operator on a ship was known as 'Sparks'. Even today, the 'funk' in the German 'telefunk' for radio means 'spark'.

Emil Behring: The Power in the Serum

1891

Time spent as an army doctor and as a lecturer in a military college left its mark on the life of German Emil Behring. Throughout his life he rose at four in the morning and enjoyed a steak for breakfast. He had already done half a day's work in the laboratory by the time his colleagues arrived.

Such discipline brought a full and productive career, culminating in 1901 with the first Nobel Prize ever awarded for medicine. For five years, Behring worked with Robert Koch, discoverer (alone and with his students) of many of the micro-organisms that cause infectious diseases such as anthrax and tuberculosis (**1876**). Behring's main concern was with diphtheria, which causes a membrane to grow across the victim's throat and choke them.

One important factor quickly emerged: the symptoms of the disease were not caused by the diphtheria bacilli, but by a chemical toxin they produced. Perhaps such a disease could be treated by neutralising the toxin even if the germs responsible were still present in the blood. He then found that the blood of an animal immunised against diphtheria (by being exposed to weakened germs as Pasteur had pioneered with anthrax and later rabies [**1886**]) contained a chemical that countered the toxin.

Behring had made this discovery in company with Shibasaburo Kitosato, perhaps the first Japanese researcher to make an impact in European science, though he did not share the Nobel Prize Behring won in 1901. Though the discovery had involved tetanus, the notion of an 'anti-toxin' seemed to apply to other diseases.

The big step forward came with finding that blood serum from an immunised animal, that is blood with all the solid matter removed, including red and white cells, could be used to transfer the anti-toxin to another animal, which then became able to fight off the disease. By working through a series of ever larger animals—

guinea pigs, rabbits, sheep—Behring gathered enough anti-toxin by Christmas 1891 to inject a young girl who was seriously ill from diphtheria. The child recovered; the anti-toxin had done its work. 'Serum therapy' was established. Large-scale manufacture made Behring wealthy as well as famous.

Later in life, Behring worked on the challenge of tuberculosis (TB), which had also concerned Koch. He established a research institute into the disease, donating his Nobel Prize money and other property. The disease ultimately took him in 1917.

Lord Rayleigh and William Ramsay: Finding Argon

You use the gas argon everyday without knowing it. Ordinary light globes are filled with the stuff. Being 'inert' (its name in Greek actually means 'idle') and so refusing to take part in any chemical reactions, it stops the hot filament burning out just as effectively as a vacuum, and it is much easier to work with than a vacuum.

Discovering argon took two brains rather than just one. Working separately at first, and later in collaboration, the independently wealthy Englishman Lord Rayleigh and Scots-born William Ramsay, who had to work for a living, found that everyday air contained more than just the familiar nitrogen and oxygen. About 1 per cent of air had to be something else.

Rayleigh knew early in 1894 that the residue left when oxygen was removed from air was slightly heavier than nitrogen made chemically from ammonia. Ramsay suggested why this might be so. He recalled an experiment done 200 years before by the English chemistry pioneer Henry Cavendish (**1766**). Cavendish had tried to extract both oxygen and nitrogen from air by various chemical reactions, expecting to finish with nothing but ending up instead with a tiny bubble of gas that stubbornly refused to go away. He had found argon, but he did not know it. Rayleigh and Ramsay found it again (though by a different method), knew that they had found it and were honoured with Nobel Prizes in 1904, Rayleigh for physics, Ramsay for chemistry.

Rayleigh liked to point out to his audiences at lectures that though argon was the first to be found of what were commonly called the 'rare gases', it was not really 'rare'. Air contains more argon than carbon dioxide or water vapour, The total amount of argon in the air in the lecture hall would have weighed as much as any of his hearers.

Cathode Rays: The Answer Emerges

We left the cathode rays story (**1854 Geissler**) with a question. Just what were these mysterious, invisible rays? They seemed to flow from a negatively charged plate inside a tube almost exhausted of air; they made the walls of the tube glow. They travelled in straight lines like light and so could cast a shadow; but they could also make a paddle wheel turn and so perhaps were particles. Just what were they?

The Germans thought the rays were some invisible form of light, perhaps like the 'ultraviolet' rays found in **1802** by Johann Ritter. The English scientists, led by William Crookes, suspected that they were exceedingly tiny particles. Over time, the evidence built up, so that ultimately only one answer was possible.

The key discovery (and this had been known for some time) was that the normally straight path of cathode rays could be bent by magnetism. The glow in the tube walls was distorted if a magnet came near. Magnets affect only other magnets or pieces of iron (and neither was present here), or electric currents (**1820 Orsted**). So the cathode rays had to be like a current, that is, charged particles in motion. Light and similar forms of radiation are not deflected by magnets, so that ruled them out as an explanation.

For confirmation, Crookes' colleague Arthur Schuster built a cathode ray tube with a couple of extra electrodes side-on to the path of the rays. He charged one of these with positive electricity and the other with negative. Again the path was bent, and towards the positive plate. Since unlike charges attract, the answer was clear. Not only were the cathode-ray particles charged, they were negatively charged. The English were zeroing in. The same experiment had been tried by the Germans (by Heinrich Hertz, no less, the discoverer of radio waves in **1888**). He did not find any effect, which seemed to support their 'cathode rays are light' theory. But on this occasion they were wrong.

The clincher came in 1895, not from a British laboratory, but from the Frenchman Jean Perrin, who later got the Nobel Prize for it. He found a way to collect the cathode rays and measure their electric charge with an electroscope. Sure enough, it was negative. So there was the answer. Cathode rays are streams of negatively charged particles.

It was time for JJ Thomson to enter the story (**1897**).

Does Carbon Dioxide Warm the 'Global Greenhouse'?

1895

The early nineteenth century (**1827**) had seen the first stirrings of a debate that still rages today. Is human activity changing the climate of our planet?

If gases in our atmosphere trap heat and keep our planet warm, a bit like the glass of a greenhouse does (the 'greenhouse effect'), increasing levels of those gases might make it warmer. But which gas would make the most difference and could human activity increase the proportion of that gas in the atmosphere?

In 1895 Swedish scientist Svante Arrhenius (**1884**), later a Nobel Prize winner, calculated that the major impact was from carbon dioxide. He assumed (though he did not actually know) that this was increasing in abundance in the atmosphere due to the spread of the Industrial Revolution and the massive burning of coal. If it continued to increase, surely we would end up making the Earth warmer.

Arrhenius was backed by fellow Swede Nils Eckholm. In 1900 Eckholm claimed that the amount of carbon dioxide in the air could double and that that would produce a very obvious rise in the mean temperature of the Earth. Many people now believed that from time to time the Earth had passed through periods of intense cold known as ice ages (**1836**). If we could control the amount of carbon dioxide released by burning coal, Eckholm thought, the added warmth could prevent a new ice age beginning. It is an argument we still hear today.

Almost at once, Swede Knut Anderson swung the balance of the argument back in favour of water vapour as the main culprit; Irish physicist John Tyndall had argued that 40 years before. As a result, the idea that carbon dioxide was the key factor fell out of favour. We cannot control the amount of water vapour in the air; it rises and falls with the temperature of the oceans. That must mean that the temperature of the Earth's surface and any resulting climate change is something beyond our influence, and which just has to take its course. That argument is still current, too.

Guglielmo Marconi and Ferdinand Braun: Fathering Radio and Television

The remarkable discovery by Heinrich Hertz in **1888** of a new form of radiant energy sent researchers scurrying delightedly around their laboratories all over the world to find out more. Hertz himself had found that 'Hertzian waves' could be reflected and refracted with mirrors and prisms just like light. All the evidence pointed to the new waves being siblings of visible light, ultraviolet light and heat radiation, all parts of the new 'electromagnetic spectrum' (**1871**). They differed only in their wavelength; whereas visible light had waves less than one millionth of a metre long, the new waves were metres in length or even longer. Compared with light, they had very low frequencies, a few hundred thousand cycles per second (or Hertz as we say now).

1895

Clever minds were now working to improve the technology for making and detecting these waves. Edward Branley in Paris, Oliver Lodge in England and Ernest Rutherford, who had not yet left New Zealand, produced various designs for a 'detector' or 'coherer' to indicate that the waves had been received—more reliable methods than watching sparks jumping across a gap.

Hertz had been pleased to find the waves and prove Maxwell's prediction right but did not expect any useful outcome. Others disagreed. The most famous of these was the Italian Guglielmo Marconi. In 1895 he began a series of experiments to detect Hertzian waves at ever greater distances; first a few kilometres, then (with the help of friends of his English-born mother) across the English Channel, and ultimately in 1901 across the Atlantic. Stopping and starting the waves in accordance with the Morse code, he initiated 'telegraphy without wires', later simply 'wireless', and later again 'radio'. A revolution in communications was stirring.

When Marconi won the Nobel Prize for Physics in 1911, he shared it with the German Ferdinand Braun. As Marconi had refined Hertz's equipment, Braun had taken on cathode ray tubes, which had led to the discovery of X-rays (**1895**) and the electron (**1897**). Coating the inside of the tube with phosphorescent chemicals enhanced the glow caused by the electrons; magnetic and electric fields could confine the glow to a small spot and make it move around the screen. From this primitive but brilliantly conceived device descended oscilloscopes, computer screens and of course television—an appropriate device to honour alongside Marconi's radio.

Wilhelm Roentgen: X-rays by Accident?

1895 The biggest discovery in science in 1895, and there were a few to choose from, was a new form of radiation, so unexpected and mysterious that the term 'X-rays' seemed appropriate. They were also called 'Roentgen Rays', honouring their discoverer—Wilhelm Roentgen, a physics professor from the University of Munich. Roentgen would win the first Nobel Prize for Physics ever awarded.

Roentgen, like many others, was interested in cathode rays (**1854 Geissler**). They were generated by a high-voltage electrical discharge inside a glass tube almost empty of air. Various people had noted, to their annoyance, that photographic plates stored near such 'discharge tubes' were inexplicably fogged, as if they had been exposed to light. Roentgen wondered if the discharge tubes gave off some form of invisible radiation, especially since the walls of the tubes glowed where the cathode rays struck them. Perhaps the radiation could leak through the packaging surrounding photographic plates and expose them.

One evening in November, he found that this was so. His discovered that radiation easily passed through a light-proof shroud around the discharge tube, and made a plate covered with a fluorescent chemical glow several metres away. One account of the discovery has him caught by surprise. The shroud was to exclude light from the room so he could better study the faint glow on the tube walls. He glimpsed the glowing plate out of the corner of his eye.

Roentgen's X-rays were stopped by a thin sheet of metal, but could cut through human tissue and emerge with enough energy to leave an image on a photographic plate. The first 'X-ray photograph' ever taken was of Roentgen's wife's hand, clearly showing her bones and a ring she was wearing, surrounded by the shadow of her flesh. Such images were to be of incomparable value to medicine and surgery, and later to engineering and industry. More powerful X-ray tubes soon appeared. Unfortunately, the hazards of the technology were discovered only later.

Roentgen justly became one of the most famous scientists of his age. Honours and accolades poured in, culminating in the Nobel Prize in 1901. He remained strikingly reticent and modest, preferring to work alone, building his equipment with his own hands. A lover of nature and a daring mountaineer, he was finally taken by bowel cancer at the age of 78.

Henri Becquerel: Unexpected Radioactivity

French physicist Henri Becquerel's father and grandfather had been eminent researchers and professors; his own son also carried on the family business. Yet the discovery for which he is renowned, so unexpected and so momentous in its implications, was largely an accident.

1896

Wilhelm Roentgen's discovery of X-rays in **1895** had set the scientific world abuzz. These amazing new 'rays' could see through solid objects, and set many sorts of mineral glowing. Some minerals, such as those containing uranium, glowed after being exposed to sunlight. Becquerel decided to see how these minerals responded to X-rays. He had inherited a substantial collection of them from his father.

Becquerel wrapped a photographic plate and a sample of uranium salt in opaque paper, planning to expose the package to X-rays. If the X-rays made the mineral glow ('fluoresce'), the film would be fogged by the light released. He developed one plate before it was hit with X-rays, perhaps to get a 'before and after' comparison. He was amazed to find it was already fogged; a distinct shadow showed where the sample of uranium had pressed on it. So even in the total dark, and without any stimulus from X-rays, the uranium had released some 'rays' of its own. It was unprecedented.

Becquerel soon found that all compounds of uranium did the same, so the energy release had nothing to do with chemistry. Nor did physical conditions like temperature or pressure affect it. The radiation must have come from deep within the uranium atoms—this was the first glimpse of the 'atomic energy' that would become such an issue over the next century.

Fittingly, the emissions were initally dubbed 'Becquerel rays'; later 'radioactivity'. Becquerel explored their properties. He found that they caused air and other gases to conduct electricity—to become 'ionised' (**1884**). X-rays did the same. But unlike X-rays, these emissions were influenced by magnets and electric charges. So they were not a form of light. A metal cross placed between a sample of the radioactive mineral and the plate left a paler mark than the mineral itself, showing that the rays were partly absorbed by metals. Others would unravel their complexity (**1899**).

Becquerel was accorded the honours due. He shared the 1903 Nobel Prize for Physics with Marie and Pierre Curie, who built so spectacularly on his work (**1898**). Today, the strength of radioactivity is measured in becquerels.

JJ Thomson and the Electron

1897

Since its foundation in 1871, the Cavendish Laboratory at the University of Cambridge had been led by men of genius; firstly James Clerk Maxwell (**1871**), who had established the true nature of light and predicted radio waves, then John William Strutt (aka Lord Rayleigh) (**1894**), who among other things jointly discovered the rare gas argon. He was followed by JJ Thomson, of whom more in a moment, and then by the incomparable New Zealander Ernest Rutherford (**1899 Radioactivity**), whose stewardship lasted into the 1930s.

Elf-like John Joseph Thomson, universally known as 'JJ', took up the job at the Cavendish in 1884. This Professor of Experimental Physics was notoriously clumsy; his laboratory assistants tried to stop him touching the apparatus, because he often broke something. But his insights into the workings of nature were brilliant, and he was quick to follow up discoveries made elsewhere, like Heinrich Hertz finding radio waves in **1888** and Wilhelm Roentgen's X-rays in **1895**. It was Thomson's 1897 investigation of the enigmatic cathode rays that made his reputation and won him the Nobel Prize for Physics in 1906.

JJ agreed with his countrymen, such as William Crookes, that cathode rays were streams of minute, negatively charged particles; German physicists, on the other hand, thought they were some unknown form of light (**1895 Roentgen**). To prove their case, the English needed some facts about the particles' mass (m) and/or their electric charge (e), or at least the ratio of the two (e/m). JJ fired a beam of cathode rays through a magnetic and an electric field at the same time, and got the measurement he sought.

The answer was astonishing. The ratio of charge to mass of the cathode-ray particles was 2000 times greater than for the best thing to compare it to, namely a hydrogen ion—an atom of the lightest known element, given a single positive charge.

JJ's finding could mean one of two things: the charge on the electron was much higher than on the hydrogen ion, or the mass was much lower. JJ himself went for the latter, making the revolutionary suggestion that the 'carriers of electric charge are smaller than hydrogen atoms', the latter being the smallest objects then known.

This minute fragment of matter would, in time, be assigned the name 'electron', already devised by the Irish physicist Johnson Stoney in another context, but highly relevant. There was still more to discover about the electron, and JJ had a hand in that too. **▶▶1899**

Research in applied science leads to reforms, research in pure science leads to revolutions, and revolutions, whether political or industrial, are exceedingly profitable things if you are on the winning side.

JJ THOMSON

Marie and Pierre Curie: Hunting Down Radium

Most people know the name Marie Curie (née Skodowska), most likely because she was a woman in the male-dominated world of science. But Polish-born Marie was a great scientist by any standards, one of the very few to win the Nobel Prize twice. The first, in 1903, she shared with her husband, French physicist Pierre Curie. He was perhaps more eminent than she at the time, due to his discovery of the Curie Point, the temperature above which a material cannot be magnetic.

The 1903 Nobel Prize was actually split three ways, with half of the total going to Frenchman Henri Becquerel, who in **1896** had found the totally unexpected phenomenon of radioactivity in uranium, the heaviest known element. The discovery seized the imagination of the Curies, who had married in 1895, especially when Marie and a colleague found radioactivity in thorium, another heavy element. They set out to systematically explore the mineral world for other traces of radioactivity, using as a measure the power of the 'Becquerel rays' to make ordinary air conduct electricity and so cause an electroscope to lose its stored electric charge.

After a long, fruitless search, they returned to the uranium-rich mineral pitchblende. Even when all the uranium had been extracted, the mineral was still radioactive, four times more so than the uranium itself. Some other radioactive element must have still been hidden in the residue. The Curies set out to find it.

The years of toil that followed are the stuff of legend. In a makeshift laboratory, freezing in winter, stifling in summer, they slowly and laboriously reduced tonnes of pitchblende to a few fragments of the mineral they sought. It was so intensely radioactive that it glowed in the dark, so earning it the name radium. They found a second radioactive element soon after, and named it polonium after Marie's homeland. These were discoveries well worthy of the Nobel Prize.

In an enormous loss for Marie and for science, Pierre Curie was run down and killed by a cart in a Paris street in 1906 at the age of only 51. Marie was to continue the work after her husband's death, pioneering the use of radium for the treatment of diseases such as cancer, and earning a second Nobel Prize by herself. ➤➤1911

Martius Beijerinck: The First Viruses

'Tobacco mosaic disease', so called because it leaves a distinctive pattern on leaves, affects many plants other than tobacco, including tomatoes, chillies, capsicums and lettuces. So it has long been of interest to vegetable growers and their friendly scientists.

Dutchman Martius Beijerinck studied micro-organisms such as bacteria and fungi living in the soil and the impact they had on plants, including the spread of disease. In 1898 he was seeking the cause of tobacco mosaic disease. He squeezed some fluid from an infected leaf, and tried to filter out the micro-organisms he assumed it contained so he could study them. He found that the liquid which passed through the filter could still spread disease to other plants. Clearly the cause of the disease was something very much smaller than a bacteria or a fungus cell. Beijerinck said his 'living contagious liquid' contained a 'filterable virus', using a Latin word for 'poison'. The 'filterable' bit was later dropped.

His discovery was controversial. Nineteenth-century research had apparently shown that living things consist of complete cells (**1838**) and that 'cells come from cells'. Beijerinck confused the picture more when he found that he could grow these viruses only inside living plants, and not on plates of gelatin as was possible with colonies of bacteria and fungi. So his viruses were alive but perhaps only partly alive. Being so small, they would not be directly seen for many years, and it took a long time to tease out their secrets. By then, viruses (rather than bacteria) would be known to cause a number of human and animal diseases, including foot and mouth disease, polio and smallpox.

Beijerinck spent little of his illustrious career on viruses. Much more time was taken in working out how plants get the nitrogen they need to make protein. It had been known that the plants called 'legumes' (peas, beans, clover and lupins, for example) could somehow 'fix' nitrogen, that is take nitrogen from the air and make compounds called nitrates which then provide nitrogen to other plants. This 'fixing' happens in nodules of the plants' roots and it was Beijerinck that proved the nodules were home to specific bacteria (*Bacillus radiacolia*) which really did the job. Other bacteria do the 'denitrifying' task, releasing nitrogen from soil nitrates. So micro-organisms play a vital role in the nitrogen cycle in nature.

William Ramsay: The Rest of the Rare Gases

1898

The discovery in **1894** of the rare gas argon was the great joint achievement of English researchers William Ramsay and John William Strutt (Lord Rayleigh). The find won a Nobel Prize for each of them.

For Ramsay, that was not the end of the story. In 1895 he had identified a gas found by some American researchers within uranium ores. It proved to be helium, previously unknown on Earth but named because it had been found by analysing light from the Sun (**1868 Lockyer and Jansen**). Helium, like argon, was 'inert': it would not make compounds with other elements despite strong urging from red-hot metals, powerful chemicals and electric sparks.

Looking at Dimitri Mendeleyev's new Periodic Table (**1869**), Ramsay guessed that helium and argon were two members of a new family of elements. With his team at the University of London, Ramsay now painstakingly distilled liquid air, and by 1898 had found the other siblings one by one: neon (the 'new' gas), xenon ('the stranger') and krypton ('the hidden'). Together these form the 'rare' or 'noble' gases.

Each has its distinctive colours and spectral lines when made to glow in a 'discharge tube' (**1854 Geissler**). In the case of neon, this would be its source of fame, as the archetype of the neon signs.

There is a quirk in the naming of these elements. All but one have names ending in 'on'; helium is the odd one out. When it was known only by the lines it made in the Sun's spectrum, it was expected to be a metal. The names of metals usually end in 'um' or 'ium', as in sodium or magnesium (or even 'ferrum' and 'plumbum', the Latin names for iron and lead). When helium was found on Earth and proved not to be a metal, its name should have been changed to 'helion' but it was probably too late.

A few footnotes to this story. The inertness of all these elements raises the issue of their valency (**1852**). Valency means 'combining power'. Since these elements would not combine, they effectively had valency of 'zero', while for every other element, the number ran from one to four (or perhaps seven). A valency of zero would take some explaining (**1926 Bohr**).

The family of rare gases was completed a few years later by the discovery of a radioactive inert gas ('radon') given off by uranium and thorium. Then, in 1903, Ramsay found that the newly isolated element radium (**1898**) also gave off a gas. This proved to be helium again, and provided the answer to the riddle of the 'alpha particles' thrown off by a number of radioactive elements (**1899 Radioactivity**). The world within the atom was about to open.

JJ Thomson: Deeper into the Atom

The brilliant, if clumsy, English physicist John Joseph (JJ) Thomson did not rest after proving in **1897** that cathode rays were particles (electrons). There were still puzzles to solve. The ratio of the charge of the electron to its mass (e/m) was 2000 times greater than for a positively charged hydrogen atom ('ion'). This could mean that (a) the charge on the electron was 2000 times greater than on the ion, or (b) its mass was 2000 times less, or (c) some combination of the two.

1899

JJ felt sure it was (b), but needed proof. A colleague at the Cavendish Laboratory in Cambridge had made a 'condensation chamber', full of very moist air, ready to condense into 'rain drops'. JJ made drops form by passing X-rays through the chamber to give some air atoms a negative charge (that is, to 'ionise' the air). The air ions attracted water drops, so the drops were charged too. JJ worked through a series of steps to find the smallest amount of charge on any drop. He got the answer he hoped for. The smallest lump of negative electricity he could find, which he believed would represent a single electron, was the same size as the positive charge on a hydrogen ion. That cut out (a) and (c), leaving only (b). An electron really was minute compared with an atom.

Thomson was driven to an unimaginable conclusion: 'in negative electricity, we have something that involves splitting up of the atom'. If this were true, the very name 'atom', from the Greek for 'indivisible', would be false advertising.

By now electrons were turning up everywhere: as cathode rays, where they had first been found; boiling off the hot filaments of Thomas Edison's light globes (**1882**); released from metals by the photoelectric effect (**1905**); making up electric sparks and currents; and fired out of radioactive atoms as the newly discovered beta particles (**1899 Radioactivity**).

To place the ubiquitous electrons into the scheme of nature, JJ used a homely analogy, the 'plum pudding', beloved of middle-class Englishmen like himself. An atom was like the whole pudding, composed of positively charged 'stuff', with electrons stuck in like the plums. The electrons could be liberated from the pudding in many ways: by friction (as in static electricity), by chemical action (in batteries), by light (the photo-electric effect), by heat (in light globes) and by the power of radioactivity.

Thomson's 'plum pudding atom' explained all the known facts; it rightfully held centre stage for the moment. But things were moving fast, and by **1909** there would a serious challenger. ➤**1907**

What Is this Thing Called Radioactivity?

1899 New Zealand-born Ernest Rutherford, destined to be one of the greatest scientists of his (or any) day, was only in his mid-20s in 1897. But he was already working alongside JJ Thomson and other luminaries at the renowned Cavendish Laboratory in Cambridge, England. JJ was always quick to follow up on discoveries made elsewhere. The year before, Henri Becquerel in Paris had found unexpected rays of energy coming from uranium (**1896**). Pierre and Marie Curie (**1898**), also in Paris, dubbed the phenomenon radioactivity. Thomson set Rutherford to work to find out just what these rays were. This task took him several years.

Radioactivity proved more complex than it seemed at first. There was more than one sort of ray. Radium, which the Curies had found in 1898, gave out a 'soft' sort of ray that could be stopped by a piece of paper or a couple of centimetres of air. Uranium generated a 'hard' radiation that was much more penetrating. Rutherford called the first type 'alpha' and the second type 'beta'. Within two years a third form of radiation, even more pentrating still, had been found by the French physicist Paul Villard. These were called, naturally, 'gamma rays'.

Other researchers, including Becquerel himself, chimed in. Magnetic fields proved to be another way to sort out the rays; alpha rays were bent one way, beta rays the other. So the rays were almost certainly streams of charged particles, with the alpha particles positively charged and the beta particles negatively charged. By 1900 Becquerel had proved that the beta rays were identical to the 'cathode ray

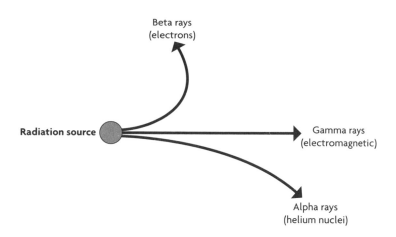

Beta rays
(electrons)

Radiation source

Gamma rays
(electromagnetic)

Alpha rays
(helium nuclei)

In this diagram, the three forms of radioactivity are represented passing through a magnetic field that runs vertically through the page. The negatively charged electrons (beta particles) have been sharply curved one way. Being positively charged and much heavier, the helium nuclei (alpha particles) curve the other way but by a much smaller amount. Gamma rays are uncharged (being similar to X-rays) and are not affected by the magnetic field. Alpha rays have the least energy and are easiest to stop; gamma rays are the hardest.

corpuscles', that is, they were streams of electrons. The gamma rays were not affected by magnetism at all, and so were clearly not charged particles. They proved to be a form of electromagnetic radiation—like X-rays, but much more powerful.

It took longer to find what the alpha particles were, but ultimately Rutherford did it. There were early clues. When helium, the 'gas from the Sun' (**1868**), was first found on Earth, it was mixed with a uranium mineral. By 1903 helium was found in every compound containing radium. Rutherford completed the hunt in 1909 (after he had already received his Nobel Prize). He collected a gas called radon that gave off alpha radiation in a glass tube, and made it glow by passing electricity through it. The spectrum of light was exactly the same as from helium.

So alpha particles were helium atoms with a couple of electrons knocked off. If JJ Thomson had thought that liberating electrons involved 'splitting the atom' (**1899**), then heaving out the 8000 times heavier alpha particles was wholesale dismemberment. We were beginning to glimpse the powerful events going on all the time, deep in the heart of matter, an understanding of which would in time have immense consequences. ➤➤**1903 RUTHERFORD**

Lord Kelvin: Getting It Wrong

The English physicist Lord Kelvin, born William Thomson, was a brilliant man and undoubtedly one of the most influential scientists of his time (**1846**). But he was not infallible or all-knowing. No scientist is. Indeed,

1900

Kelvin made some authoritative pronouncements that might well have destroyed the reputation of a lesser man.

Kelvin never lacked confidence in himself or his physics. He once famously declared: 'in science there is only physics, all the rest is merely stamp-collecting'. It is perhaps not surprising that he said at various times things like 'radio has no future', 'X-rays will prove to be a hoax', and 'we will never fly in heavier-than-air machines'. These declarations went well beyond his mistake about the age of the Earth (**1862**); that could at least could be forgiven, since radioactivity (which slows the rate at which the Earth cools) had not yet been discovered. He was not, however, alone in his forecasts. Heinrich Hertz, discoverer of radio waves (**1888**), had doubted they would ever be of any practical value.

Most memorably, Kelvin declared at a science conference in 1900 that the work of physics was done: 'There is nothing new to discover in physics. All that remains are more and more precise measurement'. The American Albert Michelson (**1887**) once said much the same thing; he thought the only progress in physics would be in refining the sixth place of decimals. It was a comment he came to regret.

To be fair, Kelvin did draw his listeners' attention to 'two dark clouds' on the horizon, clouds he probably thought would just blow away. One cloud was the failure of Michelson and his colleague Edward Morley to detect the aether (**1887**); the other was the non-appearance of the feared 'ultraviolet catastrophe' (**1900 Planck**). What a storm would break from those two clouds. The first would unleash Albert Einstein's Special Theory of Relativity (**1904**); the other would yield quantum physics (**1905**). Together these would shake the foundations of the physics Kelvin knew so well.

Max Planck: Energy Comes in Lumps

1900 Late in the nineteenth century, physicists were puzzled when they failed to find something their theories predicted should exist. The absent phenomenon was the 'ultraviolet catastrophe' and its non-appearance threw physics into turmoil. New calculations based on the accepted laws of physics demanded that any hot object, such as a piece of burning wood, should pour out most of its energy as searing ultraviolet light, blinding and burning anyone who looked at it. Yet, as we know, the energy emerges more sedately as invisible radiant heat and visible light, and careful measurement found the rules that control the release. Nothing catastrophic happens at all.

Why did the researchers expect an ultraviolet catastrophe? Think of an oven or a furnace with only a very small door to look through (physicists would call this a 'cavity'). As the temperature rises, energy starts washing back and forth in the form of 'light waves' of various kinds and wavelengths, from long wavelength infra-red light (**1800 Herschel**), through the various colours of visible light and on even to very short ultraviolet waves (**1802**).

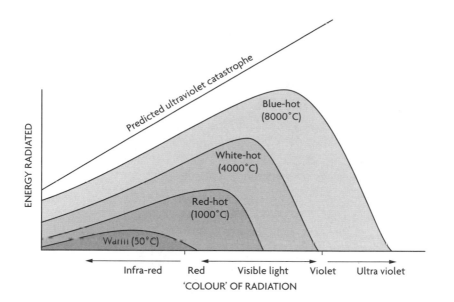

The lower four of these 'radiation curves' sum up what was first measured late in the nineteenth century. As an object gets hotter, it gives off more energy, but the 'colour' of the radiation also changes. A warm object gives off mostly low-frequency infra-red energy and so is invisible in the dark. At higher temperatures, it radiates more and more at higher frequencies, becoming in turn 'red-hot', 'white-hot' and even 'blue-hot'.

The top line shows what the physicists of the time had expected to find—a relentlessly rising share of energy coming off as very high frequency violet and ultraviolet light. But this 'ultraviolet catastrophe' refused to show up. In all cases, the spectrum reached a peak (depending only on the temperature and not what the object was made from) before falling away to zero at still higher frequencies.

In the confused wash of radiation, some waves of energy will be much stronger than others. These will be the 'standing waves', whose wavelengths fit neatly into the width of the cavity a certain number of times. Consider the 'bathroom baritone' effect. When you sing in the shower, certain notes sound much louder than others, as if they are resonating with the bathroom. The sound waves for these notes fit exactly one or more times between your bathroom walls; they are your 'standing waves'.

Each standing wave is a whole number of wavelengths long. The shorter the wavelengths (or the higher the frequency) of the radiant energy (going back to the oven) the more different standing waves there will be, more for blue than for red, more still for ultraviolet. By the law of Equipartition of Energy (**1868 Gases**), each of the standing waves must carry an equal share of the available energy. So there will be much more energy being carried by the blue light waves (because there are more possible standing waves) and even more by the ultraviolet. In fact most of the energy

PLANCK'S CONSTANT

This number links the energy carried by a photon or 'light quantum' with the frequency of the associated electromagnetic radiation. This number is extremely small, so that a photon of visible light with frequency of a thousand million million cycles per second (hertz) has an energy of a million million millionths of a joule. X-rays and gamma rays with much higher frequency carry much greater energy.

should be carried by the vast numbers of standing waves associated with ultraviolet light. This was the anticipated 'ultraviolet catastrophe'.

It simply does not happen. There is no sign of it. But there was a catastrophe for physics as it was known then. Some part of nineteenth-century physics had to be wrong, or at least incomplete; researchers scrabbled for answers. The boldest suggestion came from German physicist Max Planck. The problem, he said, was with the way energy was exchanged between the standing waves to give each of them an equal share of the energy. In Planck's vision, energy could not be exchanged like water, drop by drop, but only by the bucketful. There was a smallest allowable amount of energy, a 'quantum', which could be transferred.

What's more, the size of the bucket, which had to be full to be exchanged, depended on the frequency of the radiation. The quantum of energy for blue light was bigger than for red light, and bigger again for ultraviolet. In most cases, there would simply not be enough energy about to fill the various ultraviolet buckets and so those standing waves would never receive any energy. So, no ultraviolet catastrophe. Physics was saved.

In a way, Planck was simply doing for energy what was already in place for matter and electrical charge. There was a smallest possible unit for those quantities—the atom and the electron. Now there was an 'atom of energy' too. This was a momentous step forward. Planck's quanta of energy were exceedingly tiny, and we do not normally notice that energy comes only in packets of a given size. But for very

An important scientific innovation rarely makes its way by gradually winning over and converting its opponents. What does happen is that its opponents gradually die out, and that the growing generation is familiarised with the ideas from the beginning.

MAX PLANCK

small things like atoms, the effects become noticeable. This brings us to the threshold of the bizarre world of quantum physics. Albert Einstein was the first to open the door (**1905**).

It is worth remembering that at the time the calculations were being done, ultraviolet light had the shortest wavelength among the recognised forms of radiant energy. It was soon realised that X-rays (**1895**) were radiant energy with a shorter wavelength than ultraviolet light, and gamma rays (**1899 Radioactivity**) had a shorter wavelength still. Had they known this, the physicists would have been hunting the 'X-ray catastrophe' or the 'gamma-ray catastrophe'. They would still not have found it.

Gregor Mendel Comes in from the Cold

'Simultaneous discovery' is not at all uncommon in science and technology. There are any number of incidents where researchers, working separately, have come up with the same idea or invention at pretty much the same time. For example, Michael Faraday and Joseph Henry both discovered electromagnetic induction in **1831**; Charles Darwin (**1859**) and Alfred Wallace (**1858**) both publicised their ideas on evolution by natural selection at the same time. It happens with inventions too. Alexander Bell and Elisha Gray walked into the same government office on the same day in 1876 to independently patent a telephone. It seems that there are some discoveries 'whose time has come'.

1900

The example from 1900 is special, since it involves three scientists working independently: Dutchman Hugo de Vries, German Carl Correns and Austrian Eric Tschennak. They were all breeding plants, cross-pollinating different lines to determine the laws which controlled the inheritance of various characteristics. They all came up with very similar results.

The real surprise came when they did a search of the 'literature', as any reputable scientist must do, to see if anyone else had already made the same discovery. They all found that someone had—the same someone, an Austrian monk called Gregor Mendel. In **1865** he had published results from tens of thousands of cross-breedings done in his monastery garden. Not being part of any scientific network, his findings had been ignored and then forgotten.

Now they were back on the table, and supported by three leading scientists in the field. Mendel had proposed that various characteristics, such as whether the plants were short or tall or the seeds smooth or wrinkled, were carried forward into the next generation by separate 'packages' of genetic information. Mendel had called these packages 'factors'. Sometime after 1900, they began to be called 'genes'. Various scientists have a claim to be the originator of this word: de Vries, American Walter Sutton and the Dane Wilhelm Johannsen. It really does not matter. The word came into the language; the next step was to find one (**1902 Sutton and Boveri**).

The Certainty of Uncertainty

1900

The nineteenth century began with an optimism born of the belief that nature is a machine, absolutely predictable in its behaviour, if we have enough information. The French mathematician Pierre Laplace (**1796**) was bold enough to suggest that if we knew the exact position and motion of every particle in the cosmos at any instant we could predict the future course of every particle and therefore the fate of the universe itself. This suggested that everything was knowable, not only in outline but in detail, and every question had an answer.

Optimism faded in the glare of reality. The numbers were against it. Even the smallest sample of gas contains an immense number of particles. Knowing what they will all be doing at any instant is a practical impossibility. So by the end of the century, mechanics (in gases anyway) had become 'statistical mechanics'. The talk was not of individual behaviour but of averages, of what a typical particle was doing. But even in this uncertainty there was a high level of certainty. James Clerk Maxwell (**1871**) devised rules showing how the actual velocities of the particles in a gas are distributed around the average in a predictable manner. We can know how many particles would be found in a particular range of velocities, even if we cannot identify which ones.

This approach from 'probability' or 'statistics' seemed to fit many situations. If soldiers are being issued with new uniforms, measurements of chest size show a predictable distribution around the average or most common size. A soldier chosen at random could have any chest size, but if there were 100, or better still 1000, the quartermaster stores could know with useful reliability how many of each size would be needed. As Francis Galton had found in 1883, these numbers form a normal distribution or bell-shaped curve. This started turning up everywhere—in the distribution of bullet holes around the bull's-eye on a target and the tossing of coins in a gambling game. Any event would appear random but an accumulation of events would have predictable totals.

This mindset was to fit the coming century admirably. Take the discovery of radioactivity (**1896**). As research would soon show, each burst of energy is given out by the 'decay' of a particular atom. We cannot tell when any given atom will decay but we can know with high accuracy how many will break down in a given time. Looking at it another way, we can predict how long it would take half the atoms in a sample to decay. This 'half-life' is precisely known for every sort of radioactive atom. We cannot know what any particular atom will do but that really does not matter. Only the behaviour of all the atoms together matters, and we can know that.

1901–1950

The World Stage

THE FIRST HALF OF THE TWENTIETH CENTURY saw two international conflicts (1914–18, 1939–45); these were so all-encompassing that they were called 'world wars', though the first was initially referred to as the 'Great War'. Most European nations, and many elsewhere, were drawn in. In both struggles, alliances led by Germany were defeated, ending more than 200 years of growth and military success by states grouped around Prussia. France, Britain, Russia and the United States were among the victorious powers on each occasion. The two conflicts were connected, in Europe at least, the first leaving political and territorial legacies that helped ignite the second.

The two world wars drew to an unprecedented degree on advances in science and technology, with the first military use of tanks and other powered vehicles, aircraft, radio and poisonous gases in World War I, and of radar, rockets, jet engines, code-cracking machines and nuclear weapons in World War II.

At the same time, they greatly stimulated science and technology. For example, the antibiotic penicillin was rushed into production to meet military needs, a decade having lapsed since its discovery. The use of radar generated both the technology and some of the observations that led to radio astronomy. The quest for the atomic bomb gave rise to the first of the great national laboratories, employing thousands of researchers in the United States. The need to crack the enemy's secret codes spawned the first massive, if primitive, computers.

Many other conflicts and military adventures embroiled European nations. Britain waged the Boer War in South Africa up to 1902, and other European powers consolidated their grip on African colonies. Russia duelled with Japan (1902–04),

Britain had troubles in Ireland and Spain endured civil war in the 1930s. Between the two world wars there was a struggle of a different kind. Overspending and financial speculation led to a collapse of stock markets in 1927, initiating a global economic depression that ended only with the outbreak of World War II.

Autocratic governments rose to power in several states. By the 1930s both Germany (under Adolf Hitler) and Italy (under Benito Mussolini) were fascist dictatorships; they were allies in World War II. Persecution of Jews caused many people, including leading scientists, to escape to Britain or the United States, strengthening the scientific enterprise of those nations. Austria, too, had a dictator before the nation was swallowed up by an expanding Germany. Russia, which was now part of the Union of Soviet Socialist Republics (USSR, or Soviet Union), was under the dictatorial rule of Joseph Stalin, who listed among his favourites the geneticist Triofim Lysenko, whose Lamarkian ideas on evolution suited the dictator's political purposes. More orthodox science withered.

ARTS AND IDEAS

The twentieth century in Europe began in the glow of nineteenth-century certainty, born of prosperity and colonial expansion. However, this was soon shattered by the cataclysm of World War I. Between the end of that torment and the beginning of the Great Depression, life in Europe and the United States grew more uncertain and frenetic, characterised by the music and life styles of the 'jazz age' of the 1920s.

The outpourings of creative minds—in music, literature, art and ideas—were immense, with the growth of new forms and experimental styles, many reflecting the uncertainty of the times. In painting, France remained prominent, building on the impressionist tradition, with Maurice Utrillo, Henri Matisse, Camille Pissarro and Claude Monet among others.

Elsewhere, artists like Marc Chagall and Wassily Kandinsky of Russia, and Salvador Dali and Pablo Picasso in Spain were defining a more radical position (Picasso's *Guernica* expressed the horror of the Spanish Civil War). Later came the American Jackson Pollock. Germany discovered the 'expressionists', typified by Ernst Kirchner. In totalitarian states like Nazi Germany, and especially the new Soviet Union, art was made subservient to ideology and political ends.

Music saw a flow-on of nineteenth-century romanticism, often blended with folk idioms and traditions: Britain's Ralph Vaughan Williams, William Walton, Frederick Delius, Gustav Holst and Edward Elgar; France's Maurice Ravel and Claude Debussy; Russia's Sergei Rachmaninov; Germany's Richard Strauss; and Italy's Giacomo Puccini. Other composers broke strongly, even violently, with the past: Russia's Igor Stravinsky and Sergei Prokofiev; and Germany's Arnold Shoenberg and Alban Berg.

The United States was finding its own distinctive musical voice through Scott Joplin and Irving Berlin, and later George Gershwin and Jerome Kern, who wrote not only 'popular' songs but also music for the stage. But the rising force in music in

America and elsewhere was jazz, grown from roots in African music and transforming rapidly through early improvised forms to the sophistication of the 'swing era' of the 1930s. Many 'serious' musicians took jazz seriously. All forms of music were lifted by the explosive growth of the recording industry.

Novelists continued to explore new forms to chronicle their changing times: Americans Scott Fitzgerald, Sinclair Lewis, Ernest Hemingway, John Steinbeck and Gertrude Stein; from England, Somerset Maugham, E. M. Forster, Virginia Woolf, Graham Greene, Evelyn Waugh, D. H. Lawrence and John Galsworthy (and in a lighter vein, P. G. Wodehouse); Frenchmen Marcel Proust and Jean Cocteau; Russian Maxim Gorky; Germans Thomas Mann and Franz Kafka; and Ireland's James Joyce.

There were poets aplenty: Robert Frost and Ezra Pound in the United States; T. S. Eliot, W. H. Auden, Wilfred Owen and Walter de la Mare in Britain; Rainer Maria Rilke in Germany, and Ireland's W. B. Yeats.

Theatre-goers could see for the first time James Barrie's *Peter Pan*, J. M. Synge's *The Playboy of the Western World*, Arthur Miller's *Death of a Salesman*, Eugene O'Neill's *Long Day's Journey into Night*, Bernard Shaw's *Pygmalion*, Noel Coward's *Private Lives*, Jean Anouilh's *Antigone* and Tennessee Williams' *A Streetcar Named Desire*. But the most influential culture of the day was the cinema, especially with the advent of the 'talkies' in 1927.

There was an abundance of provocative new thinkers: psychologist Carl Jung, William James (*The Varieties of Religious Experience*), Max Weber (*The Protestant Ethic and the Rise of Capitalism*), Karl Popper (*The Open Society and its Enemies*) and Simone de Beauvoir (*The Second Sex*). Aldous Huxley (*Brave New World*) and George Orwell (*1984*) provided disturbing visions of the future, as did later dictator Adolf Hitler (*Mein Kampf*).

THE QUANTUM, THE GENE AND THE UNIVERSE

During the first half of the twentieth century, science advanced on all fronts, but progress in three areas was particularly spectacular. Nuclear physics and quantum physics opened up the world of the very small—atom-sized or smaller. At the other end of the scale, we found our place in the pattern of the cosmos, within our own galaxy and among the countless other galaxies that fill the expanding universe. Between these extremes, we came closer to solving the riddle of heredity—the structures and chemicals that control inheritance.

Albert Einstein is perhaps the most familiar name; his major discoveries include his two theories of relativity. But brilliance was everywhere, and even Einstein drew on the work of others and needed other scientists to confirm his predictions.

By mid-century, nuclear fission had been discovered and the first atomic reactor had been built. Antimatter had been predicted and found. The source of the Sun's energy had been revealed, and light shed on the intertwined issue of the origin of elements. Radio signals and 'cosmic radiation' had been detected falling to Earth from regions beyond. Order had been established among the variety of stars, and the ultimate fate of very large stars deduced.

We knew that motion changed our measurements of length, time and mass, and that gravity bent light. We had realised the dream of the alchemists and turned one element into another many times. At the same time, we realised our inability to measure accurately everything about the motion of tiny particles.

Major medical discoveries included the vital role of vitamins, the existence of insulin, blood groups, penicillin, a chemical cure for syphilis and evidence that 'conditioning' can affect human behaviour. Radioactivity was shown to have the power both to heal and to kill.

In the Earth sciences, two theories were advanced, both unpopular to begin with but later vindicated: the notion of continental drift, and the theory that the pull of the other planets on the Earth brings on the ice ages. Radioactivity gave us the first reliable estimates of the age of the Earth. The Richter Scale provided a measure of the energy of earthquakes.

Underpinning all of this was a growing transformation in the way the enterprise of science was organised and supported. Governments took more interest, especially where the new knowledge supported national defence or economic growth. Some set up 'national laboratories' (such as Britain's Department of Scientific and Industrial Research) to explore areas of concern. Grandest of these was the huge Manhattan Project established by the United States in the 1940s to refine the science and technology of nuclear weapons. This set a pattern for the future.

Spurred by the example set by Thomas Edison (**1882**), business and industry were beginning to use science to develop new products or find solutions to production problems. Many more 'scientists' (the word was now in common use) were active; the universities, hospitals and research centres where most of them worked were more numerous and better endowed. Being a scientist became an established career and profession. Few important discoveries were now made by amateurs and part-timers.

As in previous centuries, the 'technology of knowledge' was a driving force, particularly the new wizardry of electronics, used to detect and amplify faint signals. Cloud chambers in nuclear physics, big new telescopes in astronomy and the electron microscope in biology continued to illuminate the previously undetected and unknown, and the pace of discovery was quickening.

Hugo de Vries: The Reality of Mutations

Hugo de Vries of Holland was one of three scientists who in **1900** independently rediscovered the almost 40-year-old insights of Gregor Mendel (**1865**) into the laws of biological inheritance. A year later, de Vries added something more to the mix, something that responded to one of the biggest objections to Charles Darwin's theory of evolution (**1859**). Darwin claimed that in any generation of living things, some vary from the rest in a superficial way. They have some characteristic in their bodies or behaviour that makes them better able to survive and bear young to carry forward that trait. Darwin summed up this process in the phrase 'survival of the fittest'. But where did that advantage come from? What is the source of the 'variation'? Darwin had no answer, though his own experiments, such as with breeding pigeons, showed how advantage could be taken of such differences when they occurred.

French biologist Jean Lamarck (**1809**) had argued that organisms gained these new capabilities by actually adapting to their environment and then passed those on to their offspring. Wading birds had developed longer legs through venturing into deeper waters in search of food. In nearly a century of observation, no one had actually seen that happen. Mendel had looked for it. He found that setting different varieties of a plant side by side so they shared the same environment had no observable effect on succeeding generations.

Now it appeared that 'Lamarckism' could not happen anyway. German biologist August Weismann had discovered that at a certain point in the very early development of a plant or animal (around day 59 for a human embryo) certain cells are set aside and given a special role; they are to become the gametes, the 'half-cells' in the form of eggs, ovules, sperm and pollen that will transmit genetic information to the next generation. After that day, nothing that happens to the rest of the organism can influence what the cells contain, and so what is handed on.

So 'variation', these small differences among individuals descended from the same stock, still needed explaining. In the 1880s de Vries had been experimenting with evening primroses. He noted from time to time sudden changes in some feature of the plants from one generation to the next, changes that then bred true. Perhaps the new 'fitness for survival' emerged not as a series of imperceptible changes but in large and decisive shifts, big enough to be acted on by Darwin's natural selection.

De Vries devised the term 'mutation' for such changes, from the Latin for 'change'. His primroses proved to be not typical of living things in general—most mutations being very small—but the idea and the name stuck. The challenge now was to find out how these mutations came about. If inheritance was bound up with the chemical nucleic acid in the heart of every cell, as was now widely accepted (**1884 Hedwig and Weismann**), mutations must somehow occur there too (**1926**).

Oliver Heaviside and Arthur Kennelly: How Marconi Bridged the Atlantic

1902

When Guglielmo Marconi (**1895**) first planned to send a radio signal across the Atlantic in December 1901, many doubted he would succeed. The physicists argued that if radio waves were like light, they would travel straight out into space. A signal sent from Cornwall would never reach Newfoundland, since it was way below the horizon. But Marconi was confident; he persevered. History records that the signal did reach Newfoundland, so the wise men of physics had to think again.

A solution was put up in 1902 by the American Arthur Kennelly, and again a few months later by Englishman Oliver Heaviside. They argued that a reflecting layer

THROUGH THE ATMOSPHERE		
250 km		240 km: Space shuttle orbits
200 km		Aurora visible (from 240 km down to 110 km) 200 km: 'F' layer of ionosphere
150 km	THERMOSPHERE/IONOSPHERE	150 km: sounding rockets (gathering information) reach here 120 km: 'E' layer of ionosphere. Aurora visible (up to 230 km) 100 km: 'D' layer of ionosphere
90 km	Mesopause: temperature −100°C	
	MESOSPHERE	80 km: meteors ('falling stars') burn out
50 km	Stratopause: temperature 0°C	Air pressure: 1 millibar
	STRATOSPHERE	40 km: weather balloons reach here 30 km: spy planes reach here 20 km: 'ozone layer' filters solar UV
15 km	Tropopause: temperature −50°C	Air pressure: 10 millibars
Surface	TROPOSPHERE	The 'weather zone' Air pressure: 1000 millibars

The table shows a slice through the lower 250 kilometres of our atmosphere, according to modern studies. Air pressure falls rapidly with altitude. At the top of the stratosphere, air pressure is one-thousandth of its sea-level value; 99.9 per cent of our air lies below this level. Air temperature falls through the troposphere, rises through the stratosphere, falls through the mesosphere and rises through the thermosphere. The boundaries between these layers are known as 'pauses', as in 'tropopause', to indicate that temperature has ceased falling and is starting to rise, or vice versa.

*The ionosphere is thickest at three different levels, each layer reflecting and absorbing different frequencies of radio waves. The ozone layer (**1974**) is thickest in the lower part of the strato-sphere. The ionosphere and the ozone layer are produced by ultraviolet light and X-rays from the Sun hitting atoms of oxygen and nitrogen.*

must exist somewhere in the atmosphere, like a mirror for radio waves. The signal must have bounced back and forth between the mirror and the ocean surface until it reached Newfoundland. In honour of them both, the name Heaviside–Kennelly layer was assigned to the reflecting surface.

Hardly anything was known about the layer, other than that it was there, though Marconi did notice that signals sent via the layer seemed stronger in the daytime than at night. A quarter-century passed before there was more significant information. Americans Gregory Briet and Merle Tuve measured how high the layer was by sending up a pulse of radio waves and timing how long it took to come back. This showed that the layer was about 100 kilometres above the Earth's surface, though it did vary. Later still, we found there were more than one.

By this time, we had some idea why the layer existed. Ultraviolet light and X-rays sent out by the Sun collide with atoms in the air with sufficient energy to strip away some of their electrons, leaving a mix of electrons and charged atoms (ions). Since radio waves have electric and magnetic fields associated with them, this 'plasma' of ions and electrons acts like a mirror. Ionisation decreases once the Sun goes down, which is why the layer weakens at night.

Scots-born engineer Robert Watson Watt suggested the layer be called the 'ionosphere', which was appropriate. Watson Watt would later achieve fame as the inventor of radar (at least as far as the British were concerned). Radar technology would make a lot of use of the techniques developed to explore the ionosphere, such as the way of measuring distances.

Walter Sutton and Theodore Boveri: Where Are the Genes?

When they rediscovered in **1900** the pioneering work of Gregor Mendel (**1865**), three researchers independently confirmed what he had found: various physical characteristics are passed on from one generation of **1902** living things to another as if they were controlled by separate packages of information, working in pairs. Mendel called these 'factors'; the modern term is 'genes'. But where were these genes and how did they operate?

American graduate student Walter Sutton thought they were tangled up in some way with the chromosomes, the thread-like structures that appear and disappear inside the nuclei of cells at the time cells divide in two. In what Walter Flemming (**1869**) had called 'mitosis', the chromosomes form, double in number and then separate, with each of the new cells getting a copy of each chromosome.

Working with grasshoppers, Sutton was first to notice (in 1902) that chromosomes occur in pairs, so that there is always an even number of them. The members of each pair are (apparently) identical but noticeably different from those in other pairs. Thinking about Mendel's pairs of 'factors' (genes), he wondered if each chromosome in a pair carried one member of each pair of genes. For Mendel's laws of inheritance to work, one gene would have to come from each parent. How could that be?

Sutton saw the answer in the less common sort of cell division often called 'reduction division', but which he called 'meiosis'. Instead of creating complete cells ('diploid') as in mitosis, this creates half-cells ('haploid'). These are the sex cells or gametes, such as the sperm and egg, which unite in fertilisation to form a new complete cell. In meiosis the chromosomes are not first duplicated, so that each gamete gets only one chromosome from each pair, not both.

So when a sperm and egg, or a pollen grain and ovule, unite in fertilisation, the resulting whole cell receives one chromosome in each pair from each parent, and so one gene for a particular characteristic from each pair. So chromosomes are equipped to be the repositories of Mendel's genetic factors, needing to interact in pairs to control inheritance. Despite the importance of all this, Sutton did not pursue a career in genetics, but instead ran a surgery practice in Kansas until he died, aged 39.

The German scientist Theodore Boveri reached similar conclusions independently. He made an important observation while working with the eggs of sea urchins. An organism needs its full quota of chromosomes (36 in the case of sea urchins) to operate or to develop normally. So the various chromosomes must be each responsible for different characteristics, and need to be paired to do their job.

Ivan Pavlov: At the Sound of a Bell

1903   Reading Charles Darwin's *The Origin of Species* turned Russian theological student Ivan Pavlov away from the Church and towards a long career in science (he was active right up until his death at the age of 87). This brought him high scientific acclaim, including the 1904 Nobel Prize for Medicine, but also frequent conflict with the totalitarian governments of his homeland, especially once it became the Soviet Union. It was his reputation that kept him free from persecution.

Pavlov went into medical research. He was particularly interested in digestion; his studies with dogs showed that the stomach goes into action only when stimulated by the production of saliva in the mouth. Clearly there is some connection between these two organs through the nervous system. Pavlov made his dogs salivate by presenting them with food, but soon found that other things could

> *While you are experimenting, do not remain content with the surface of things. Don't become a mere recorder of facts, but try to penetrate the mystery of their origin.*
>
> IVAN PAVLOV

start them drooling, even the sound of his footsteps as he entered the room. So he decided to see if other sounds and actions could get the saliva flowing.

Most famously, he rang a bell when he presented food, and soon saw that the dogs would salivate at the mere sound of the bell, whether food was presented or not. So the 'drooling reflex', over which the dog had no control at the sight of food, had become 'conditioned' and now responded automatically to a very different stimulus. This opened up many more questions, which Pavlov investigated by collecting and measuring the amount of saliva produced in various circumstances.

Once established, did a conditioned reflex stay for ever? No. He found it could be eliminated over time by not reinforcing it—by ringing the bell and not giving food. The disappointed dogs would soon stop drooling at the sound of the bell. Yet an 'extinguished' conditioned reflex is quite easily reset by once again linking the food and the bell. It can even reappear spontaneously after some hours.

He also saw that anything that sounded like the bell could set the dogs off—a telephone for example. The stimulus became 'generalised'. Yet by rewarding some bell sounds with food and not others, Pavlov could teach his dogs to discriminate between different sorts of sound.

Pavlov doubted that his studies with dogs, published in 1903, meant anything for humans, but it did lead to the development of the 'behaviourist' school of psychology. The American John Watson showed that conditioning could influence emotions such as fears and phobias. More darkly, Pavlov's approach also provided a mechanism for 'brainwashing'.

Ernest Rutherford and Frederick Soddy: Unravelling Radioactivity

Radioactivity, discovered so unexpectedly by Henri Becquerel in **1896**, proved more and more complex the longer it was studied. For example, it was soon found that three different sorts of rays came from radioactive minerals: alpha, beta and gamma (**1899**). The first two were particles (helium nuclei and electrons respectively) and the third were like X-rays but much more penetrating. At the same time, the list of identified radioactive elements grew. Soon, so many were known that the hunt was on for some sort of order, some 'family relationships' among them.

1903

New Zealand-born Ernest Rutherford and his colleague Frederick Soddy were in the lead here. They found that many radioactive elements were linked in chains of transformation. One chain started with uranium and 14 links later ended with non-radioactive lead. Each link was a different element, either lighter or the same weight as the previous one. Some links gave off strong radiation, but tended to go quiet quite quickly. At each transformation, alpha, beta and gamma rays were given off. The chains were complex but orderly and intelligible. Rutherford secured the 1908 Nobel

Atomic number	Element	Atomic weight	Sort of decay	Half-life
92	Uranium	238	Alpha	4.5 billion years
90	Thorium	234	Beta	24.5 days
91	Proactinium	234	Beta	1.14 minutes
92	Uranium	234	Alpha	230 000 years
90	Thorium	230	Alpha	83 000 years
88	Radium	226	Alpha	1590 years
86	Radon	222	Alpha	3.8 days
84	Polonium	218	Alpha	3 minutes
82	Lead	214	Beta	27 minutes
83	Bismuth	214	Beta	29 minutes
84	Polonium	214	Alpha	One ten-thousandth of a second
82	Lead	210	Beta	22 years
83	Bismuth	210	Beta	5 days
84	Polonium	210	Alpha	140 days
82	Lead	206	Stable	

An example of natural alchemy. The many steps in the radioactive decay of uranium 238 to lead, passing through several other elements: thorium, proactinium, radium, radon, polonium, bismuth and a couple of radioactive isotopes of lead. Each alpha decay reduces atomic number by 2 and atomic weight by 4. Beta decay increases atomic number by 1 but does not change atomic weight (a neutron in the nucleus has decayed into a proton and an electron). Note the huge range of half-lives.

Prize for Chemistry for this work. Some thought that Soddy should have shared it; he was rewarded with his own prize in 1921.

Rutherford and Soddy looked hard at the way the elements in these chains 'decayed', changing step by step into other elements. A simple rule seemed to apply. In a given time, the intensity of radiation in any sample fell to half its original level, as if half the atoms had changed. This 'half-life' varied enormously from element to element (most elements, of course, are not radioactive at all). The half-life of radium was 1600 years, for radon it was four minutes, but it was as much as 4.5 billion years for uranium and perhaps only a hundred millionth of a second for other elements. Usually, the more intense the radiation, the shorter the half-life. The patterns of decay would soon provide a way to determine the age of rocks and the Earth itself (**1921**).

It also posed a puzzle. While it was certain that half the atoms in a sample would decay within the half-life, there seemed no way of knowing which atoms would actually break down. All seemed equally likely to transmute, no matter what had happened before. So while we knew with accuracy the outcome with a large number of atoms, if we had only one atom there was no certainty at all, just a probability. For nineteenth-century physicists, what happened at any time was precisely determined by what had happened before. Now, that confidence in cause and effect was draining away. Into this uncertain and changing climate, 'quantum physics' was already being born. ➤➤**1909 RUTHERFORD**

Heinrich Lorentz and George Fitzgerald: The Prehistory of Einstein's Relativity

In **1887** Americans Albert Michelson and Edward Morley tried to detect the 'aether', the medium that was supposed to carry light through space as a series of ripples. The physics of the day said that light should travel at different speeds in different directions because the Earth moves through the aether (a bit like a boat against or with the current). But Michelson and Morley found that the speed was always the same; the aether could not be detected.

1904

Dutch physicist Heinrich Lorentz had an explanation; Irish physicist George Fitzgerald had much the same idea at much the same time. They were probably both trying to save classical physics, for which the failure to find the aether had been a disaster.

Michelson and Morley had raced light rays over paths set at right angles to each other, one track in line with the supposed movement of the Earth through the aether, the other across it. Lorentz (and Fitzgerald) suggested that the path parallel to the aether was shortened by the aether as it streamed past, by just enough so that even though light had been slowed down by battling against the aether, the time taken for the journey was unchanged. It was ingenious. Light would always appear to travel at the same speed. The aether might still exist but it would be forever hidden from view.

The two men wrote the idea down in a mathematical form, called the Fitzgerald–Lorentz Transformation (FLT for short). Imagine a train passing very fast (meaning close to the speed of light) as you stand on the platform. You and your surroundings are one 'frame of reference', and the interior of a carriage on the train is another against which you can compare measurements. FLT links the two frames, provided the train is going at a constant speed, but the effects are only noticeable when the

> *Creating a new theory is not like destroying an old barn and erecting a skyscraper in its place. It is rather like climbing a mountain, gaining new and wider views, discovering unexpected connections between our starting point and its rich environment.*
>
> ALBERT EINSTEIN

train is travelling close to the speed of light. At our everyday speeds, any effect is all but impossible to detect.

If you were able (by some ingenuity) to measure the length of a rod in the train as it went past, or mark the ticking of a clock or find the mass of an object, you would find (by comparing them with the metre rod, the clock and the kilogram weight on the platform beside you), that the rod in the passing train looks shorter, the clock appears to tick more slowly and the object seemed heavier. If someone in the carriage did the same they would find that your rod is shorter, your second longer and your mass greater compared with theirs.

It does not matter that you might think you are standing still. Only the relative movement between you and the train matters. You and the person in the carriage will disagree about these vital measurements, yet objects and light will behave and move according to the same laws in the carriage and on the platform (those laws are 'invariant'). So there is nothing you can do on the platform or in the carriage to tell you how fast you are travelling. (We all know this from travelling in a lift; once it has started moving, only the slight shaking tells us we are not standing still.)

Einstein built on these ideas in a very important paper he published in 1905, one of three he put out that year. He called the paper *The Electrodynamics of Moving Bodies*, but it sets out what is usually called the Special Theory of Relativity ('Special' because his 'General' theory came later, in **1915**). Because he added more ideas of his own, it is usually called Einstein's Theory of Relativity, but he may not have done what he did without Fitzgerald and Lorentz. Einstein started out by accepting that light always appears to travel at the same speed (as Michelson and Morley had shown), and did the maths to see what followed from that. The outcome was essentially what Fitzgerald and Lorentz had stated in the FLT.

The predictions of the FLT, bizarre as they seem, have actually been confirmed. Experiment has shown that a moving clock (for example, in a satellite orbiting the Earth) does seem to tick more slowly than one on the ground; mass does increase with speed, say for electrons nearing the speed of light (that was shown within a few years). Distances do shorten. We also find that light sets a speed limit for the rest of the natural world. Nothing can go faster than light, nothing material can even go as fast as it. Einstein, with a little help from others, fundamentally changed the way we see the world. ➤➤**1905 EINSTEIN**

Albert Einstein: Particles of Light

1905 In 1905 a little known 26-year-old German-born patent clerk working in Switzerland published three scientific papers in a leading physics journal. One paper explained a puzzling phenomenon called Brownian motion (**1827**). Another initiated what we now call the Special Theory of Relativity (**1904**). Important, indeed revolutionary, as those papers were, it was the third that would win Albert Einstein a Nobel Prize in 1920. That third paper unleashed perhaps the most important and puzzling idea in twentieth-century science—quantum physics.

The subject of the paper was the photoelectric effect, found by Heinrich Hertz, the discoverer of radio waves (**1888**). When light fell on a metal surface, it released negative electricity, which could be gathered by a metal plate nearby. Hertz did not take much notice of it, but 10 years later his countryman Philip Lennard did. He set up an experiment to measure how much energy the electrons carried away as they were bumped out of the metal under the impact of light.

Lennard expected that the more intense the light, the more energy the electrons would carry, but it was not so. More light released more electrons, making a bigger current, but did not change their energy. Very strange. What mattered was the colour of the light. Blue light produced more energetic electrons than red light; if the light was too red, no electrons came off at all.

Classical physics, so successful in the nineteenth century, was in the dark here; it had no answer. But at the start of the new century, Einstein did. He knew that his compatriot Max Planck (**1900**) had explained mysteries in the radiation from hot objects by assuming that energy could be transferred only in packets ('quanta') of a certain size, and that the amount of energy in a packet depended on the frequency (colour) of the associated light or infra-red radiation.

In a radical move, Einstein gave these energy bundles an independent existence. Light, he said, can behave as if it is a steady stream of these energy packets, or 'light quanta', later to be known as 'photons'. According to Planck, the bluer the light (that is, the higher its frequency) the more energy each light quantum would carry. A certain amount of energy would be needed to extract the electrons, so only light that was not too red would do. If the light quanta carried more energy than the minimum, it ended up with the electron, so the energy of the photons would depend only on the colour of the light. More quanta (more intense light) simply meant more electrons set free. All the evidence seemed to fit.

The 'classical' physicists, for whom light was indisputably a wave, and had been for a hundred years (**1801 Young**), were outraged. One of the most outspoken was the brilliant American experimenter Robert Millikan. For 10 years he laboured unsuccessfully to prove Einstein wrong. All his experiments supported Einstein's vision of light-as-particles. But few were on Einstein's side as yet.

Crucial new evidence arrived in 1922, around the time that quantum physics, which owes so much to Einstein's ideas (but which he ultimately rejected), was being firmed up. American Arthur Compton set up an experiment in which X-rays or gamma rays (now seen as closely related to light) were bounced off slow-moving electrons. The frequency of the radiation was decreased by the encounter. This strongly suggested that the X-rays and gamma rays were indeed behaving like particles, and they had given up some of their energy to the electrons. This lowered their frequency. Again Einstein's answer was the best.

The battle was not over. Light still behaved as waves on many occasions, say in interference and diffraction (**1801 Young**), and X-rays had been found to do the same (**1915 Braggs**). So at most Einstein was saying that what are usually thought of as waves can also be thought of as particles. Frenchman Louis de Broglie would soon claim the reverse is also true (**1924**). ➤➤**1907**

Frederick Gowland Hopkins: Vital Amines

1906 Today we all accept the importance of vitamins and the health consequences of going without them, but the name is only a century old and the concept not much older. The name was coined (as 'vitamine') in 1912 by the Polish scientist Casimir Funk. He combined 'vita' (for 'life') with amine, a common chemical grouping containing nitrogen. He had already isolated one, which he called thiamine, from rice husks.

The rice husk story is a little older. In 1905 English doctor William Fletcher had found that a serious tropical disease called beriberi could be prevented if people ate unpolished rice, that is, rice with the husk on, rather than 'polished rice'. Fletcher believed that the husk contained some special nutrient to ward off the disease. Seven years later, Funk had found it.

The name most associated with the vitamin story is English doctor Frederick Hopkins, even though it formed only a small part of his illustrious career. Hopkins, related to the poet Gerard Manley Hopkins, had most of the honours: a knighthood, Order of Merit, President of the Royal Society (immediately after Ernest Rutherford) and the Nobel Prize for Medicine in 1929. Drawn into science by playing with his father's old microscope, he truanted school a lot. He worked in an insurance office and then as a forensic assistant before turning to medicine.

Around 1906 Hopkins became convinced that small quantities of specific complex chemicals ('growth factors') were essential to health. He began to feed mice diets limited to sugar, starch, salt and milk protein. They fell sick and recovered only when given whole milk. Clearly the milk contained some of these essential complex chemicals. Hopkins isolated two: one was later called vitamin C and the other proved to be a mixture of vitamins A and D. Funk's thiamine was the first of the B group of vitamins. By 1940 others had been identified: B2 (riboflavin), B3 (niacin), B4 (folic acid), B6 (nicotinic acid) and B12. Enthusiastic researchers had given the names B5 and B7 through B11 to other chemicals, but they later proved not to fit the definition; either they were not really essential for health or they could be made in the body.

The role of vitamin D in preventing the crippling bone deformation rickets was settled by Edward Mellanby in 1922, as was the way in which it is synthesised in the skin under the action of ultraviolet light from the Sun (**1802**) (which strictly speaking means D is not a vitamin). Vitamin E was found in leafy green vegetables in the same year. The story of vitamin C (ascorbic acid) goes back to **1747** when James Lind found that citrus juices prevented scurvy. The vitamin was identified in 1912, and in 1935 became the first vitamin to be made artificially.

Albert Einstein: The Most Famous Equation

1907

There is one equation in science that just about everyone knows, even if they are not quite sure what it means: $E = mc^2$, where E is energy, m is mass and c is the velocity of light. The equation proclaims that matter can be converted into energy and vice versa, a mind-stretching concept. The speed of light is a very big number, 300 000 kilometres per second. The equation has this number multiplied by itself, so a small amount of matter will make an enormous amount of energy. If totally converted, 1 kilogram of matter, equal to the mass of 1 litre of water, would yield as much energy as 10 million tonnes of oil.

Matter and energy had long been seen as being quite separate, though each was thought to be separately 'conserved' and unable to be created or destroyed. The Conservation of Matter and the Conservation of Energy were in their time 'laws of nature'. Now it appears that it is 'mass plus energy' that is conserved. In the never-ending drive for unity among the phenomena of nature, matter and energy could be seen as two sides of the one coin.

This new vision flows from Albert Einstein's Special Theory of Relativity (**1904**). This said, among other things, that moving objects get heavier as they approach the speed of light, a speed they can never actually reach. This gives us a handle on this 'energy/mass equivalence'. The energy of motion of an object depends on both its mass and its speed. Suppose we had a rocket that could approach the speed of light. If we keep firing the engine and providing energy, more and more of that energy will have to go into increasing the mass of the rocket, since there is a limit to how much we can speed it up. So in a sense, energy has been converted to matter. As Einstein said, the mass of a body is a measure of its energy content.

A more complete explanation of this concept, of the kind Einstein wrote down in 1907, is mathematical, but evidence that the relationship holds is everywhere. Conversion of matter into energy powers the Sun and the stars (**1920**); four million tonnes of the substance of the Sun vanishes every second to be replaced by the energy that keeps the Sun shining. The nuclear fission (**1938**) that powers nuclear power plants and nuclear weapons likewise turns mass into energy. The production

> *Most of the fundamental ideas of science are essentially simple, and may, as a rule, be expressed in a language comprehensible to everyone.*
>
> ALBERT EINSTEIN

of pairs of 'mirror particles' from pure energy (**1928 Dirac**) is a less everyday example. So E = mc² is rightfully the best known equation in science. The relationship it summarises is one of the most fundamental in nature. ➤➤**1915**

JJ Thomson: Weighing Atoms

1907

Cathode ray tubes had preoccupied physicists in Germany and Britain for many years (**1854 Geissler**), and interest did not cease once the English physicist JJ Thomson used one in **1897** to prove that cathode rays were particles called electrons, which went from the negative cathode to the positive anode in the discharge tube. German researchers had already found something else moving—positively charged particles (or ions) much heavier than the electrons and going the other way. These could be concentrated by boring a hole through the cathode to capture them; they were called for a time 'canal rays'.

JJ knew these 'rays' were actually streams of atoms without one or more of their electrons. He had run electrons through magnetic and electric fields to find the ratio of their charge to their mass. From around 1907 he did the same for the canal rays. He already knew the charge on the ions was one or two or three times the charge on the electron (but positive, not negative), so by getting these canal rays to bend in a magnetic field, he was effectively 'weighing the atoms', getting much better results than the nineteenth-century chemists had done in measuring atomic weight.

It was painstaking work but atoms of many different elements were weighed directly for the first time. There were surprises, of course. When the gas neon was put into the machine, out came not one weight but two, 20 units (oxygen being 16) and 22 units, with the lighter atoms being much more plentiful. The old chemists had obtained the average value (20.2), which had of course puzzled them. Now here was an answer. There were in fact two sorts of neon, chemically the same, one a little heavier than the other.

JJ called this new technology 'mass spectroscopy', likening it to spectroscopy, in which a prism or grating separated light into colours and measured its wavelength. Cambridge University colleagues took over. Frederick Soddy coined the name 'isotopes' for the versions of an element with different masses. Francis Ashton made it his life's work (Nobel Prize for Chemistry, 1922). Of 300 known isotopes he weighed 200. He cleared up the long-running mystery of the mass of chlorine. By the old methods chlorine had an atomic weight of 35.5. Ashton proved it has two isotopes, Cl35 and Cl37, the first isotope being three times more plentiful than the second.

Just about all the elements tested had isotopes. Krypton had five. Uranium had two, U235 and U238, a very important distinction as later events proved (**1938**). The challenge would come in trying to separate isotopes in more than the minuscule amounts handled by mass spectroscopes, given that isotopes differ only in mass.

Kamerlingh Onnes:
Liquid Helium and Superconductivity

In the mid-nineteenth century, soon after Lord Kelvin (aka William Thomson) announced that there was a lowest possible temperature, which he called absolute zero (**1848**), a serious drive to reach even lower temperatures got under way. A few years later, working with James Joule, Kelvin uncovered the Thomson–Joule Effect (**1852**), which showed how gases could be cooled by being repeatedly compressed and then expanded. Once below a 'critical temperature', increased pressure could crowd the gas particles together to form a liquid.

One by one the major gases were liquefied: air first of all, then oxygen and nitrogen separately at around 100° below zero or 170° on the absolute scale (170K). By now the quest had its own title, 'cryogenics', which literally means 'making ice'. It was helped by the invention of the Dewar flask by English physicist James Dewar, to keep things very cold. This was essentially the 'thermos flask' as we know it today, using a vacuum to limit the flow of heat. Carbon dioxide behaved differently from other gases. At everyday pressures, it would not form a liquid at all, going straight from the gas to a solid ('dry ice').

Hydrogen resisted becoming a liquid until the temperature was down to only 20° above absolute zero (20K). The Dutch physicist Kamerlingh Onnes led the team that achieved that in 1906, and then conquered helium at 4K in 1908. Later work pushed the temperature still lower, to less than one degree above absolute zero, prompting the joke that the coldest place on Earth was in the physics department at the University of Leyden. Onnes won the Nobel Prize for Physics in 1913.

Discoveries did not cease there. If a loop of tin, lead or even frozen mercury was immersed in liquid helium, an electric current, once started, continued flowing, apparently indefinitely. All electrical resistance had been lost; this was the first example of 'superconductivity'. At the time, this needed such low temperatures it was hard to imagine any use for it.

Liquid helium itself did some odd things. It seemed to lose all resistance to movement (viscosity), becoming a 'superfluid'. It would not stay in a flask, but flowed up and over the sides. Superfluids and superconductors defied explanation by the old 'classical' physics. Only the new 'quantum physics' (**1927**), with special rules for the behaviour of the very small, could do so, and ultimately did.

Karl Landsteiner: The Varieties of Blood

In 1818 English obstetrician James Blundell was attending a patient who had bled severely following childbirth. In fear for her life, he extracted blood from her husband's arm with a syringe and injected it into the woman. She survived, the first successful blood transfusion recipient on record, and most of Blundell's later patients did too. Perhaps he was lucky, because transfusions often went wrong and patients died.

Doctors began to suspect that not all types of blood were compatible, that the patient's blood could cause the donated blood to clot, sometimes with disastrous consequences. The first to look at this matter systematically was Austrian Karl Landsteiner, then working in a hospital in Vienna. By 1909 he was able to divide blood into four major groups, which he called A, B, O and AB. Finding the blood type could ensure compatibility, though it was shown soon after that type O blood can be given safely to anyone ('universal donor'), and an AB patient can take blood from anyone ('universal recipient'). These developments made blood transfusions much safer.

Further research gave a reason. Each red blood cell has a marker on its outside surface, an 'antigen' which is either A or B, though some people have both antigens (AB) and some have neither (O). Our defence system is primed to detect and reject blood with the wrong sort of marker. Type A patients reject anything carrying B; B patients reject A.

These blood types are inherited; they involve genes for A and B, though neither is 'dominant' (**1865**). Landsteiner saw a further use for blood types—as 'admissible evidence' in the courts, for example, to help determine paternity. Landsteiner also helped find the Rhesus factor in blood (so called because it was identified first in Rhesus monkeys). It is either positive or negative.

Though his blood group discoveries won Landsteiner a Nobel Prize in 1930, he had many other discoveries to his credit, including the virus that causes polio, and a test for syphilis. He helped build up a new area of study called immunology, which would become very important in future decades as we understood more about our natural defences against disease.

Ernest Rutherford, Hans Geiger and Ernest Marsden: Every Atom Has a Core

1909

In **1899** the English physicist John Joseph 'JJ' Thomson startled the world with his 'plum pudding' model of the atom: the newly discovered, negatively charged electrons (**1897**) were studded through a mass of positively charged matter to make each atom. A mere decade later that image was under heavy challenge.

Ernest Rutherford, the New Zealander who had worked under JJ at the Cavendish Laboratory in Cambridge and who already had the Nobel Prize for discovering what radioactivity was, now ran physics at the University of Manchester. He planned to use the alpha particles coming out of radioactive elements as probes to examine the contents of JJ's plum pudding. Two research students, German Hans Geiger (**1912**)

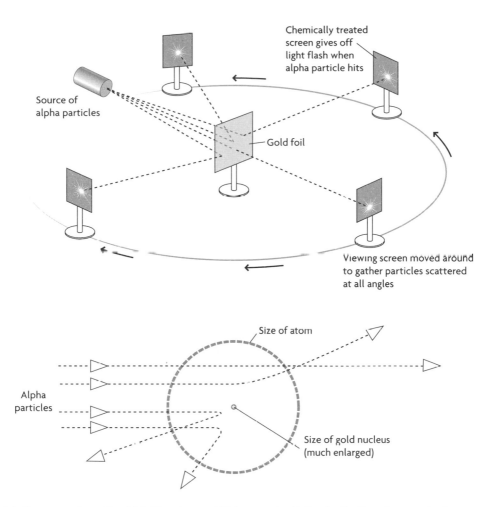

This is the apparatus with which Marsden and Geiger, under Rutherford's direction, made the profound discovery of the atomic nucleus. The lower diagram shows how the amount of deflection of the alpha particles depends on how close they come to the minute nucleus.

and Englishman Ernest Marsden, later a professor in New Zealand, set up a piece of gold foil; gold had the heaviest easily available atoms and could be hammered into a very thin sheet. On one side stood a source of the alpha particles; on the other side, a screen that glowed when struck by an alpha particle. They counted the 'hits' through a microscope.

At first, all was as JJ's model predicted. Most of the alpha particles barely changed direction at all as they went through the foil. However, a small but significant number were turned from their path by 10 degrees or more. With Rutherford's encouragement, Geiger and Marsden looked for even larger deflections and found those too, including (and this really shocked them) a few alphas sent back where they came from. Rutherford is supposed to have wept when he heard this; it was, he claimed, as if you had fired a cannon ball at a piece of tissue paper and it came back and hit you.

Rutherford had at once grasped the reason. The lump of positive charge in the atom had deflected the positively charged alpha particles. Since nearly all had gone straight through the detecting screen and only a few went off in other directions, that lump had to be very small, in fact 10 000 times smaller than the atom as a whole. If the atom could be enlarged to the size of a cricket ground, the core of the atom would be only the size of a cricket ball. The plum pudding was actually mostly empty space. In Rutherford's words: 'this is the most incredible event that has happened to me in my life'. **➡➡1911 RUTHERFORD, 1912 GEIGER**

Ejnar Hertzsprung and Henry Russell: Classifying the Stars

1910

A quick glance at the night sky might suggest that 'a star is a star' but there is, in fact, a great variety. Some are bright, some are faint. Some may be white or whitish, some have a blue tinge, others are orange or even red. Colour reflects the surface temperature of the star, at least relative to our star, the Sun. Blue stars are hotter than our Sun, red stars are cooler, white stars are about the

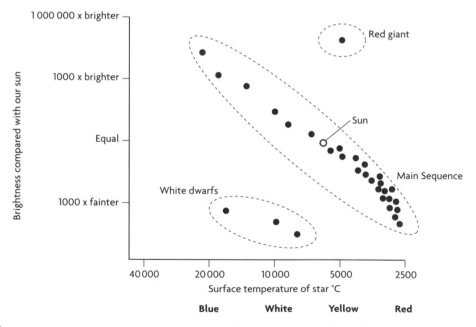

The stars nearest to us in space are arranged on this diagram according to their colour or temperature (horizontal axis) and their brightness compared with the Sun (vertical axis). This shows the three main groupings—Main Sequence, white dwarfs and red giants—and that our Sun is a little brighter and hotter than the average.

same. Brightness is more complex. Some stars are further away than others; we know this from measuring parallax (**1838**). But even when this is taken into account, some stars are intrinsically brighter than others. In general, brighter stars are bigger than faint ones.

Around 1910 two researchers, the Dane Ejnar Hertzsprung and the American Henry Russell, independently at first, started plotting these variables out on a chart, soon to achieve fame as the Hertzsprung–Russell (HR) diagram. The biggest and brightest stars were marked at the top, the smaller and fainter ones at the bottom. Hot blue stars were on the left, cool red stars on the right.

The stars line up in a striking fashion. Nearly all the stars, at least 90 per cent, are on a long thin line from the top left (very hot and very bright) down to the lower right (cool and dim). Our Sun is somewhere in the middle of this line, which is called the Main Sequence. Over time, two other groups shaped up: one on the upper right, red but bright and therefore big (the 'red giants'); the other towards the bottom, hottish (white) but dim and therefore small. These were dubbed the 'white dwarfs'. Having pegged out this classification of the stars, the big challenge was to explain it. Why were red giants and white dwarfs relatively so rare? What was so special about being on the Main Sequence? At the time it was not even certain how stars sustained their prodigious outpourings of energy. An understanding of that would have to come first (**1920 Eddington**); more detail emerged in **1938** (**Bethe and von Weiszacher**).

Paul Ehrlich: Chemicals Against Disease

The discovery by William Perkin in 1856 of a beautiful purple dye in coal tar, previously though of as a waste (**1856**), started a 'gold rush' among chemists. They began to track down and identify all sorts of chemical compounds, many of them 'organic' (containing carbon). As methods improved, hundreds or even thousands were extracted and named. Our understanding of chemistry increased enormously as a result, but the main agenda was to find chemicals that were useful in some way, say in industry or medicine: dyes, drugs or synthetic materials that could be manufactured and sold.

1910

The vision that some of these new chemicals might help to control disease and promote health was the principal motivation of Paul Ehrlich, a German doctor who spent his early career working with Robert Koch, discoverer of the organisms causing anthrax and tuberculosis (**1876**). Early in his career, Ehrlich had found Perkin's purple dye very useful in staining animal tissues and micro-organisms, so that features that were usually invisible became obvious. Walter Flemming had found the same (**1869**). (Bacteriologists today still use methods derived from those Ehrlich pioneered).

By the early twentieth century, Ehrlich was deeply involved in 'chemotherapy', the search for chemicals that could be targeted against identified disease organisms—'magic bullets', as he liked to call them. This involved testing hundreds of chemicals in the search for those that worked; the hundreds had been chosen from many thousands on the basis that they were constructed in a way that made them more likely to succeed.

Erlich's public fame rests mostly on salvarsan, the arsenic-containing chemical that killed the organism (the 'spirochete') that caused syphilis. The first successful drug was number 606 in his exhaustive series of trials (though it had been initially judged ineffective); later, number 914 (neosalvarsan) proved to be even better. Ehrlich, who ate little, was sustained through these labours by a diet of 25 strong cigars every day; he always had a cigar box under his arm.

Like many before and since, Ehrlich had to battle opposition to his new ideas and methods, but he persevered, so benefiting public health and cementing his own reputation. Ehrlich shared the Nobel Prize for Medicine in 1908.

Thomas Hunt Morgan: Genes, Chromosomes and Fruit Flies

1911

In **1900**, as the long-forgotten findings of genetics pioneer Gregor Mendel were being rediscovered, American biologist Thomas Hunt Morgan was studying fruit flies in New York in an attempt to understand their reproductive development. Morgan was fascinated by Mendel's experiments breeding peas in his monastery garden (**1865**), though a bit sceptical about the laws of inheritance he had uncovered. He was keen to see if those rules applied to animals. Fruit flies could go through 30 generations in a year; they took up little room and needed little care. Morgan decided they were ideal subjects.

He began cross-breeding them (ultimately by the millions) as Mendel had done with peas, looking for the same patterns. After some years without success, he had a big break when some flies with white eyes turned up in one batch. Fruit flies generally have red eyes. He crossbred red- and white-eyed flies, finding the first generation all had red eyes but the 'recessive' white eyes turned up in the second generation, with the numbers supporting Mendel's laws.

All the white-eyed flies were males; it appeared that 'white-eyedness' was 'sex-linked'. By now the word 'gene' was being used to describe a file of inherited information about a certain characteristic, and most biologists agreed that genes were somehow tied up with chromosomes, which appeared when cells were dividing (**1902**).

Here Morgan had some good fortune. The salivary glands of fruit flies have very big chromosomes that are easy to see and study. There are eight chromosomes in each cell (humans have 46); six of them match up into three identical pairs, the other two are different, as Morgan found. They are called X and Y chromosomes because of their shape, and they determine the sex of the fruit fly. XX means a female, XY a male (it is the same with us).

Morgan argued that the gene for white eyes had to be on the Y chromosomes (so that only males had it). This suggested that the genes controlling all the various characteristics of a fruit fly were on particular chromosomes. By staining the chromosomes with dyes, he could make out a great number of stripes. Around 1911 Morgan announced that these stripes of tissue *were* the genes, lined up along the chromosomes like beads on a string. His student Alfred Sturtevant made the first 'maps' of chromosomes, showing where the genes for the various characteristics were located.

Other findings followed. You would expect all the genes on a particular chromosome to be inherited together; a fruit fly with any of those characteristics would have them all. This commonly happens, but not always. Morgan found that 'crossing-over' occurs: bits of chromosomes are exchanged during the formation of the sex cells (ovum and sperm) that unite to form the next generation. Crossing-over adds a lot of variety to the offspring by mixing up the characteristics, producing more differences for Charles Darwin's natural selection (**1859**) to work on. It is a powerful force for diversity in nature.

By the time Morgan came to write his influential *Mechanism of Mendelian Inheritance* in 1915, his scepticism about Mendel (and indeed about Darwin's natural selection) had gone. He had worked through the evidence and was convinced; he had added more evidence of his own. In 1933 he became the first non-physician to be awarded the Nobel Prize for Medicine.

Marie Curie: The Promise and the Peril

The life and death of Marie Curie was a testament to the power of radioactivity, both creative and destructive. During World War I, she pioneered the use of radon gas—given off by the radium that she and her **1911** husband Pierre had discovered (**1898**)—as a treatment for cancer. Glass vials filled with the gas were inserted directly into tumours.

She had discussed such possibilities with Pierre many years before when they had both seen what radioactivity could do to body tissues. Their hands were scarred from the effects of handling radioactive chemicals day by day. They were both constantly tired, partly from the gruelling nature of the work, but at least partly from the cumulative effects of radiation exposure. Pierre had once bound a tiny sample of radium to his arm for 10 hours; the resulting wound, which resembled a burn, took three months to heal and left a permanent scar.

Marie Curie died in 1934 of leukaemia, most probably caused by a lifetime of radiation exposure, from both radium and X-rays. During the war she took charge of delivering X-ray machines to help the treatment of the wounded in the trenches, and often took the photographs herself without adequate protection. She had kept a small sample of radium by her bed as a sort of nightlight, since it glowed in the dark.

> *We must not forget that when radium was discovered no one*
> *knew that it would prove useful in hospitals. The work was one of*
> *pure science. And this is a proof that scientific work must not be*
> *considered from the point of view of the direct usefulness of it.*
> *It must be done for itself, for the beauty of science, and then there*
> *is always the chance that a scientific discovery may become,*
> *like the radium, a benefit for humanity.*
>
> MARIE CURIE

Curie's life had been turbulent since the accidental death of Pierre in 1906. She took over his professorship at the Sorbonne, the first ever held by a woman there or perhaps anywhere in science. Her second Nobel Prize, in 1911, and an invitation to a conference with the cream of the world's (male) physicists was offset by failure to be elected to the French Academy of Science, though the vote was close. There were also allegations of impropriety in her private life, and two men fought a duel over some of the rumours.

Since Curie had shared the 1903 Nobel Prize for Physics, some have argued that the 1911 Nobel Prize for Chemistry rewarded the same work. Her major new achievement since 1903 was to isolate the pure metal radium. She was able therefore to define its properties in detail. Given the growing importance of radium, particularly in medicine, this was crucial work.

Ernest Rutherford and Niels Bohr: The Solar System in the Atom

1911

In **1909** the world of physics was shaken up by the discovery by Ernest Rutherford and some colleagues at the University of Manchester that atoms were not solid balls of mixed positive and negative matter, as JJ Thomson had argued (**1899**), but mostly empty space with a tiny lump of positive matter at the centre. This discovery raised more questions than it answered. Where were the negatively charged electrons in all this? If they were drifting in the region around the core, what stopped them falling into it under the pull of electrical attraction? What stopped the core, stuffed full of positive charge, from blowing itself apart?

By 1911 Rutherford had some preliminary answers. He had already settled on the name nucleus for the central lump of charge, likening it to the tiny spot at the heart of every plant and animal cell (**1838**). He proposed that the electrons went around and around the nucleus, just like the planets around the Sun. They do not fall into

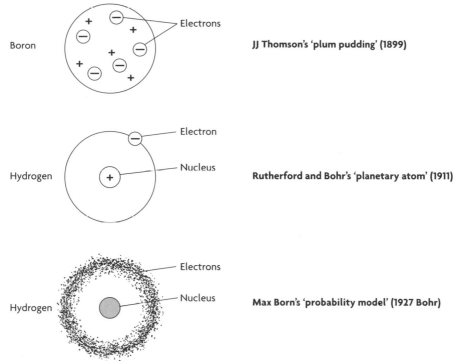

Boron — JJ Thomson's 'plum pudding' (1899)

Electrons

Hydrogen — Rutherford and Bohr's 'planetary atom' (1911)

Electron
Nucleus

Hydrogen — Max Born's 'probability model' (1927 Bohr)

Electrons
Nucleus

The diagrams show a quarter-century of new ideas about atoms. JJ Thomson (1899) thought that the electrons in an atom were stuck into a mass of positive charge like plums in a pudding. Rutherford and Bohr likened an atom to the solar system, with electrons moving in precise orbits around the nucleus like planets around the Sun.

But the growth of quantum physics and its 'indeterminacy' required precision to be replaced by probabilities. By the mid-1920s Max Born, for example, visualised electron 'clouds', with the darkness of the cloud indicating the chances of an electron being in one place rather than another.

the nucleus for the same reason that the Earth does not plunge into the Sun; it is moving and falls 'around'.

But another problem came up at once. When a charged particle, like an electron, is 'accelerated' (made to change its speed and/or direction), it gives off energy. That is how Heinrich Hertz had first made radio waves back in **1888**; the sparks he used were in fact electrons made to jump rapidly backwards and forwards. If electrons were orbiting the nucleus, they were constantly changing direction, so they should have been pouring out energy all the time, losing speed and falling into the nucleus in a split second.

Obviously, this did not happen. Rutherford was stuck for a solution at first, and a bit annoyed that everyone expected all the answers to come at once. An answer did come, in the form of a new research student. Niels Bohr from Denmark arrived in

Manchester in 1912 after failing to get on with JJ Thomson (**1907**) at the Cavendish Laboratories in Cambridge. Bohr liked Rutherford because he was willing to praise the work of others; Rutherford liked Bohr because, among other things, he played football.

Bohr also had the most extraordinary mind, innovative and iconoclastic. Why did the orbiting electrons not radiate energy? Well, said Bohr, they just don't, not as long as they travel in 'allowed orbits'. They give up (or take in) energy only when they transfer from one 'allowed orbit' to another.

And what makes an orbit 'allowed'? Bohr's answer would make him the creator of quantum physics. As with matter (the atom), electric charge (the electron, **1897**) and energy (the quantum of Max Planck, **1900**), something about the orbits had to be 'quantised', that is, only certain precise values would be permitted by nature. Bohr argued that what was quantised was the angular momentum of the electron, a combination of its mass, its speed and its distance from the nucleus. Allowed orbits have one or two or three or more quanta of angular momentum. So one, two, three and so on are the 'quantum numbers' for these orbits.

And the difference in the allowed amounts of angular momentum involves the same number that Planck used to make his quantum of energy, Planck's constant. It was an amazing development, and the beginning of a whole new way of thinking about physics. It would explain so much, beginning with the colours of light that various chemical elements emit (**1913**) and then the Periodic Table of Elements (**1924**). ➤➤**1913 BOHR, 1919 RUTHERFORD**

Henrietta Leavitt:
The Standard Candle in the Heavens

1912

In 1902 Henrietta Leavitt, then aged 34, secured a permanent position at the observatory at Harvard University in the United States, after working as a volunteer for a number of years. Her work analysing images of stars on photographic plates was meticulous and she was given a key role in a major project, to develop a precise standard for the brightness of stars as determined from photographs.

Around 1912 Leavitt worked on a group of variable stars (stars that fluctuate in brightness over time) called Cepheid variables. The first of these had been found in 1784 in the constellation of Cepheus, hence the name. Cepheids are very regular in their light changes over days or weeks and she could easily pick them out on photographic plates exposed at different times and compared with each other. She concentrated on the two great collections of stars in the southern sky known as the Magellanic Clouds; they had been first sighted by the Spanish navigator Ferdinand Magellan in the sixteenth century.

Leavitt found 25 Cepheids among these stars and noticed an interesting fact. The brighter a star appeared to be at its peak brightness, the longer was its 'period', the time it took to go though its cycle of change. She realised that since these stars were clustered together and so all roughly the same distance from the Earth, stars that appeared equally bright *were* equally bright. So the time taken for a Cepheid variable to complete a cycle of change was a measure of how bright it actually was.

The Dane Ejnar Hertzsprung, co-author of the soon-to-be famous Hertzsprung–Russell diagram for classifying stars (**1910**), took this up. He found the distance to some Cepheids much closer to Earth by measuring their parallax (**1838**). This allowed him to link the actual brightness of a Cepheid precisely to its period. So if the period of a Cepheid variable anywhere in the cosmos could be measured, its apparent brightness was a sure clue to how far away it was. It was a major step forward in charting the cosmos. Henretta Leavitt's Cepheid variables became a 'standard candle' for measuring distances out into the deep universe. This would prove vital in settling the big questions about the size and shape of our star system and its relationship to others.

Henrietta Leavitt was a prominent scientist at a time when few women had an opportunity to shine. Her success came through diligence and dedication; in the same way she also overcame recurrent illness and deafness, as well as the all-too-common disdain of male colleagues. During her career she found more than half the variable stars ever discovered up to the time of her death. She died from cancer, aged 53.

Victor Hess: Radiation from the Cosmos

Austrian physicist Victor Hess sometimes put his life on the line for his science. In 1912, to investigate a phenomenon that puzzled him, he went aloft in a balloon, without oxygen tanks, to a height of 5000 metres. His reward was confirmation of a theory and ultimately a share in the Nobel Prize for Physics in 1936.

1912

To understand what puzzled Hess, we need to go back a bit. The discovery of radioactivity by Henri Becquerel in **1896** had been a shock, coming at a time when science was thought to have run out of surprises. Radiation was found seeping from naturally occurring minerals containing elements like uranium and thorium. This radiation was ionising (as were the equally new X-rays): it could give atoms of air an electric charge. Ordinary air, which normally is an insulator, would then carry electricity. So if some electric charge is stored (in an instrument called an electroscope) it will slowly leak away through the air. The more radiation there is about, the faster the stored charge will disappear.

Researchers found that their estimate of the total quantity of radiation that they thought should have been coming from radioactive minerals in the ground—rocks, stone buildings and so on—was well short of what they actually measured. Furthermore, screening out the background radiation, which messed up delicate measurements, proved difficult. Some of it got through 10 centimetres of lead, which should have stopped any known radioactive rays. Somewhere there was a source of powerful penetrating radiation. Where was it?

Victor Hess, then teaching at the University of Graz, wondered if the unaccounted-for radiation came not from the ground but from higher up, from the atmosphere or even space. This led to his daring balloon flights. The rate at which electric charge was lost from an electroscope showed that the radiation got stronger as he ascended; it was nearly 10 times stronger at the top of his flight than on the ground. Later it was called (not by Hess) 'cosmic radiation'.

Hess devoted most of the rest of his career to unravelling its secrets. In 1939, for example, he found that the amount of radiation arriving from space went up and down over a 27-day period, which matched the rotation of the Sun; the Sun must therefore generate or at least influence much of the background cosmic radiation.

To escape persecution by the Nazis, Hess left Austria in 1937 and settled in the United States. He was one of many who did so, in the process greatly strengthening American science.

Piltdown Man: The Man Who Never Was

1912 Outright fraud is rare in science. While many researchers, even the great ones, have made mistakes, few have tried to claim discoveries they have not actually made, and if they do, they are usually exposed quickly. This transparency, while not perfect, is what makes science work. It comes from a tradition of openness before the evidence and of publication of results so that others can repeat the experiments and check the findings. If results cannot be replicated, they are commonly set aside.

In 1912 Charles Dawson, a solicitor and amateur hunter after bones, dug up a curious skull in a gravel pit at Piltdown in Sussex, England. He reported the finding to the British Museum. Over coming years, more bones were found, including a jaw. When assembled, the head of 'Piltdown Man' posed a challenge to existing theories of human origins. Its modern cranium, combined with an ape-like jaw, was quite different from the look of *Homo erectus* (**1856**), who supposedly lived during much the same period, around 500 000 years ago, judging from the depth at which the Piltdown remains had been uncovered.

Despite the conflict, and even though Dawson died in 1916, others continued to promote the find, known formally as *Eoanthropus dawsoni* or 'Dawson's dawn man'. It fitted the expectation of a 'missing link' between apes and men, and it gave the British a 'fossil man' of their own; the other important finds of Cro-Magnon man and Neanderthal man had been made in Germany and France (**1856**).

For four decades, the bones were locked away from study and scrutiny, apparently for security reasons, though there may have been other motives. In 1953 testing with radioactive fluorine showed that the skull and the jaw were very different in age; the skull was maybe 50 000 years old, the jaw only a few decades. Other tests proved that the jaw had been chemically stained to make it look old. While the skull was human, the jaw proved to be from an orang-utan, though the teeth were from a chimpanzee, filed to make them fit.

Someone had taken a lot of trouble to make the bones look authentic and to bury them where they would be found. Arguments continue to this day as to who perpetrated the hoax and why, though Dawson himself was an early suspect. Opponents of evolution have seized upon the scandal to brand other vital evidence as equally bogus, though we can also claim the case shows that it is relatively easy to expose fraudulent evidence, given the chance.

Vesto Slipher: Red Shifts and Blue Shifts

When Christian Doppler discovered his effect in **1842**, he recognised that it would apply not only to sound waves but also to light. The pitch of a siren gets lower (longer wavelength) as it moves away and goes up (shorter wavelength) as it comes closer. Similarly, he theorised, a receding light source would get redder (longer wave length), an approaching source would become bluer (shorter wave length). Of course, the speed of light is enormously greater than the speed of sound, and light sources would have to be moving very fast indeed for 'red shifts' and blue shifts' to show up.

1912

And show up they did, not on Earth but out among the stars or more precisely the 'nebulas'. These small, variegated patches of light had been much debated for centuries. Many tens of thousands were known. Mostly they had ill-defined shapes, but some looked like spirals (**1847**). Some had evidence of stars in the midst; others were merely gas.

Around 1912 American astronomer Vesto Slipher examined light from some of these nebulas through his spectroscope. This instrument broke light into its colours (**1814 Fraunhofer**) and revealed the light and dark lines that showed the presence of various elements. The two yellow lines generated by sodium vapour in the nebulas were always distinct and easy to recognise, but they were not usually in the place expected or showing their yellow colour. In nearly every case, these lines were 'red shifted'—moved into the orange or even the red—indicating that the nebula was travelling away. The amount of 'red shift' was at least three times greater than that of any known star, so these nebulas were travelling much faster than any star was. Only one case of a 'blue shift' turned up, and that was the largest and brightest of the nebulas, the one in the constellation of Andromeda. So if Doppler was right, that one was getting closer.

Slipher noted something more. Among the spiral nebulas, the more red-shifted a nebula was, the fainter it appeared to be. If fainter nebulas were further away, that would mean nebulas moving fastest are furthest away. Slipher was a cautious man and did not make that connection. Nearly two decades later Edwin Hubble did (**1929**).

Alfred Wegener: Continents Adrift

1912

If you look at a world map, you will soon see how similar the coastlines are on either side of the Atlantic. If the ocean were not there, South America would fit into Africa very neatly. Francis Bacon (**1626**) was one who noted this but it was usually dismissed as a coincidence.

German weather researcher Alfred Wegener was in his 30s when he first took the observation seriously. Researching through scientific journals, he found other evidence that a link had once existed between Africa and South America. Many of the fossil remains of plants and animals seemed very similar, but that was usually explained by imagining a 'land bridge' that had once connected the two continents but was now sunken below the sea.

Wegener proposed something much more radical: that the two landmasses had been physically joined at a time when the Atlantic Ocean did not exist. There was more: he argued that all continents had once been grouped together in a supercontinent stretching from pole to pole. He called this Pangea, meaning 'all lands'. It had broken up 200 million years ago and since then the continents had drifted to their present positions.

Few new ideas have received so hostile a reception as Wegener's 'continental drift'. The scientific world condemned it, with few exceptions, as absurd and without foundation. Some of the vitriol came from resentment that Wegener, who specialised in meteorology, was theorising about geology. But the critics had some good points. Most importantly, what possible forces could make these vast masses of solid rock plough through the equally solid rock on the ocean floor?

Wegener had some suggestions, involving the pull of the Sun and the Moon or the rotation of the Earth, but these were clearly inadequate. How the continents had moved was a question without an immediate answer, but the case that they had moved was already quite impressive: the fit of the continents; the distribution of plants and animals, both now and in ancient times; the presence of very similar rock types and mineral deposits on landmasses now widely separated; and evidence that the various continents had once had very different climates. Wegener thought this last point particularly persuasive.

Wegener endured the often personal criticism of his grand vision stoically, and never ceased to improve and promote it. His book *The Origins of the Continents and Oceans* went through four editions, the last just before his death in 1930, aged 50. Interest in climate and the weather had taken him to Greenland several times; on his last expedition he died there, apparently from a heart attack. Three decades later, most geologists began to accept that he had been right after all (**1953**).

Charles Wilson and Hans Geiger: Counting the Unseen Particles

Nuclear physics was a booming industry in the early twentieth century. Stimulated by the discovery of radioactivity (**1896**), physicists were dismantling the atoms previously thought untouchable, revealing (at least in their mind's eye) the tiny nucleus and the orbiting electrons (**1909**). In time, the nucleus, too, would be taken apart (**1932 Cavendish Laboratory**).

But no one had ever seen an electron, and no one ever would. Even atoms would not be glimpsed for many decades. The researchers had to rely on other methods to locate and count the fragments of matter. These techniques gave the same benefit to nuclear physicists that telescopes and microscopes had provided for centuries to other scientists; they made the previously unseen (and unknown) visible.

Early on, physicists found that the new 'rays' released by radioactive atoms

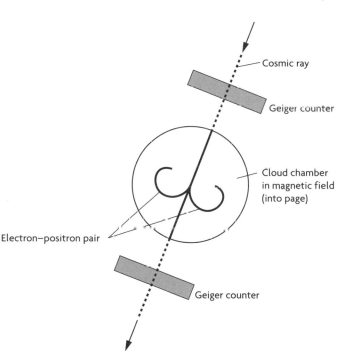

Sophisticated machinery is needed to detect fleeting fragments of subatomic matter and make their paths visible. Here a high-energy cosmic ray (proton or gamma ray) passes through a 'cloud chamber' full of air saturated with water vapour. The ray triggers the Geiger counters above and below, and so activates the chamber. The air is expanded and cooled, and water condenses as tiny drops on the trail of the ionised air atoms created by the ray.

*In this case, it also catches the tracks of an electron and a positron (**1928 Dirac**) created from the energy of the cosmic ray. The tracks curve in opposite directions in the magnetic field running through the chamber.*

(**1899 Radioactivity**) were actually streams of particles and made bright spots on a chemically treated screen (**1909**). Counting the spots by eye was tiring work and prone to human error. Machines to do the job relied on the fact that the rays could ionise air, making it able, if briefly, to conduct electricity. This was the basis of German-born Hans Geiger's counter, more formally (and later) the Geiger–Muller counter. Each particle passing through triggered a pulse of current; many of these together drove a needle to measure the amount of radiation and produced the easily recognised storm of audible 'clicks'.

Making the paths of the particles visible needed the 'cloud chamber', a marvellous machine perfected by the Nobel Prize–winning Scotsman Charles (CTR) Wilson at Cambridge. It too relied on radiation ionising the air, leaving a trail of charged air particles in its wake. Expanding the air in the chamber a little cooled it—the Thomson–Joule effect (**1852**)—and allowed water vapour to condense as droplets on the trail of air ions, making a visible track. The track a particle left said a lot about it. If the track was thick, the particles making it were travelling slowly or were heavily charged; if the particles were sent through a magnetic field, the track bent one way or another, a lot or a little, depending on the charge and mass of the particles.

The two technologies could be combined to study cosmic radiation (**1912 Hess**)—streams of charged particles arriving from space. Geiger counters above and below a cloud chamber sounded the alarm that a cosmic ray had passed through: the chamber expanded immediately and the cosmic-ray track was captured on film. This is how Patrick Blackett produced electron–positron pairs in **1932** (**Cavendish Laboratory**), proving that antimatter existed.

Niels Bohr: Explaining the Spectra

1913   For 100 years physicists had struggled to explain what is plainly visible when light from a heated gas is broken up into its various colours by a prism or a diffraction grating (**1814 Fraunhofer**). The 'spectrum' that results does not have all the colours that Isaac Newton (**1672**) had first extracted from sunlight, but only a few lines of very pure colour on a black background.

Likewise, when light from a star like the Sun was broken down, the full spectrum could be seen, but crossed by many dark lines, as if very specific colours were

> *When it comes to atoms, language can be used only as in poetry. The poet too is not nearly so concerned with describing facts as with creating images.*
>
> NIELS BOHR

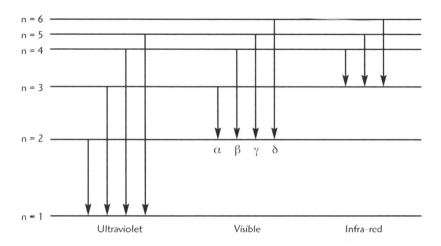

n = 6
n = 5
n = 4

n = 3

n = 2

α β γ δ

n = 1

Ultraviolet   Visible   Infra-red

This chart shows how Niels Bohr explained the characteristic lines in the spectra of hot atoms. It shows the first six energy levels (orbits) in a hydrogen atom. When an electron falls from a high-energy level to a lower one, it releases energy. The more energy is released, the higher the frequency of the radiation generated.

*Electrons falling to the ground state (n = 1) make spectral lines in the ultraviolet part of the spectrum. Falling to the n = 2 makes lines in the visible spectrum (these are the lines in the spectra shown in **1814 Fraunhofer**). Transitions to the n = 3 level generate lines in the infra-red spectrum. Electrons jumping up from lower to higher levels absorb equivalent amounts of energy.*

missing. Why was that so? The lines, both bright and dark, had been carefully measured and listed. Every element had its own particular set of lines, and they were linked by certain mathematical formulae. But that was still not an explanation.

The 'Great Dane' Niels Bohr had the answer. It followed from the 'planetary atom' idea he had worked out with Ernest Rutherford (**1911**). In this model, electrons travel around the nucleus of an atom in certain allowed orbits. The energy of an electron in any orbit is precisely determined by its distance from the nucleus. Electrons could lose energy only by dropping down from one orbit or 'energy level' to a lower one, with each level identified by a quantum number n. So only precise amounts of energy were involved in these transitions and those generated the thin lines of pure colour seen in the spectra.

It was all quite straightforward, at least at first, but the thickets soon grew. There were more spectral lines, even for the simplest element, hydrogen, than Bohr's model could explain. So the model needed some refinements. One was to allow the orbits to be elliptical rather than just circular (as Kepler had done with the orbits of the planets in **1609**). Another was to let the elliptical orbits point in different directions in space. Of course, not all shapes or directions were allowed—only some. 'Quantisation' was the name of the game. A third was to think of the electron spinning either clockwise or anticlockwise.

Bohr ended up with a set of four quantum numbers for each energy level, rather than the one he began with. But the scheme worked. He could explain the bright lines in the spectra of at least the lighter elements, something no one else had done. Yet from the start there was a worrying issue. Bohr knew he was mixing together old and new ideas. Were the electrons really racing around the nucleus like planets around the Sun, or was that just a useful way to picture them? Physics could not really leave the nineteenth century until Bohr answered that question (**1927**). ➤➤**1926**

The Two William Braggs: The Architecture of Crystals

1915 Only one father–son pair has shared a Nobel Prize, though other parent–offspring couples have won different prizes. William Henry Bragg was English, but William Lawrence Bragg was born in Australia when his father was Professor of Physics at the University of Adelaide. Both had distinguished careers in England later, being successive directors of the Royal Institution (**1798 Thomson**) and sharing the 1915 Nobel Prize for Physics. At only 25, W. L. was the youngest laureate ever, and later succeeded the mighty Ernest Rutherford as head of the Cavendish Laboratory in Cambridge.

Their mutual interest in crystals was sparked in 1912 when German physicist Max von Laue shone a beam of X-rays into a crystal of a mineral. He was more interested in the X-rays than the crystal, wanting to see if X-rays could be considered as waves rather than as particles. A crystal, with its regular arrangement of atoms in parallel planes, was like a diffraction grating (**1814 Fraunhofer**), which can break light up into its various colours because light behaves like waves. Von Laue's experiment proved that in this case X-rays, too, were 'wave-like', though that argument was only just beginning (**1924**).

The Braggs turned the experiment on its head. Knowing that X-rays were waves of a given wavelength, they shone them through crystals of different materials. They wanted to know how the atoms in the crystals were arranged. Depending on the spacing and grouping of the atoms, the emerging X-rays made different diffraction patterns, with bright or dim beams that could be found by turning the crystal in all directions in front of a detector. It was laborious work, as was analysing the results to find the arrangements and spacings of the unseen atoms. This second task was highly mathematical, with complex calculations all done before computers were

> *The important thing in science is not so much to obtain new facts as to discover new ways of thinking about them.*
>
> WILLIAM LAWRENCE BRAGG

invented. It often took years, even for the highly skilled W. L.

The early work was done with simple crystals like common salt, but the real power of the technology was revealed when X-rays illuminated crystals of organic compounds, such as the disease-fighting drugs being developed by Paul Ehrlich (**1910**). Understanding the architecture of such chemicals made it possible to deduce how they worked, and perhaps to design and make better ones. This sort of analysis, now highly automated, is crucial to the modern pharmaceutical industry.

Later again, photography was used to record the diffracted X-rays (as Wilhelm Roentgen had found in **1895**, X-rays leave bright and dark spots in photographic film, as light does). This speeded up the experiments, if not the calculations. In this way, Linus Pauling (**1928**), Rosalind Franklin and others found the structure of proteins, and then of the genetic material DNA, making possible the watershed discovery about DNA by James Watson and Francis Crick in **1953**.

Albert Einstein: Bending Space and Time

A total eclipse of the Sun on 29 May 1919 made German Albert Einstein the most famous scientist in the world. On the basis of a new theory, published in 1915, Einstein had predicted that rays of light passing though the powerful gravitational pull of the Sun would be minutely bent from their usual straight-line path. This would change, ever so slightly, the apparent positions of these stars in the heavens.

1915

A total eclipse of the Sun was an ideal opportunity to test this radical idea. With the light of the Sun blacked out, nearby stars were visible and their positions could be measured. Astronomers led by Englishman Arthur Eddington (**1920**) recorded observations during the eclipse. They found that the positions of the stars were shifted by tiny amounts that appeared to correlate with Einstein's predictions. The figures later proved not quite as precise as first thought, but the event was widely reported at the time and Einstein's public reputation was established.

Einstein's General Theory of Relativity, on which this prediction was based, was, at heart, a new vision of gravity. According to Isaac Newton, one lump of matter pulled on another, even through empty space, with a force that he admitted he could not explain (**1687**). So, for example, the Moon was pulled into a curved path around the Earth.

In Einstein's vision, inspired to some extent by his countryman Ernst Mach (**1877**), the matter in the Earth 'bends' the space through which the Moon moves. The Moon, seeking as always the shortest path through space, finds that path is now curved into a closed loop, its orbit of the Earth. Wherever there is matter, said Einstein, space is no longer flat but curved. The more matter is present, the more space curves, even perhaps curving back on itself (**1916**).

According to Einstein's theory, matter also affects the flow of time. A clock runs more slowly close to a large mass like the Earth than further away. This was much harder to prove than the bending of space, but in the 1960s clocks sent up in rockets (and later in satellites) to where gravity was weaker, ran minutely faster than on the ground. Contrary to Newton (**1716**), there is no absolute time as there is no absolute space. Space and time are relative.

In the meantime, something else had been better explained by Einstein than by Newton. Astronomical observations over many years had shown something odd. about the planet Mercury. The point in its orbit that is nearest the Sun (its perihelion) is not fixed in space but moves extremely slowly around the Sun. In other words the whole slightly elliptical orbit of Mercury is rotating. Einstein explained this 'advance of the perihelion of Mercury' as another consequence of the bending of space by the immense mass of the Sun. The other planets are further away and do not feel the effect. Newton had predicted some advance too, but only half as much as Einstein had predicted and as was actually found.

We should realise that Einstein did not render Newton obsolete. Einstein's is a different way of thinking about gravity and in some extreme circumstances gives better predictions. But Newton's 'law of gravity' with its 'action at a distance' is still perfectly adequate for nearly all circumstances. In the late twentieth century space probes were sent on journeys to the fringes of the solar system, often arriving within a few kilometres and a few seconds of their due time and place after journeys of many years and billions of kilometres, with their controllers relying on Newton alone.

Karl Schwarzschild: Black Holes and White Holes

1916 Albert Einstein had recast a lot of physics in his General Theory of Relativity in **1915**, which argues that matter curves space. A year later, some of its many implications began to emerge. The German physicist Karl Schwarzschild imagined a huge star, many times the mass of our Sun, with its fuel exhausted at the end of its life, collapsing under its own weight. As its matter began to pile up in an ever-decreasing volume, gravity there would intensify or, as Einstein put it, space around the star would become more and more bent. Rays of light trying to leave the shrinking star would follow increasingly curved paths.

At a critical point, the light paths would close back on themselves and the light would no longer be able to get away. The collapsing star would have become what we now call a 'black hole', surrounded by an 'event horizon'. Everything going on inside the horizon would be forever hidden from view. Nothing, not even light, would ever emerge. Black holes are minute compared with the stars from which they form; perhaps a few hundred metres across instead of millions of kilometres.

Black holes were not a new idea. The eighteenth-century clergyman John Michell had visualised a 'dark star' in **1783**, based on Newton's ideas about gravity. Schwarzschild's proposal, based on Einstein, had a lot more detail and some startling conclusions. He saw no reason why the collapse would cease once the black hole had formed. It would continue, out of sight, until the density of matter in the heart of the hole, the force of gravity there and the curvature of space all became infinite. This cataclysm was a 'singularity', a point at which the laws of physics as we know them would cease to apply. This was such a bizarre suggestion that for decades Schwarzschild's projection was not taken seriously. The term 'black hole' was not even used until 1968.

In theory at least, the whole process could run backwards, with space–time un-curving, and matter and energy emerging from a singularity that has been dubbed appropriately a 'white hole'. This would resemble in miniature the vast outpouring of matter and energy popularly called the Big Bang, and which many now see as the way our universe began (**1950**).

Less well accepted is the notion that black holes and white holes can be connected by 'worm holes'—tunnels through space and time. Even if they exist, using worm holes as high-speed transit routes to other times and places in the universe means first surviving 'spagettification', the process of being torn into ever finer strips by the overwhelming force of gravity inside the black hole.

Harlow Shapley: Where in the Galaxy Are We?

In **1543** the radical new theory of Nicholaus Copernicus had removed the Earth from the centre of the universe. Instead of the Sun, the planets and the immense sphere of stars revolving around a stationary Earth, the Earth **1918** now revolved around the Sun. Perhaps, though, the Sun was still at the centre of the universe. After all, the stars seem to spread out in all directions.

There was early evidence to the contrary from looking at the Milky Way. Most astronomers now believed that this band of light across the sky indicated that the Sun and its family resided in a vast assembly of thousands of millions of stars, called a galaxy, shaped like a wheel (**1750**). Since the Milky Way appeared wider and denser on one side (that is, towards the constellation of Sagittarius) than on the other, the Sun must sit somewhere off centre, nowhere near the heart of our galaxy.

In 1918 American astronomer Harlow Shapley found more evidence of this. He concentrated on the distribution of dense balls of stars called 'globular clusters'. He

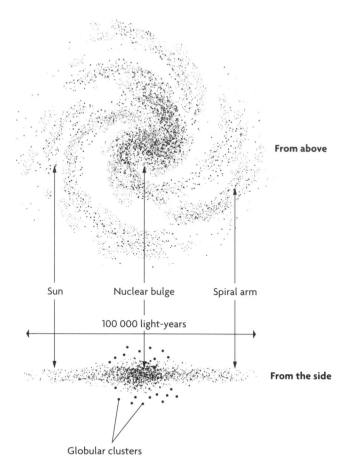

From above

Sun   Nuclear bulge   Spiral arm

100 000 light-years

From the side

Globular clusters

This is how astronomers today visualise our galaxy, like the firework called a 'Catherine wheel'. About 100 000 light-years across, and 10 000 light-years thick in the middle, the Milky Way holds about 400 million stars in its central bulge and in four or five spiral arms. If our galaxy were the size of a dinner plate, the nearest companion galaxy would be 4 metres away.

used the new method of Henrietta Leavitt (**1912**) to work out how far these clusters were from the apparent brightness of Cepheid variables within them. He found that hundreds of these clusters were arranged in a huge ball around the core of our galaxy. They were certainly not centred on the Earth or the Sun. His measurements indicated that we are placed more than halfway from the centre of the Milky Way towards the edge.

Shapley also provided the first estimates of the size of our galaxy: 30 000 light-years across, and perhaps 6000 light-years through the middle, tapering towards the edge, with the Sun about 10 000 light-years from the centre. Modern estimates inflate these numbers by about 300 per cent, but we still live in the 'outer suburbs'.

Shapley was able to do much of this work because he had a big new telescope, a monster with a 2.5-metre mirror. This was noticeably bigger than the previous record-holder, built by the Earl of Russe (**1847**) and, unlike his, fully steerable. To underline how far technology had progressed, the new telescope was more than twice the diameter, with four times the gathering power of William Herschel's biggest-in-the-world of **1789**. The new telescope sat atop Mount Wilson, north of Los Angeles, and was the first telescope to be deliberately located away from the growing lights and pollution of major cities. It saw first light in 1917 and would not be topped in size and power for 30 years.

Ernest Rutherford: The 'New Alchemy'

The old alchemists had dreamt that they might be able to transmute a base metal like lead into a noble one like gold. That seemed impossible once John Dalton (**1808**) had established that chemical elements were made up of indestructible, immutable atoms. These could rearrange themselves into molecules of different compounds, but lead into gold was not an option.

1919

In 1919 New Zealand-born Ernest Rutherford, then at the University of Manchester and already a Nobel Prize winner, reported that he had become the first alchemist of modern times. He did not change lead into gold but nitrogen into oxygen, which was no easier. He was still working with the alpha particles that had been so crucial in proving that every atom had a tiny nucleus (**1909**). These particles, which were thrown out by radioactive elements like polonium, had been proven (by Rutherford himself) to be the nuclei of helium atoms, with a charge of two and a mass of four (**1903**).

Rutherford and his team fired these particles into gases like hydrogen, oxygen and nitrogen to see what would happen. The outcomes with nitrogen were the most surprising. The detection screen showed some powerful particles had come out. These were not nitrogen atoms knocked out of the gas by alphas, nor even stray alpha particles, but hydrogen atoms (missing only an electron). If a helium atom going in had become a hydrogen atom coming out, then the nitrogen must have become oxygen. Otherwise the masses and charges of the various particles would not add up. The nuclei of the helium and nitrogen had been broken down and rearranged. Rutherford had indeed transmuted one element into another. Alchemy was possible.

Rutherford predicted that if he could fire in particles with much greater energy, the nuclei of many elements could be broken open to reveal what was inside. This

> *It is important for men of science to take an interest in the administration of their own affairs, or else the professional civil servant will step in. And then the Lord help you.*
>
> ERNEST RUTHERFORD

$$He_2^4 + N_7^{14} \longrightarrow O_8^{16} + H_1^2$$

This equation sums up Rutherford's 1919 achievement: he converted one element into another, the dream of the alchemists. Helium atoms (alpha particles from a radioactive element) were made to collide with atoms of nitrogen gas. Very energetic hydrogen atoms were generated; this was explicable only by assuming that the nitrogen had been transmuted into oxygen.

The lower figure against each element is the atomic number—the number of protons in the atomic nucleus (which also gives the number of electrons and so determines what element it is). The upper figure is the atomic weight—the number of protons and neutrons (which had not been proved to exist in Rutherford's day). In this case, both atomic number and weight are conserved.

would not happen for another decade, but it did come (**1932 Cavendish Laboratory**). Rutherford had done much of this work during the closing years of World War I, causing him to miss meetings of an important committee on using science in anti-submarine warfare. He wrote in an apology: 'If, as I have reason to believe, I have disintegrated the nucleus of an atom, that is of greater significance than the War'.
➤➤1920

James Croll and Milutin Milankovich: What Changes the Climate?

1919 In an ice age (**1836**), our planet as a whole becomes much colder and the areas covered by sheets of ice expand. The theory that ice ages are caused by the pull of the other planets on the Earth sounds like astrology, but it is actually good science. It is usually associated with the Serbian astronomer Milutin Milankovich, but he did not think of it first.

Scotsman James Croll was largely self-educated, and never rose above a clerical position at the Scottish Geological Survey. Yet he corresponded with leading scientists like Charles Darwin and Charles Lyell and was made a Fellow of the Royal Society. Before Croll, the French astronomer Urbain Leverrier, who had predicted the existence of the planet Neptune (**1846**), had figured that the gravitational pull of the other planets, especially Jupiter, causes the orbit of the Earth to change shape from almost circular to more elliptical every 100 000 years.

In 1875, nearly 50 years before Milankovich, Croll's book *Climate and Time in their Geological Relation* proposed that these subtle changes in orbit affect climate, moving the Earth in and out of an ice age at predictable times. The science of his time was not precise enough to prove him right or wrong, so his theory was ignored.

From 1920 Milankovich added a lot of detail to the theory. He used three cycles: the 100 000 year change in the shape of the Earth's orbit, a variation in the tilt of the Earth's axis every 41 000 years, and a shift over 22 000 years in the time of year the Earth is closest to the Sun. These changes affect either the amount of energy the Earth gets from the Sun, the time of year when most energy arrives or the regions that get the most.

Collectively these can change the climate. If, for a combination of reasons, northern summers become cooler, more ice and snow will survive without melting until the next winter. Over many winters, the area under ice and snow will grow, reflecting more solar energy back into space, cooling our planet still further and hastening the onset of a new ice age. Milankovich predicted ice ages 100 000 years long, with warmer periods (interglacials) 10 000 years or so long between them. His ice ages would get steadily more severe as they went on and warm up quickly before they ended.

At the time he was writing, evidence about ice ages was still primitive, but it was known that ice sheets had come and gone several times. This was not enough to convince sceptics; the idea seemed too radical. The Milankovich Hypothesis languished out of favour for another 50 years. Like other initially rejected ideas, such as continental drift (**1912 Wegener**), the Milankovich hypothesis now enjoys wide support. Details of the waxing and waning of ice ages, gained from studying ice layers in Greenland and Antactica and sediments from the bed of the sea, match the changing pulls of the planets very closely.

It is worth noting the current pattern of the planets suggests that, human interference aside, the Earth would be heading for a new ice age.

Arthur Eddington: What Makes the Sun Shine?

Over hundreds of years, a variety of answers had been given to the question: what makes the Sun shine? Why does it continue to pour out **1920** light and heat, perhaps eternally? The suggestion that the Sun was actually on fire was an early casualty. No known fuel could keep the sun burning for more than a few thousand years, and anyway, there was no oxygen in the vacuum of space. Continual stoking by colliding asteroids and comets would not have helped much.

The remains of heat from the initial collapse of the nebula from which the Sun and planets had formed (**1796**) would have gone cold after perhaps 100 million years. Radioactivity was already proving that the Earth was much older than that (**1921**) and billions of years of sunshine seemed necessary for the slow process of

> *Life would be stunted and narrow if we could feel no significance in the world around us beyond that which can be weighed and measured with the tools of the physicist or described by the metrical symbols of the mathematician.*
>
> ARTHUR EDDINGTON

biological evolution to create the current variety of plants and animals.

Cambridge astronomer Arthur Eddington was already noted for his observations of the 1919 eclipse of the Sun that confirmed Einstein's General Theory of Relativity (**1915**). Eddington drew upon the work of another Cambridge man, Francis Ashton. Using the mass spectroscopy devised by JJ Thomson (**1907**), Ashton had carefully weighed hydrogen and helium atoms. He showed that a helium atom weighed a little less than four hydrogen atoms. Eddington seized on this, suggesting that inside the Sun and other stars, hydrogen atoms, four at a time, are being 'fused' together to make helium atoms.

In **1919** Ernest Rutherford had found that 'alchemy' was possible with modern science: one element could be transmuted into another, such as nitrogen into oxygen. Eddington was proposing alchemy for another purpose. The tiny loss in mass when four hydrogen atoms were transmuted into one helium atom would reappear as energy, as Albert Einstein had allowed for in his famous equation (**1907**). Calculations showed that the steady conversion of even a small amount of the Sun's immense stores of hydrogen into helium would power it for billions of years. The crucial details would be worked out later (**1938 Bethe and von Weiszacher**), but as Eddington said, 'we need look no further for the source of the energy of the stars'.

Ernest Rutherford:
The World within the Nucleus

1920

The word 'atom' comes from the Greek for 'indivisible'. By the early decades of the twentieth century, the term no longer fitted. It was clear that atoms had hidden components, and these could be separated. JJ Thomson had found a place for the negatively charged electrons inside atoms (**1899**). The remaining positive charge was shown to be in a tiny central lump or nucleus (**1909**), around which the electrons appeared to orbit (**1911**). The question was now: what is the nucleus made of? Can that also be taken apart?

A century earlier, English chemist Edward Prout had suggested that all atoms are built up from atoms of hydrogen, but measurements of atomic weight at that time did not support that idea (**1808 Dalton**). No atoms appeared to weigh as much as an exact number of hydrogen atoms; chlorine, for example, weighed 35.5 times as much. But the discovery of isotopes, varieties of an element with different atomic weights (**1907**), solved that problem; something very like the Prout Hypothesis was back in favour. Ernest Rutherford, the great pioneer in all this new knowledge, argued that every nucleus is made up of identical particles he called 'protons' (from the Greek meaning 'first'), the number being equal to the element's atomic weight. So an oxygen nucleus had 16 protons, carbon 12, chlorine either 35 or 37 and so on.

But there was a problem. A new characteristic about atoms was gaining prominence. This was 'atomic number', which counted the number of electrons in an atom or the number of positive charges in its nucleus. For carbon this was six, for oxygen eight, for chlorine always 17. Atomic number matters. Elements in the periodic table (**1869**) are better arranged by atomic number rather than atomic weight.

Perhaps the nucleus also contained electrons, light enough to make little difference to the weight, but able to neutralise some of the positive charge and so make the difference between atomic weight and atomic number. So the common form of carbon would have 12 protons and six electrons, chlorine either 35 or 37 protons balanced by 18 or 20 electrons. If any of these electrons were ejected from heavy radioactive atoms, they could be the source of the beta particles that Rutherford had first explained (**1899 Radioactivity**).

In 1920 Rutherford himself had a better idea. He visualised a companion particle for the proton, a particle of the same mass but with no electric charges. He called this the 'neutron'; it would not be detected for another 12 years (**1932 Cavendish Laboratory**). The neutron hypothesis explained isotopes; same number of protons, different number of neutrons. So the two forms of chlorine both had 17 protons but with either 18 or 20 neutrons. There was an explanation, too, for beta particles. Perhaps, in radioactive elements, neutrons spontaneously changed into protons, firing out electrons to keep the total electric charge constant.

Rutherford also speculated that the presence of the neutron helps to keep the protons together; otherwise their mutual repulsion would have blown the nucleus apart. He was postulating a new force of attraction, later to be called the 'strong force', that would act only between protons and neutrons (**1936**). Here again he was ahead of his time; this force would one day power nuclear reactors and the atomic bomb (**1938 Nuclear Fission**).

Arthur Holmes: Dating the Earth

The discovery of radioactivity in **1896** transformed the long-running debate about the age of the Earth (**1862**). Radioactivity in rocks provided a long-lived source of internal heat, allowing the Earth to be hundreds of

1921

millions or even billions of years old, as the geologists and the supporters of evolution required. And the slow transformations of radioactive minerals gave hope that we could determine just how old the Earth is.

In the vanguard of this development was, once again, the energetic Ernest Rutherford. With Frederick Soddy, Rutherford had teased out the life stories and family relationships of the various radioactive elements (**1903**) as they passed through generations of decay, ultimately ceasing to be radioactive at all. Some of these elements had offspring that decayed quickly, while the parents, like uranium and thorium, could hang around for millions, even billions, of years. Rutherford thought some sense of the passing of the ages could come from comparing quantities of these long-lived and short-lived elements in a mineral sample.

Era	Period	When it began	Reason for name	What happened
Cenozoic ('recent life')	Quaternary	1.8 my BP*	Fourth division of rocks as named by Arduino (1759).	Mostly taken up with a series of ice ages. Evolution of modern humans.
	Tertiary	65 my BP	Third division of rocks as named by Arduino (1759).	Rise to dominance of mammals and birds.
Mesozoic ('middle life')	Cretaceous	145 my BP	From Latin for 'chalk'; the White Cliffs of Dover were laid down at this time.	Greatest dominance of dinosaurs; they were wiped out at the end of the period (1980). Early mammals and birds.
	Jurassic	213 my BP	From the Jura Mountains in Switzerland, where rocks of the period were first studied.	Early swimming and flying reptiles. First flowering plants.
	Triassic	248 my BP	From the threefold division of the rocks of this period in Germany.	Beginning of the Age of Reptiles ('dinosaurs'). Supercontinent Pangea (1992) begins to break up.
Palaeozoic ('old life')	Permian	286 my BP	After Perm region in Russia, where rocks of this period were first studied.	Palaeozoic Era ends with largest 'mass extinction' (1860) on record.
	Carboniferous	360 my BP	After the major coal beds laid down at this time.	Supercontinent Pangea forms as continents collide. The earliest reptiles appear. Huge swamp forests later become coal beds.
	Devonian	410 my BP	After Devonshire, where rocks of this period were first studied.	Major diversification of fish. Insects and spiders colonise land. The first seed–bearing plants grow.
	Silurian	440 my BP	After the Silures, an ancient Celtic tribe.	The earliest fish appear. Life begins to colonise land. The first plants with stems appear.
	Ordovician	505 my BP	After the Ordovices, an ancient Celtic tribe.	Rapid evolution of invertebrates. Modern ozone layer forms, screening out harmful UV light.
	Cambrian	544 my BP	After the Latin name for Wales, where rocks of this period were first studied.	The first complex (multicelled) life forms evolve (sponges, jellyfish).

*my BP = million years before present

This table, with the dates established by studies of radioactivity in rocks, covers only the last 540 million years of Earth's history. In Precambrian times, life was confined almost entirely to single-celled organisms living in the sea.

It was the dogged Englishman Arthur Holmes who persisted with this technique through to ultimate triumph. Rutherford and Soddy had worked out that radium decays into non-radioactive lead in ten steps, each step creating a distinctive new element. Holmes argued that since each step happens at a known rate, comparing the amounts of the various 'daughter' elements (including the lead) would indicate how long the process had been going on, or rather, how long since the parent mineral had hardened (since that prevents any of the daughters from subsequent decays escaping).

Making these measurements was complex and time consuming, so much so that many geologists thought it not worth the trouble. Neither did they care for a physicist like Holmes intruding into their domain. But Holmes battled on against this indifference, constantly refining his techniques and reducing the errors.

By 1921 he was winning. The speakers at a meeting of the British Association for the Advancement of Science that year agreed that radioactive dating was valid and that it indicated that the Earth was 'several billion' years old. This left plenty of time for evolution and slow geological change to generate the look of the Earth and its inhabitants. In 1926 the United States Academy of Sciences set up a committee to study the matter further. The final report from that committee was written mostly by Arthur Holmes. You can guess what it said.

Modern studies using radioactivity indicate that the crust of the Earth first hardened more than four billion years ago. The first living things appeared quite soon after (**1993**).

Frederick Banting and Charles Best: The Hunt for Insulin

Canadian doctor Frederick Banting worked a lot with children, including some who suffered from diabetes. This is a serious condition, especially **1922** for children; they cannot control the level of sugar in their blood. It fluctuates wildly, affecting many of their body functions, including the blood vessels, the kidneys and the eyes; this can cause serious illness or death in later life.

Banting knew that the gland called the pancreas was somehow involved. Like others, he suspected that it produced a chemical that controlled the sugar level. Though it had never been seen, this chemical had a name—'insulin'; the region of the pancreas where it was supposed to be made was called the 'Islets of Langerhans', after the German anatomist who had first found it. Diabetics presumably did not make enough insulin. Feeding them extracts of pancreas taken from pigs did not seem to help this. The pancreas has a double function: it also makes an enzyme called trypsin that helps digest proteins (**1851 Bernard**). Presumably, in making an extract of the pancreas to give to patients, the digestive juices had destroyed the insulin.

Banting had the idea of tying off the duct that carries trypsin into the bowel. This would cause the trypsin-making cells to wither but leave the insulin-making cells intact. After a time, insulin could be extracted without fear of it breaking down. In 1922, working with student Charles Best, Banting followed this course with the pigs, extracting the insulin which, when injected, proved effective in controlling diabetes.

He saved many lives and much suffering and was rewarded with the Nobel Prize for Medicine only a year later.

Insulin is one of the 50 or so hormones in the human body. The name hormone comes from the Greek for 'I excite', because these chemicals basically 'get things going'. They can raise blood pressure and heart rate (adrenaline), bring on labour (oxytocin), stimulate sexual desire (testosterone) and cause allergic reactions (histamine). Others keep things ticking along. Insulin and some related hormones control the storage of sugar (glucose) in the liver as glycogen. Dopamine and melatonin influence the functioning of the brain. The human hormone system is complex and sophisticated. Knowledge of it has grown immensely since the first hormone (adrenaline) was identified in the 1890s.

Vilhelm Bjerknes: The War of the Winds

1922 The invention of the electric telegraph around 1840 had transformed the study of the weather. For the first time, 'synoptic charts' could be drawn, providing a 'synopsis' of the weather over a large area at the same time. American Joseph Henry (**1831**) had been one of the first to construct such a chart.

These charts basically showed the distribution of air pressure, with small, intense regions of low pressure (cyclones) and larger regions of high pressure (anticyclones). Most of the populations and industries in Europe and North America lay close to a low-pressure trough between a high over the polar region and another further south. This trough was the path for a steady stream of cyclones that pushed through from the west, bringing major changes in the weather, and sometimes very destructive winds and storms, particularly in winter.

Better weather forecasting for Europe and North America clearly needed an understanding of why these mid-latitude cyclones develop, and how they might grow and move. The crucial new ideas came out of Norway soon after the end of World War I. Meteorologist and physicist Wilhelm Bjerknes and his son Jakob spent part of the war setting up a network of observing stations across Norway. The weather information would always be sparse, and needed a theory of the mid-latitude cyclones to fill in the gaps.

By 1922 Bjerknes was ready to publish. He argued that the cyclones arise from a clash between two large-scale wind streams: north-easterlies that bring cold, dry air down from the high-pressure system that lies around the north pole; and south-westerlies that carry moister, warmer air up from the high-pressure belt to the south. These winds come together along a narrow zone Bjerknes called the 'polar front', lying on average about 60 degrees north of the equator. He chose the term deliberately; with World War I barely over, the image of struggle between competing forces seemed appropriate. As the opposing winds try to push past each other, the turning of the Earth helps create great eddies spinning anticlockwise; these are the

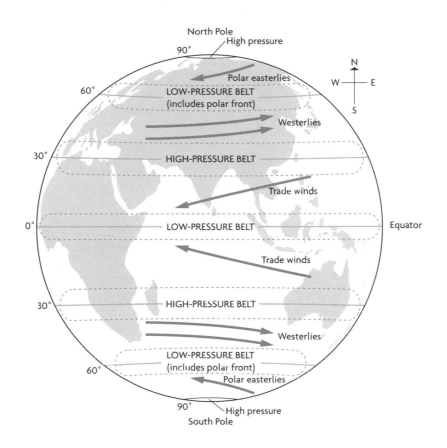

Winds across the surface of the Earth blow from places where the pressure is higher than average (over the poles and in regions around 30 degrees north and south latitude) to places where it is lower than average (along the Equator and in regions about 60 degrees north and south latitude).

*Because the Earth spins (**1835 Coriolis**), the winds do not blow directly north or south. Winds blowing towards the equator (the trade winds and the polar easterlies) appear to come mostly from the east; winds blowing away from the equator become westerlies (as in the 'Roaring Forties').*

Where the westerlies and the polar easterlies collide at about 60 degrees north and south latitude, the polar front occurs with its many travelling low–pressure cells (cyclones) that influence weather in Europe and North America.

mid-latitude cyclones that so affect the weather at those latitudes.

This graphic image was backed by good science and sound mathematics, and remains well accepted today as the basis of weather forecasting in these regions. Embedded in the cyclones are fragments of the major fronts—cold fronts bringing dry air south, warm fronts leading moist air north. One story says that the usual symbols for the fronts, an arrowhead for cold fronts and a semicircle for warm fronts, are based on the shapes of the German and British helmets seen in the trenches of World War I.

Bjerknes and his son left another profound legacy. They were the first to try to put the weather into mathematical terms, writing equations to express the interactions between key factors like pressure, temperature, matter and energy flow, and humidity. Predicting the weather involved solving these equations, dozens at a time, many times over. It was very slow work until the arrival of computers, but that process is the basis of all forecasting today.

Edwin Hubble: Galaxies Beyond Our Own

1924

The Hubble Space Telescope, launched into Earth's orbit in the 1990s, has transformed our view of the cosmos both near and far through its stunning images. It is a fitting memorial to the American astronomer Edwin Hubble, whose ideas and discoveries had comparable impact in their time. He solved forever the riddle of the nebulas and gave us a better (and more humble) understanding of the place of our planet and Sun in the grandest order of nature.

Hubble had access to the new 2.5-metre telescope on Mount Wilson in California, the most powerful in the world. His colleague Harlow Shapley (**1918**) had used it to gauge the size and shape of our star system, the Milky Way galaxy, with its billions of stars, by monitoring a special class of stars called Cepheid variables (**1912 Leavitt**).

In 1924 Hubble went looking for similar stars among some of the faint smudges of light called nebulas. Evidence was building inexorably that at least some nebulas were not among the stars, as for example William Herschel had believed (**1789**), but far beyond them. For example, new stars ('novas', later called 'supernovas') had appeared in some nebulas, brightening and fading like those seen near at hand by Tycho Brahe in **1572** and Johannes Kepler in **1604**. But the supernovas in the nebulas were faint, meaning they were very far away. Thanks to Vesto Slipher (**1912**), we knew that some nebulas were moving away from us at speeds exceeding 1000 kilometres a second, far faster than any star.

Hubble ended any debate. He compared the apparent brightness of the Cepheids he found with the time they took to go through their cycle of changes (which was a measure of how bright they actually were). His figures showed that the nebulas that hosted those Cepheids had to be hundreds of thousands of light-years away, at least 10 times further than any star in our galaxy.

The conclusion was clear. Many nebulas are immense, independent congregations of stars, siblings to our own galaxy, each carrying many billions of stars. They are the 'island universes' that the German philosopher Immanuel Kant prophesied nearly 200 years earlier (**1750**). Dotted through space as far as our telescopes can see, they are separated from each other and from us by vast gulfs of barren space. If each galaxy were the size of a bicycle tyre, the nearest companion galaxy would be 30 metres away.

➡**1929**

Louis de Broglie: A Particle Is a Wave

Royalty has been rare in science. Frenchman Louis de Broglie was a prince, though the nation he lived in was now a republic. He was also brilliant, putting forward a startling new idea in the thesis he submitted for

1924

his doctorate at the age of 24. Revolutionary ideas were becoming commonplace in the physics of the day, but his examiners were so dumbfounded they called in Albert Einstein to give an opinion. Einstein agreed wholeheartedly with the young man's proposal and he got his doctorate.

The essence of de Broglie's theory destroyed the traditional distinction between waves and particles. Einstein had already begun this process: in **1905** he had shown that light, long thought to be waves, could also be thought of as particles, which he called 'light quanta'. Greatly impressed by what Einstein had done, de Broglie took the reverse path. He said that particles such as electrons could be thought of as waves. He even had a formula for the wavelength of the 'associated wave', the wave that he imagined accompanied a particle as it moves through space.

Einstein's opinion notwithstanding, many physicists thought the whole thing outrageous and unacceptable. Particles are particles and waves are waves; it has ever been so. But de Broglie soon had some strong support. For one thing he could give another and less arbitrary explanation for the very important 'allowed orbits' that Niels Bohr had visualised for electrons orbiting an atom (**1913**), and which explained both the line spectra of gases and the Periodic Table of the Elements. De Broglie showed that the circumference of one of these allowed orbits was exactly a whole number of electron wavelengths long, just enough room for an electron 'standing wave'. His electron wavelengths flowed naturally from formulas already set down, by Max Planck (**1900**) and Einstein (**1907**). It all hung together.

Most impressively, electrons were soon shown to do what waves do, for example be diffracted like light through a grating. Electrons were sent through crystals, with their regular arrangement of atoms. When the electrons came out again, they made regular light and dark circles on a photographic plate, just like the 'fringes' Thomas Young had seen with light (**1801**) and Heinrich Hertz with radio waves (**1888**). Here was powerful evidence indeed of the reality of de Broglie's wave–particle duality, the idea that waves and particles are two ways of looking at the same events.

Ironically, among the physicists who proved that electrons can be waves was the

> *Two seemingly incompatible conceptions can each represent an aspect of the truth. They may serve in turn to represent the facts without ever entering into direct conflict.*
>
> LOUIS DE BROGLIE

British physicist George Thomson. In **1897** his father JJ Thomson had first proven that electrons ('cathode rays') were particles, not waves. George Thomson and American colleague Clark Davisson won the Nobel Prize for Physics for this work in 1937. De Broglie already had one from 1929, aged only 30—he was the youngest ever solo prize-winner.

Wolfgang Pauli: Only One Electron Allowed

1925

The theory that the daringly innovative Danish physicist Niels Bohr had put forward to explain the pattern of spectral lines seen in the light of heated gases (**1913**) was a great success up to a certain point. It handled the spectrum of hydrogen very well and helium adequately but it struggled to explain the precise positions of spectral lines for more complex elements.

In addition, there was a new problem. Each atom had a 'ground state', the lowest energy level an electron could occupy, and the one closest to the nucleus. Why did not all the electrons simply give up their energy and fall into the ground state? The orbits of the outermost electrons took up most of the space in an atom. (As Hans Geiger, Ernest Marsden and Ernest Rutherford had shown in **1909**, the nucleus of an atom is almost vanishingly small.) So having every electron in the ground state would make atoms collapse and destroy the organisation of matter. Of course this did not happen. But why not?

An answer came in 1925 from German physicist Wolfgang Pauli. Pauli was a brilliant thinker but a hopeless experimenter, even more so than most theoreticians. He was so inept that apparatus were reputed to stop working when he did no more than walk into the room. This entered folklore as the Pauli Effect.

Pauli's explanation to the riddle was simple but profound. Not all electrons could go to the ground state because there was only enough room there for two. More generally, only two electrons can have the same set of three quantum numbers (**1913**). This rule is the Pauli Exclusion Principle. Why it was so, he could not say. He was not even sure if electron 'orbits' were real or just a useful image. No matter. Thanks to the Pauli Exclusion Principle, the world around us is not in a continual state of collapse. Thanks to the same principle, Pauli won the Nobel Prize for Physics in 1945. ➤➤**1932**

Herman Muller: What Causes a Mutation?

1926

The American biologist Thomas Hunt Morgan (**1911**) studied fruit flies for decades and so greatly increased our knowledge about inheritance. But he did not breed his millions of flies unaided. Like any significant researcher in the twentieth century, he had lots of students, nicknamed the 'fly squad', many of whom became significant researchers in their own right.

Among these was Herman Muller. He joined the team at Columbia University in the United States in 1912, soon after the first white-eyed fruit flies suddenly appeared, apparently from nowhere, among hordes of red-eyed ones. The Dutch biologist Hugo de Vries (**1901**) had given the name 'mutations' to these sudden changes. Muller wondered what caused them, and if they could be made to happen.

Over the next decade, Muller sought a cause in changes in the environment. Temperature seemed important: there were more mutations if the flies' environment was warmed up. But the breakthrough came from using X-rays. Muller figured that these might deliver a blast of energy large enough to disrupt the complex chemicals that he believed made up the genes carrying vital information about inheritance. He was looking for 'an ultramicroscopic accident'.

This approach certainly worked: 30 minutes under X-rays caused 100 times more mutations than normally appeared in a week. Breeding the irradiated flies with untreated ones produced all sorts of visible changes in the offspring: wings could be broad or bumpy or curly, eyes could be purple or yellow, bulging, flat or indented. Some flies became sterile, many were so damaged they died (the mutations were lethal). The longer the exposure to radiation or the more intense it was, the more mutations occurred.

As a result of this work Muller was later outspoken about the dangers of increasing the amount of radiation that people experienced, say from increased use of nuclear power or atomic weapons testing. Muller received his Nobel Prize for Medicine in 1946, just as the atomic age was dawning.

Other researchers continued what Muller had begun. We now know that many things in the environment cause mutations, in addition to X-rays: other sorts of radiation, including ultraviolet light and radioactivity, and many different forms of chemicals. They also appear to happen spontaneously. Where Muller thought that very few mutations were helpful in moving evolution on, and most were harmful, it now seems that most mutations are 'silent', making no difference at all.

Niels Bohr: Explaining the Periodic Table

Dimitri Mendeleyev's classification of the known elements into the rows (periods) and columns (groups) of his Periodic Table (**1869**) was one of the triumphs of chemistry in the nineteenth century. Every known element

1926

had its place; initial gaps were filled later with newly found elements. The patterns and trends in chemical and physical properties were all represented. But why did it work? Did the structure of the table reflect some hidden order within the atoms of the various elements?

Danish genius Niels Bohr thought so. A desire to explain the table was the main reason he put forward his vision of the arrangement of electrons in an atom in **1913**, though it did other things too, such as account for the spectra of the various

WHY THE PERIODIC TABLE OF THE ELEMENTS WORKS

First period			Second period			Third period			
Element	n = 1		Element	n = 1	n = 2	Element	n = 1	n = 2	n = 3
1 Hydrogen	1		3 Lithium	2	1	11 Sodium	2	8	1
2 Helium	2		4 Beryllium	2	2	12 Magnesium	2	8	2
			5 Boron		3	13 Aluminium	2	8	3
			6 Carbon		4	14 Silicon	2	8	4
			7 Nitrogen		5	15 Phosphorus	2	8	5
			8 Oxygen		6	16 Sulphur	2	8	6
			9 Fluorine		7	17 Chlorine	2	8	7
			10 Neon		8	18 Argon	2	8	8

The Danish physicist Niels Bohr explained the structure of the Periodic Table by arguing that the electrons in each atom fill the empty spaces in the various 'shells' around its nucleus, starting with the lowest (n = 1). The last element in a period has its outermost shell full or carrying a 'preferred' number of electrons, for example, eight. Bohr's rules showed that the n = 1 shell can hold only two electrons. The first period therefore has only two elements. The n = 2 shell can hold eight electrons, so the second period has eight elements. The n = 3 shell can hold 18 electrons; filling the first eight of those places generates the eight elements in the third period. The same basic pattern holds for the heavier elements and the higher numbered shells, but the actual order of filling of the shells becomes more complex.

elements. He had developed a set of quantum numbers to represent the different positions or 'energy levels' an electron could occupy as it went around an atom.

German physicist Wolfgang Pauli and his Exclusion Principle in **1925** had decreed that no energy level could hold more than two electrons. Bohr's principal quantum number n designated a group of energy levels called a shell. But shells could have various numbers of energy levels, each slightly different. According to Bohr's quantum rules, the lowest shell (n = 1) had only one energy level so could hold only two electrons. There were four energy levels in the next shell (n = 2), so it could hold eight electrons; the n = 3 shell held 18 electrons, and so on.

At this point Bohr had two crucial insights. The first was obvious enough. Under normal circumstances, the energy levels would be filled up with electrons starting at the bottom. So in the case of oxygen with eight electrons, the n = 1 shell would be full, leaving six electrons in the n = 2 shell. For sulphur with 16 electrons, both the n = 1 and n = 2 shells would be full, leaving six electrons in the n = 3 shell.

The second insight was that an element's chemical properties, particularly its 'combining power' or valency (**1852**), depended on how many electrons there were in the outermost shell. If there was only one electron there, the element would be an alkali metal, with valency 1, like lithium, sodium or potassium. If there were two electrons, the element would be an alkaline earth metal, like magnesium, calcium or strontium (valency 2); if seven, a halogen (chlorine, bromine, iodine). If the outer shell was full, with two or eight or 18 electrons, the element was a rare or noble gas (helium, neon, argon). Carbon and silicon, with four outer electrons, had valency 4.

This provided a fundamental and comprehensive explanation for the periodic trends in the behaviour of elements. They simply reflect the progressive filling up of the energy levels and shells within the atom. Bohr's drive to explain the Periodic Table was satisfied. A few years later, American chemist Linus Pauling would use the same reasoning to explain how atoms bind together to form molecules (**1928**).
➤➤**1927**

Werner Heisenberg: The Uncertainty Principle

The Uncertainty Principle is a sort of a zero-sum game. The more you know about something, the less you can know about something else. The idea arose in 1926 in the fertile brain of the German physicist Werner Heisenberg, one of a half-dozen brilliant minds of the 1920s who decided what quantum physics is all about. So it is usually called Heisenberg's Uncertainty Principle, though he was quite certain about it.

1926

Heisenberg said that if you try to measure the speed of something exceedingly small, say an electron, with very high accuracy, its position in space becomes fuzzy. Find out where it is as precisely as you can, and you will be uncertain how fast it is going. One view of this is that trying to measure one of these, say position, disturbs the whole system and makes the measurements of the other less certain.

More fundamentally, uncertainty seems to come inevitably from one of the facts about the 'quantum world': everything comes in packets of a certain size—energy, electric charge, mass and so on. Some physicists suspect that even space and time can be infinitely divided, though the building blocks there would be very small indeed. One thing that does have a small available package is 'action', first described by Pierre de Maupertuis in **1746**. Action is equal to momentum times distance (where momentum is speed times mass). So a smallest measurable unit of action means a smallest measurable unit of distance times momentum.

This 'uncertainty' has odd consequences. For example, we cannot know precisely where a proton or an electron is. If you think you have one in some sort of a box, there is a finite chance that it is actually outside the box, because its position is fuzzy. This is known as 'quantum tunnelling'.

> *Since the measuring device has been constructed by the observer, we have to remember that what we observe is not nature itself but nature exposed to our method of questioning.*
>
> WERNER HEISENBERG

Action is also equal to energy multiplied by time. The Uncertainty Principle therefore limits the accuracy with which you can concurrently measure both energy and time. Measuring energy accurately needs a long period of time. If you have only a little time to measure, your estimate of energy will be woolly.

There is a fascinating consequence of this: 'virtual particles'. These live for a fleeting amount of time, so small that the error in measuring their energy is greater than the energy itself. So we cannot be sure whether these particles exist or not. The modern image of a vacuum, which older physics says is just empty space, has it seething with virtual particles, created and then destroyed before we are able to measure their energy and confirm their existence.

Of course, this bizarre behaviour is found only among atoms and 'subatomic particles'. At our level of nature, with mice and men and mountains and moons, huge in size compared with Planck's constant, all seems predictable and in order.

Niels Bohr: Is the Quantum World Real?

1927

Niels Bohr once said: 'If quantum physics doesn't profoundly shock you, you haven't really understood it'. Though often hard to understand, the Dane Bohr was far from unworldly. Like so many physicists of his generation, he was caught up in the building of the atomic bomb in World War II, and led the opposition to its military use. Yet his greatest legacy is a radical theory that often seems contrary to common sense and everyday experience.

Quantum physics (QP) is a product of the twentieth century. It was born in **1900** with German physicist Max Planck, who argued that energy is transferred from place to place only in very precise (and very small) amounts, which he called 'quanta'. A few years later (**1905**), his countryman Albert Einstein associated those quanta with light particles (later called photons), reviving the old notion that light should be considered as a stream of particles not a series of waves. All the available evidence seems to fit Einstein's idea.

Two decades on, Frenchman Louis de Broglie (1924) argued that a beam of electrons, previously thought of as particles, could also be regarded as waves. Here was one of the early platforms of QP—'wave–particle duality'. Light rays or electrons could usefully be seen either as particles or as waves, depending on the circumstances. This was a long way from the physics of the nineteenth century, where particles were particles and waves were waves.

Since we could not say for sure what these things were, we could not describe them with old-fashioned precision. According to the Uncertainty Principle of the German Werner Heisenberg (1926), the more we know about, say, the speed of a particle, the less we know about where it is. Traditional certainty was steadily crumbling. Heisenberg was one of a small group of brilliant young men who through the 1920s dug deeply to see what the mathematics of QP said about physics. Germans Edwin Schrödinger and Max Born were in there, as was the Englishman Paul Dirac (1928).

But the leader of the pack was the Dane Niels Bohr. Early on he had used some of the ideas of QP to explain the structure of the atom (1911), the origins of spectra (1913) and the reason the Periodic Table of the Elements works (1924). But those insights, brilliant as they were, depended on thinking of electrons as particles located in particular positions around the atom. Already that model was being replaced by one in which we were not quite certain what an electron was doing or where it was but we could state with precision the probability of it being in one place or another. Instead of precise electron orbits, we had diffuse electron clouds.

By 1927 Bohr had distilled some of the key factors into the 'Copenhagen Interpretation' of QP (that's where he was working at the time). One key concept was 'complementarity'; different ways of looking at events or phenomena can be equally valid, though we cannot say which is 'true'. Then there was the 'correspondence rule'. QP applies to exceedingly small things seen up close, but if we look at them from far enough away we have to get measurements like those from classical physics. That's why we can get by in the everyday world without worrying about QP, just as we survive without regard for Einstein's relativity (**1904**, **1915**). In most circumstances, the physics of Galileo (**1604**) and Newton (**1686**) is quite adequate.

Hardest for most people to accept is Bohr's notion that making an observation affects what is observed; no observer can be really objective. Indeed, some say that nothing really exists until it is observed. Before that time we have just probabilities. What is more, what you go looking for influences what you observe. Set up an experiment to see waves, you will see waves; look for particles, you will find particles.

Bizarre as some of these ideas seem, QP is very powerful; it lets scientists calculate with precision interactions between matter and energy in chemistry and elsewhere. Einstein, who did so much to start QP off, acknowledged that. Yet he never really accepted Bohr's point of view with its fundamental 'indeterminacy'. He devised 'thought experiments' against it. 'God does not play dice', he once said. That argument continues.

Frederick Griffiths and Oswald Avery: The 'Transforming Factor'

1928 The decades-long search for the secret of inheritance began a new stage in 1928. American biologist Frederick Griffiths wanted to understand why some forms of the microbe that causes pneumonia (*Diplococcus pneumoniae*) caused much worse cases of the disease than others.

Under the microscope, he saw a difference. The virulent strain (the 'S strain') had an outer casing made of many sugar molecules, which gave it a smooth appearance. The harmless form had no such coating and looked much rougher (the 'R strain'). Griffiths was looking for a pneumonia vaccine, and was trying out the principles of vaccination, as developed by Louis Pasteur (**1886**). A weakened form of the disease-causing organism can give a patient a mild dose of the disease but also generate immunity against the full-blown disease if it strikes.

Griffiths thought that a mixture of the virulent S strain, killed by heat, and the non-dangerous R strain might make an effective vaccine. Unfortunately, the laboratory mice died when he injected the mix, which had somehow become virulent. How could that be? Under the microscope, he saw why; the living R strain now had the smooth coats. It seemed some 'transforming factor' was at work, transferring the smooth coat, and the associated virulence, from the dead bacteria to the live ones.

Griffiths was busy with other matters and did not pursue his 'transforming factor'. In 1943 Canadian scientist Oswald Avery and some colleagues took another look. At that time, there were two candidates for the factor, two types of chemicals that might carry genetic information in some sort of code.

The popular choice was a protein. That was a long chain of 20 different kinds of amino acids, making an 'alphabet' with 20 letters. The other candidate, with much fewer votes, was DNA (deoxyribose nucleic acid). This too came in long molecules but with only four different 'letters'. Surely, for complex genetic messages 20 letters were better than four. The computer age would teach us that any amount of information can be transmitted with just two letters (0 and 1), but at the time the reasoning seemed sound.

Avery and his colleagues soon proved it wrong. They went back to Griffiths' R and S microbes and extracted the 'transforming factor'. Using various enzymes, they removed first any proteins in the substance, and then any DNA. With the proteins gone, the transformation from harmless to dangerous still took place, but not without DNA. The case was solved. From now on, DNA was the place to look for the reason we resemble our parents.

Alexander Fleming: Chance and Penicillin

The discovery of penicillin is often quoted as an example of serendipity in science. Discoveries can be made by chance, but it takes a keen eye and mind to turn a chance observation into discovery.

1928

Alexander Fleming, a Scottish doctor working at St Mary's Hospital in London, had a great interest in finding ways to prevent infection. His experiences in the trenches of World War I showed how deadly infections can be, often killing more soldiers than bullets or shells.

Fleming took a holiday in 1928, leaving unwashed a number of glass dishes in which he had been growing the sometimes deadly bacteria *Staphylococcus aureus* ('golden staph'). On returning, he began to clear up but noted that blue–green mould was growing on one dish. More interestingly, the growth medium on the dish was free of bacteria around each cluster of mould. It seemed that something seeping from the mould had stopped the bacteria owing.

The mould had apparently come from soil in the garden, blown in through an open window. Fleming identified the mould as *Penicillium notatum*, so he called the yet to be identified bacteria-killer 'penicillin'. There was scant evidence of its ultimate value. Fleming reported that it was unstable and difficult to purify. As a doctor, Fleming was not the man to do the job. So after a few years, the work petered out.

A decade later, German-born chemist Ernest Chain, working in Oxford with Australian Howard Florey, came across a report of Fleming's discovery. The pair decided to find a way to produce it in larger amounts, and the rest, as they say, is history. Penicillin cured bacterial infections in mice, then in human patients, and by the end of World War II was saving the lives of wounded soldiers. Half a ton of penicillin was being produced every month. Fleming, Florey and Chain shared the Nobel Prize for Medicine in 1945.

Penicillin was the first of the 'antibiotics'; different moulds would later produce other drugs such as streptomycin. Along with the 'sulpha' drugs that had come into use before penicillin was mass produced (1932), penicillin and its kin gave doctors powerful new weapons against serious infections. Ominously, strains of bacteria not killed by these 'wonder drugs' began to emerge within a few years. Drug resistance was an unwelcome example of Charles Darwin's natural selection.

Linus Pauling:
Bonding in Chemistry and the World

1928 American chemist Linus Pauling won two Nobel Prizes, a very rare feat. They were not both for science, as Marie Curie's were. Pauling won one for chemistry and another, more controversially, for peace. Politically very active, he strongly opposing the testing of nuclear weapons and United States involvement in the Vietnam War during the 1970s. He was enormously productive in chemistry, working in many fields, making major contributions to the understanding of proteins and ultimately the structure of DNA (**1953**), but his reputation was later clouded by his advocacy of 'megadoses' of vitamin C (**1906**) as a protection against the common cold.

Early in his career Pauling tackled one of the oldest questions in science: what holds molecules together in chemical compounds? A century earlier, Swedish chemist Jöns Berzelius (**1814**) had theorised that every molecule has a positively and a negatively charged part, and that mutual attraction did the job. This worked well for what were called 'ionic' compounds, compounds that conduct electricity when dissolved or melted. This covered acids, bases and salts but not, for example, water itself, nor the vast range of organic compounds. It did not explain how two oxygen atoms or two nitrogen atoms can form a molecule; that issue caused Berzelius to reject Avogadro's Hypothesis (**1811**) and set some of chemistry back 50 years.

The answer lay with electrons, which Berzelius did not know about. Ernest Rutherford and Niels Bohr had led the way in visualising atoms containing electrons (**1911**). Pauling knew that in ionic compounds such as common salt (sodium chloride), electrons are actually exchanged. The sodium atom gives an electron to

> *Science is the search for truth—it is not a game in which*
> *one tries to beat his opponent, to do harm to others.*
>
> LINUS PAULING

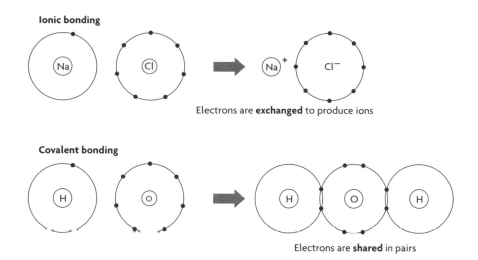

Ionic bonding

Electrons are **exchanged** to produce ions

Covalent bonding

Electrons are **shared** in pairs

Most bonds holding molecules together involve electrons. In the first example, the sodium atom donates one outer electron to the chlorine atom, which then has a full outer shell of eight electrons. The positively and negatively charged 'islands' then attract to form molecules of the ionic compound sodium chloride (salt). The ions can separate ('dissociate'), so if the salt is dissolved in water or melted the resulting liquid conducts electricity.

In the case of water, both the hydrogen and oxygen atoms seek a full outer shell of two (for hydrogen) or eight (for oxygen) electrons and gain that state by sharing pairs of electrons. Each electron pair forms a 'covalent bond'. Covalent compounds do not dissociate and so are poor or non-conductors of electricity.

the chlorine atom, so the sodium atom now has a positive charge and the chlorine is negatively charged. Clearly this did not happen in water or hydrogen or lots of other substances. Pauling pursued the idea that the electrons in those compounds are shared. Others, such as his countryman Gilbert Lewis, had mused on this idea before, but Pauling worked out the details.

The fundamental idea goes back to Neils Bohr (**1913**) and goes like this. In bonding with other atoms to make compounds, atoms try to reach a stable arrangement of their orbiting electrons. The preferred situation is to fill up their outer shell, say by having two or eight electrons there. An atom of hydrogen has a single electron. In a hydrogen molecule, the two atoms can each have a claim on two electrons, a sort of 'time-sharing'.

In water, each hydrogen atom can contribute one electron and the oxygen atom can contribute six electrons. This gives each hydrogen atom access to two electrons and each oxygen atom access to eight electrons. So both the hydrogen atom and the oxygen atom have full shells and are therefore satisfied. The diagram above explains this further.

We have here an explanation of the fundamental concept of valency or 'chemical combining power' (**1852**). With a valency of two, oxygen can bond with two hydrogen atoms (as in H_2O). Each of these bonds is a shared pair of electrons. Nitrogen needs three electrons to put eight electrons in its outer shell. So it can establish three shared-electron bonds with hydrogen, giving a valency of three and compounds like NH_3 (ammonia). Carbon needs four electrons to make up the comfortable eight, and so will bond with four hydrogen atoms in CH_4 (methane).

Pauling called these common linkages between atoms in compounds 'covalent bonds' and set all this out in his *The Shared Electron Chemical Bond* in 1928. It made his reputation and pushed chemistry ahead several vital steps.

Paul Dirac: Predicting Antimatter

1928

One measure of the stature of Paul Dirac, English-born of a Swiss father, was that he held the Lucasian professorship of mathematics at the University of Cambridge for nearly 40 years. This was the chair once filled by Isaac Newton and as of 2005 occupied by Stephen Hawking.

Dirac was legendary for his terseness. It was said (in jest) that he had only three words in his vocabulary: 'yes', 'no' and 'I don't know'. Dirac was not an experimenter; he worked with pencil and paper and mathematical equations to predict things for the laboratory people to look for. The most famous of these predictions was the 'positive electron'. Around 1928 Dirac summed up contemporary understanding of the electron by combining equations from the two major ideas in physics of his time: quantum physics (**1927**) and special relativity (**1905**). In 1931 he noted that if his equations were true, the electron should have a 'mirror image', the left/right reversal of your image in a mirror. He thought the electron's mirror image particle would be positively charged, rather than negatively, so he called it the 'positron'.

More startling predictions followed. The positron could be thought of as an anti-electron: if a positron and an electron collided, they would annihilate each other and both would disappear in a flash of energy. Running that backwards, energy could be turned into matter provided a pair of 'mirror particles' was the result. So under certain conditions a positron and an electron could be created simultaneously, an example of 'pair production'.

All this was mathematics, but it proved to be true. In 1932, while photographing the tracks of cosmic rays (then the only source of energy great enough for pair

> *It is more important to have beauty in one's equations than have them fit experiment.*
>
> PAUL DIRAC

production), American Carl Anderson found some streaks that might be caused by electrons, but which bent the 'wrong way' when passed through a magnetic field. They were positrons. Soon after, Patrick Blackett (**1932 Cavendish Laboratory**) in England photographed pair production as it happened, with two tracks appearing out of nowhere and curving in opposite directions, as Dirac had predicted.

It now seems that all particles have 'antiparticles'. The antiproton, for example, is negatively charged; it was not found until 1952. We can visualise whole assemblies of 'antimatter'. Bring matter and antimatter together and nothing would remain, just energy.

Edwin Hubble: The Universe Expands

American astronomer Edwin Hubble provided our first clear image of how the Earth and the Sun are placed in the large-scale order of the universe. Others had shown that the Sun is only one star among billions in the 'Milky Way' galaxy. In **1924** Hubble showed that our galaxy is merely one among many others, perhaps millions, scattered through space, separated by distances on average at least 30 times greater than the size of the galaxies themselves.

1929

In 1929 Hubble had further news: the galaxies are on the move. Using an equivalent of the Doppler Effect for light (**1842**), astronomers such as Vesto Slipher had proved that most of the galaxies were receding from us (**1912**). Only the nearest of them, the Andromeda Galaxy, was getting closer. Hubble systematically plotted the speed of recession, obtained from the 'red shift' of the galaxy, with its distance away, gained by studying Cepheid variables (**1912 Leavitt**). He found a surprisingly simple relationship. The further away a galaxy was, the faster it was departing. The relationship between speed and distance became known as the Hubble Constant.

The conclusion was very clear. We live in an expanding universe. Hubble showed that galaxies 135 million light-years distant were racing away at 20 000 kilometres per second. Both those numbers would have been considered fanciful only a few

---

HUBBLE CONSTANT (H)

This number indicates the rate of expansion of the universe, since distant galaxies are moving away from us faster than those close by. Recent measurements of H show that the velocity of recession increases by about 18 kilometres per second for each additional million light-years in distance. Consequently, galaxies and quasars about 12 billion light-years away are leaving us at nearly the speed of light. This distance marks the edge of the observable universe and indicates the time since the 'Big Bang' (**1950**).

decades before. If nothing intervenes, the spaces between the galaxies will grow over time. Conversely, there must have been a time in the past when they were all very much closer together.

That gives the Hubble Constant another significance. Looking out into space means looking back in time. We see the Sun not as it is 'now', but as it was eight minutes ago, since light takes eight minutes to cross space to reach us. Light from the nearest stars is four years old by the time we see it, that from the Andromeda Galaxy two million years old. The most distant galaxies we can see therefore represent the oldest events of which we could have any knowledge.

How far back we look in time depends on how distant those objects are; that depends on the rate of expansion, on the Hubble Constant. It therefore binds together the age of the universe, its size and its rate of growth. Finding the exact value of this important number would be a challenge for the next 60 years (**1992 COBE**).

Wolfgang Pauli: 'The Little Neutral One'

1932 By the 1930s physicists had been studying the 'rays' from radioactive elements for three decades. The beta particles, which were streams of electrons (**1899 Radioactivity**), had been clocked travelling at a great range of different speeds. In fact, any speed seemed possible. In the age of quantum physics, that was a problem. According to the new philosophy, everything in the tiny world of the atom, including the energy (and therefore the speed) of particles, came only in parcels of a certain size. A continuous range was simply not allowed.

German physicist Wolfgang Pauli, whose agile mind had created the 'no-more-than-two-electrons-in-one-place' rule (**1925**), had a solution here, too. The energy packages that drove the beta particles from radioactive atoms were indeed of certain sizes, depending on the allowed 'energy levels' deep within, but the electron carried away only part of that. The rest had been carried off by something else, something that escaped detection by the instruments gathered around.

To be so elusive, the other 'particles' could have no charge, no mass, no interest in interacting with other matter. They were barely 'there'. Pauli called his predicted particle the 'little neutral one'. What that was in German is not recorded, but the Italian physicist Enrico Fermi called it the 'neutrino', and the name stuck.

Neutrinos were indeed slippery. They would not be proven to exist until 1955, when they were found only in the exhaust of a nuclear reactor in 1955.

> *I do not mind your thinking slowly; I mind you publishing faster than you think.*
>
> WOLFGANG PAULI

The initial concept gave neutrinos no mass, but later researchers were not so sure. Even if neutrinos have only a tiny mass, they are so numerous (much more plentiful than atoms) that they could make up a major part of the total mass of the cosmos. This made neutrinos early candidates to explain the 'missing mass' of the universe, once we knew that it was missing (**1936 Zwicky**).

Karl Jansky: Radio Signals from Space

One of the major advances in twentieth-century science and in the history of astronomy began by pure chance. In 1932 Karl Jansky was an engineer at the Bell Telephone Company in the United States. The company's radio

1932

links across the Atlantic and Pacific carried telegraph messages and phone calls. Both were plagued by static, which crowded in on the traffic, making the telegraph messages illegible and the phone calls very hard to understand. Jansky was given the task of improving the 'signal to noise ratio'.

Jansky built a large receiving antenna on a rotating platform at a field station in New Jersey. He could turn this to locate the source of troublesome interference. Obvious sources included thunderstorms, other transmitters, noisy electrical equipment in factories and even aircraft. Yet there remained a faint hiss that seemed to come from nowhere in particular.

The unexplained source seem to get stronger and weaker throughout the day, and for a time Jansky thought it came from the Sun. But an eclipse of the Sun had no impact and in any case the source seemed to come up over the horizon four minutes earlier every day, so getting ahead of the Sun by two hours a month. This suggested that something out among the stars made the noise. Jansky soon found that the static was strongest when the Milky Way, with its billions of stars, was overhead, particularly the thick wide part in the constellation of Sagittarius. Here was evidence that celestial objects emit not only light but also radio waves.

Understandably, the discovery was of no interest to the Bell Telephone Company. There was nothing they could do about static coming from space; they closed the project down and moved Jansky onto other duties. After he published his findings in technical journals the story made the front page of the *New York Times* in 1933. He had to fend off suggestions that extraterrestrials were sending messages.

Astronomers took little interest and frowned on the suggestion that radio waves might come from the Milky Way. What could cause them? The only follow-up was by a radio amateur called Grote Reber, who built the first radio telescope in his backyard and drew the first maps of the strength of radio noise across the sky around 1937. Again the astronomy community ignored the work. Discoveries during and after World War II changed their attitude (**1950 Big Bang**).

Jansky's path of discovery was replicated in remarkable fashion in **1965** by two other Bell Telephone employees, but their radio waves came from far beyond the Milky Way.

A 'Year of Wonders' at the Cavendish Laboratory

1932

A year in which a number of amazing things happen is sometimes called an *annus mirabilis*, a year of miracles. The year 1932 at the Cavendish Laboratory at Cambridge has been so described. Under the unsurpassed leadership of the ebullient New Zealander Ernest Rutherford, who had taken over the laboratory from JJ Thomson in 1919, a powerful group of physicists and technicians, many of them young, had been assembled in the wake of World War I and were hacking into the new sciences, notably 'nuclear physics', which had replaced 'atomic physics' as the study of the very small. In 1932 three major discoveries were made at the Cavendish; all would gain their authors Nobel Prizes; all changed the face of physics. At least two showed the influence of Rutherford.

In **1920** Rutherford had the idea that particles with no charge combined with protons to make up the nuclei of atoms. In 1932 his reserved (and never very healthy) second-in-command James Chadwick found these 'neutrons' after 10 years of frustrating work. When the nucleus of the light metal beryllium was hit with alpha particles, it released an intense penetrating radiation that other researchers thought was gamma rays (**1899 Radioactivity**). Chadwick did some tests and found they were actually the long-sought neutrons. As we were to find out in the next decade, neutrons play a significant role in reactions in the atomic bomb (**1938 Nuclear Fission**), and Chadwick himself had a major part in the Manhattan Project, which produced the first weapons.

In the same year, another of Rutherford's suggestions came to fruition. He had argued that really fast particles such as protons could break into atoms more efficiently than the now venerable alpha particles and perhaps reveal much more. The dour, nuggetty Englishman John Cockcroft, and the younger Ernest Walton from Ireland built a machine to provide such a bombardment. They proceeded to break open the nuclei of many different elements, producing sprays of alphas and other fragments that confirmed what the nuclei contained. This has been called the first 'splitting of the atom', though Rutherford might have claimed to have done that in **1919**. Certainly it was part of a big new move in physics to replace existing equipment with machines that depended more on electrical engineering than the traditional 'string and sealing wax'.

The third Cavendish triumph of 1932 belonged to Patrick Blackett—handsome, charming, radical in his politics and a major scientific statesman in the years to come. Blackett was a master of the cloud chamber (**1912 Wilson and Geiger**), which made visible the tracks of fragments of matter such as those involved in Rutherford's **1919** bit of alchemy. He later adapted these techniques to photograph the paths of cosmic rays, energetic particles from space (**1912 Hess**). Before the Cockcroft/Walton and similar machines, cosmic rays were the highest energy particles available for study. Some had enough energy to create new particles out of nothing, such as the electron–positron pairs whose production had been predicted by Cambridge man Paul Dirac (**1928**). Some of Blackett's pictures showed the exact moment of this act of creation, with the tracks of the two particles curving away in opposite directions.

Charles Richter: How Big Was that Earthquake?

Whenever an earthquake occurs, the first question is usually 'How big was it?' Easy to ask; not so easy to answer. If you are asking how much damage it did, how much it shook the ground, you are asking about the 'intensity' of the earthquake, given by a number on the Mercalli Scale (named after an Italian geologist). That is like asking how strong is the signal from a radio station at the location where you are listening. A reading of one on that scale means the quake is barely felt even by instruments; a reading of ten means total destruction.

1935

More fundamental is the 'magnitude' of the earthquake: how much energy it released at its point of origin, how much the rocks moved there. The Richter Scale developed by American Charles Richter in 1935 answers that question. Initially the scale applied only to earthquakes in southern California (of which there are plenty), near Los Angeles, where he worked. Now it is used around the world, and Richter's name is part of our everyday language.

Earthquakes vary enormously in magnitude; destructive ones release millions of times more energy than small ones just large enough to be measured. So the scale is logarithmic: each step means a tenfold increase in energy released. An earthquake of magnitude seven has ten times more destructive potential than one of magnitude six, a hundred times more than a five, a thousand times more than a four.

The scale is open-ended; the biggest earthquakes ever recorded are around nine on the scale, such as the earthquake that caused the disastrous tsunamis in Southeast Asia in December 2004. It is unlikely that a 10 will ever occur, but the scale could handle that and even more.

In 1954 Richter noticed something else about the scale: it reflects how likely quakes of different magnitudes are to occur. Magnitude six earthquakes are 10 times more likely to occur than magnitude seven and 100 times more likely than magnitude eight.

Stations are now in place around the planet to watch for earthquakes. Seismographs pick up the earthquake energy waves, often after they have travelled many thousands of kilometres through the Earth. Relying mostly on the times the waves reach the various stations, experts can figure out just where on Earth the quake occurred, how far underground and how powerful it was. Plotting these on a map of the world shows that most earthquakes occur in narrow zones, especially in the 'ring of fire' around the Pacific and along a long spur running through Asia, the Himalayas, the Middle East and on to Europe. At the far end of this spur sits Lisbon, all but destroyed by a disastrous earthquake in **1755**.

The pattern of earthquakes (and volcanoes) gives powerful support to the theory of plate tectonics, which says that large fragments of the Earth's crust are in continual motion relative to each other (**1992**).

Franz Zwicky: New Stars and Super New Stars

1936

Over many centuries, watchers of the heavens have seen many things change. Some changes are regular and predictable, like the phases of the Moon and the movements of the planets. Others occur without warning, such as new stars or 'novas' suddenly appearing where no star has been seen before. Tycho's Star of **1572** and Kepler's Star of **1604** (**Inconstant Heavens**) were on the list, bright enough to be seen in the daytime before they faded, and Chinese astronomers had seen a very bright one in 1052.

Once it was understood that our own collection of stars was not the only one in the universe (**1924 Hubble**) and that other galaxies existed, astronomers went looking in them for novas, though to be seen at such distances, the outbursts would have to be very bright, worthy of the name 'supernovas'. The stars 1052, 1572 and 1604 were of that class.

An assiduous searcher for supernovas was the brilliant but eccentric American astronomer Franz Zwicky, known to accost students whom he did not know with the question 'Who the hell are you?' Beginning in 1936, he found 20 supernovas in other galaxies. Some of them were as bright as the whole of the rest of the galaxy for a few months, before they faded away. Clearly something cataclysmic had occurred to release such energy.

Zwicky devised a scenario. A big star, he argued, would consume its energy very quickly (by which he meant in a few million years). With its fuel suddenly exhausted, and with no energy from within to keep it puffed up, it would collapse suddenly and catastrophically, shedding its outer layers as it did so. The collapse would liberate immense energy, but only for a time; the dying star would fade to nothing in a year or two.

When all went quiet again, two things would remain: a cloud of debris, gas and dust, now known as a 'supernova remnant' (the Crab Nebula in the constellation of Taurus shows where the supernova of 1052 occurred); and an ember of the giant star, squeezed so tight by the collapse that it would be as dense as an atomic nucleus. Each cupful of such matter would weigh millions of tonnes; such an object 10 kilometres across would be as heavy as the Sun. Under such intense gravity, protons and electrons would be squeezed together to make neutrons (**1920 Rutherford**). Zwicky called these 'neutron stars'. Thirty years passed before one was discovered (**1967 Bell and Hewish**).

Zwicky left something else for us all to ponder. He studied the way in which galaxies pull on one another through their mutual gravity. According to Newton

(**1686**), that attraction affects the way the galaxies move. Yet adding up the mass of all the visible stars and clouds of gas did not seem to make the galaxies heavy enough to behave as they did. Much of the mass of the galaxies (later estimated to be 90 per cent) seemed to be missing, that is, it was in the form of 'dark matter' that could not be seen. He reached a similar conclusion by studying how galaxies rotated. Nearly 60 years later, we knew for sure how right he was (**1992 COBE**).

Hideki Yukawa: What Carries the Strong Force?

The Nobel Prize for Physics awarded in 1949 to Hideki Yukawa was a milestone in the spread of 'western' science outside its homeland in Europe and North America. Yukawa spent most of his career in Japan, though he moved to the United States after World War II. Another threshold was crossed at the same time. Yukawa had put forward a startling new idea to explain what was happening at the very heart of matter.

1936

Here was the question: what holds the nucleus of every atom together? A nucleus contains protons, all with a positive charge, and neutrons that have no charge. The mutual repulsion between the protons should shatter any nucleus into bits. This does not happen; some force of attraction must hold together the mix of protons and neutrons, collectively called nucleons.

Physicists had already worked out some of the features of this force. It was about 100 times stronger than the electromagnetic force that works between electric charges, so it was called the 'strong force'. It operates only between nucleons; electrons do not feel it, and it works only over short distances, about the size of the nucleus. Beyond that short range it has no influence. Scientists later realised that the breaking of this force liberates the energy of nuclear fission (**1938**).

Yukawa had a novel approach to explaining this force. He likened it to two people with a basketball. If one throws the ball at the other, they will move apart, if they both grab the ball at once, it pulls them together. Yukawa proposed that the strong force involved the exchange of a new particle that he called a 'meson', because he thought it would have a mass somewhere between that of a proton and an electron.

The most likely place to find Yukawa's meson was in cosmic radiation (**1912 Hess**), where it might be created out of pure energy. So a search was organised. A likely candidate soon turned up in cloud chamber pictures (**1912 Wilson and Geiger**), but it was not the particle everyone sought. It had the right mass but it was not attracted to nucleons, so could not have carried the strong force. The particle, soon dubbed the 'mu-meson', was rather like a heavy electron: totally unexpected and apparently unnecessary. Isador Rabi, a leading American physicist, is reported to have said, on hearing of the discovery, 'Who ordered that?'

Yukawa's particle, now called the 'pi-meson', was finally found in 1947, but by then the proliferation of 'fundamental particles' had begun. In addition to the proton, neutron and electron, the positron (**1928 Dirac**), the neutrino (**1932 Pauli**) and now the two mesons, new particles began turning up in cosmic radiation and, after the war, in the new 'atom-smashing' machines (**1952 Cosmotron**), at an exciting but puzzling rate.

The sub-nuclear 'zoo' was getting very crowded, housing dozens of fragments of matter, some of whom survived for mere fractions of a second. The question was asked, as it had been about the elements 100 years before: why are there so many? Was there any limit to the number? Was there any order? There certainly was (**1963**).

Hans Krebs: The Energy Cycle of Life

1938

Early experimenters like John Mayow (**1674**) likened respiration, by which animals and plants get the energy they need to live and grow, to the burning of a candle. Oxygen was needed, energy was released, and carbon dioxide and water were the wastes. Not until the twentieth century did scientists understand how complex the process of respiration actually is. The eyes of generations of biochemistry students have glazed over at diagrams of the whole journey, with dozens of steps and multiple transformations of compounds.

So it was an immense achievement by Hans Krebs, a German-born chemist who spent most of his working life in England, not only to visualise the intricacies of the process but by careful and elegant experimentation to demonstrate that it worked the way he thought it did. The core of the three-part process of respiration is still usually called the Krebs Cycle. Krebs was Jewish and, like so many, fled Germany with the rise to power of the Nazis. He came to England at the invitation of Frederick Gowland Hopkins, whose personal fame rests mostly on his work with vitamins (**1906**).

Cycles are common in nature. Many substances and processes go around and around, through intricate patterns of change, adding and shedding complexities along the way, to end up where they began. Water, carbon and nitrogen all have their cycles, part of the big picture of the natural world. Krebs' achievement was to find a cycle within each cell, the cycle that makes energy available for use. Others had found many of the steps in the chain of transformation; it was Krebs who closed the chain and made it a cycle.

The first stage of extracting energy from glucose needs no oxygen. That process extracts only one-third of the energy available in glucose or other energy sources. The second stage, the one Krebs untangled, is a cycle because molecules of one compound (citric acid, which we find in lemons and oranges) pass through many stages, stimulating many transformations, taking in water, sending out carbon dioxide and hydrogen ions (which after another stage become energy-rich packages called ATP) and end the circuit unchanged, ready to start again. Only the third stage of the process (phosphorylation) needs oxygen. The diagram opposite shows the detail.

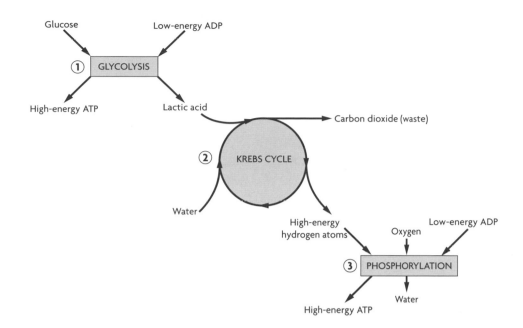

This may look complex, but it is a greatly simplified view of how we extract energy from foods like glucose. In stage 1, the glucose passes through glycolysis, which can happen anywhere in the cell and needs no oxygen. Some of the energy in the glucose is handed on to molecules of the 'energy carrier' called ADP, turning it into the more energetic ATP, which then powers everything else that happens in the cell.

The glucose emerges from stage 1 as lactic (or pyruvic) acid (it is the build-up of these wastes that makes our muscles feel tired), which becomes the fuel for stage 2, the Krebs Cycle. This stage (and stage 3) takes place in the mitochondria, specialised sites for energy release. This cycle also takes in water and churns out carbon dioxide (as a waste) and hydrogen atoms carrying extra energy.

In stage 3, phosphorylation, the hydrogen atoms hand on their excess energy to more ADP molecules (making much more ATP than the first stage does) but must then combine with oxygen to form water. This reaction is the fundamental reason we need oxygen to live.

Hans Bethe and Carl von Weiszacher: The Ovens in the Stars

English astronomer Arthur Eddington had worked out what keeps the Sun and other stars shining (**1920**). Four atoms of hydrogen, the most abundant element, are 'fused' to form one atom of helium. The helium atom weighs a little less than the four hydrogen atoms, and the lost mass (which amounts to four million tonnes every second in the case of the Sun) reappears as energy.

When it came to the details, American Hans Bethe and German Carl von Weiszacher came up with much the same scheme independently. They visualised two processes. In the first, which happens in the smaller stars like our Sun, hydrogen atoms, or rather the protons in them, are added together one by one to form a helium nucleus of mass three, which was not known to exist until a few years before. Two helium 3 atoms together make a helium 4 atom and release two protons to start the process again. This is sometimes called (confusingly) 'hydrogen burning', though the hydrogen is not actually on fire.

$$H_1^1 + H_1^1 \longrightarrow He_2^2 \longrightarrow H_1^2 + \text{electron}$$

$$H_1^2 + H_1^1 \longrightarrow He_2^3$$

$$He_2^3 + He_2^3 \longrightarrow He_2^4 + H_1^1 + H_1^1$$

These three equations summarise the process of nuclear fusion in stars, whereby hydrogen becomes helium, releasing energy. It involves two isotopes of hydrogen and three isotopes of helium, one of which is radioactive by beta decay. The upper of the two numbers beside each element symbol is the atomic weight (number of protons plus neutrons). The lower number is the atomic number (number of protons only).

The other process needs higher temperatures and so happens only in the bigger stars. It begins with a carbon nucleus, formed by three helium atoms getting together. This steadily picks up protons, generating more helium and releasing energy, but also making heavier elements like nitrogen and oxygen. This 'nuclear cooking' is the only way these vital elements are made anywhere in the universe. All the nitrogen and oxygen in the air and in our bodies has come from such stellar ovens.

Over many millions, even billions of years, helium, the 'ash' from these 'fires', builds up and finally puts them out. Soon after, helium 'burning' starts. As helium atoms are linked together, even heavier elements are cooked up: the calcium and phosphorus in our bones, the iron in our blood, the magnesium in chlorophyll, the sodium and potassium that send signals along our nerves.

Understanding how stars make energy also helps explain their life stories and the distribution of the various sorts of stars, as shown on the Hertzsprung–Russell diagram (**1910**). Hydrogen 'burning' powers stars for most of their lives, so 90 per cent of stars, including our Sun, are Main Sequence stars. When helium burning starts, vast new energy is released within; the star puffs up, becoming perhaps 100 times larger, its surface cooling and becoming redder. The star is now a red giant. This stage of a star's life does not last very long (in astronomical terms), so red giants are relatively rare. Many such stars are unstable, fluctuating in brightness. Some become the Cepheid variables that have been so important in our understanding of the larger universe (**1912 Leavitt**).

Billions of years into the future, our Sun will pass through this change of life, most likely engulfing the Earth as it does. So, life and death are linked together in the life cycles of stars. The process that creates the building blocks of plants and planets and people can also destroy them. Once the helium fuel too is exhausted, the central fire goes out and the remaining heat leaks away into space. The star ends its life as a white dwarf.

Bethe and von Weizsacher spent the years following their discoveries on opposite sides of World War II, working on their countries' programs to make nuclear weapons.

The Work of Many Hands: The Discovery of Nuclear Fission

In terms of political and military consequences, no discovery of the twentieth century surpasses nuclear fission. For 40 years, physicists had worked to understand radioactivity (**1896**) and the energy it releases from

1938

deep within atoms. At the end of the 1930s it seemed possible to some people that such energy could be released explosively and destructively, as a bomb. If such outpourings could be controlled, there was also the prospect of an almost inexhaustible source of energy for peaceful purposes. The eminent scientist Ernest Rutherford had famously called such thinking 'moonshine', but the evidence was otherwise.

The trail leads back to 1934. The newly discovered uncharged neutron (**1932 Cavendish Laboratory**) had become the favourite probe for nuclear physicists exploring the nucleus, which lies at the heart of every atom. Italian physicist Enrico Fermi led a team that sent neutrons into uranium, the heaviest known element, trying to make the atoms swallow the neutrons and so make even heavier elements that did not exist in nature. They appeared to succeed.

When German scientists Otto Hahn and Franz Strassman repeated the experiments late in 1938, the results were puzzling. Chemical analysis showed that rather than new elements, familiar ones had been produced, elements like barium, much lighter than uranium.

Austrians Lise Meitner and her nephew Otto Frisch grasped what had happened. Quite unexpectedly, the neutrons had split the uranium nuclei in half, releasing the energy that had held the bigger nucleus together. Compared with a chemical reaction, as in dynamite, the energy given out was immense. This nuclear 'fission' brought the prospect of nuclear power.

The findings were published; physicists around the world rushed to confirm them. In early 1939 Frederic and Irene Curie-Joliot in Paris—she was the daughter of Marie and Pierre Curie (**1898**)—added something very important: each fission released several neutrons. Under the right conditions, each of those could split another uranium nucleus, liberating further neutrons. So a self-sustaining chain reaction was possible, though it did not have to be explosive. The energy might be released fast enough to boil the kettle but not necessarily fast enough to destroy a city.

When war broke out in late 1939, Frisch, who was Jewish, escaped to Britain. In the Birmingham laboratory run by Australian Mark Oliphant, Frisch worked with fellow émigré Rudolf Peierls on a famous secret memorandum. Uranium has two isotopes: the rare U235 and the abundant U238 (**1907**). Research showed that neutrons split only U235. To make an 'atomic bomb', the memorandum said, the uranium must be processed ('enriched') to greatly increase the proportion of U235. Five years later that discovery led to the destruction of the Japanese city of Hiroshima in the first ever nuclear weapon attack.

There is an intriguing footnote to all this. Back in 1934 the Hungarian chemist Ida Noddack had expressed doubts about the new heavy elements Fermi claimed to have made. She wondered if instead the neutrons had broken the uranium nuclei up into several pieces. She did not check the idea out in the laboratory; nor did anyone else. It was dismissed as implausible. So nuclear fission could have been discovered five years before the outbreak of the war, rather than at the eleventh hour. The course of world history might have been very different.

In the early years of the war, researchers in the United States confirmed that new heavy elements could be made in the way that Fermi and his team had pioneered in 1934. One of these (element 94) was called plutonium; it too was found to split when

$$U^{235}_{92} + n^1_0 \longrightarrow St^{90}_{38} + Te^{144}_{52} + Zn^1_0$$

$$U^{238}_{92} + n^1_0 \longrightarrow Pu^{239}_{94}$$

Naturally occurring uranium (element 92) is a mixture of two isotopes. Most of it is U238, with less than 1 per cent U235. These isotopes behave differently when they absorb neutrons. U235 splits (undergoes 'fission') into two unequal fragments. There are dozens of ways this can happen; in this example, one 'fission fragment' is the intensely radioactive strontium 90, a dangerous component of nuclear 'fallout'. The fission also releases two new neutrons, able to split other U235 atoms and start a chain reaction.

U238 swallows the neutron and is transmuted into an isotope of plutonium (element 94). This goes on all the time in nuclear reactors. Like U235, plutonium can be made to split by neutrons, and so can be a source of nuclear power, both peaceful and military.

bombarded with neutrons and so was another possible nuclear explosive. The atomic bomb dropped on Nagasaki in August 1945 used plutonium. It was perhaps appropriate that the first ever nuclear reactor, built to prove that plutonium could be created in substantial quantities, was fired up in Chicago in December 1942 under the supervision of Enrico Fermi. Like many Jewish scientists, he had left Europe to escape persecution.

Karl von Frisch: The Language of Bees

Communication is crucial to humans, so much so that we tend to think we're the only creatures that do it with any real sophistication. Dogs bark **1940** and birds call, but how much information is being passed on? German zoologist Karl von Frisch concentrated on honeybees. As social insects, bees have high communication needs, especially when it comes to cooperating to find and gather food. Frisch found that they communicate by dancing.

Bees have two dances. The 'round dance' is performed when foragers have found food within 75 metres. The more complex 'tail-wagging dance' means food is further away. The bees move across the vertical face of the comb in a straight line, waggling their abdomens, before circling back and starting again. The dance is full of meaning. Its speed indicates how far away the food source is (slower means further); the line of the dance marks the direction.

There is impressive subtlety and sophistication here. The direction is always given relative to the position of the Sun. A line of dance vertically upward means the food lies in the direction of the Sun; if vertically downwards, it is in the opposite direction from the Sun. For food lying in other directions, the communicating bee dances at an angle in between. Since the position of the Sun in the sky changes throughout the day, the bees use their internal (circadian) clock to make the appropriate correction in direction. Suppose the sky is covered by cloud: Frisch found in 1949 that bees can use their capacity to detect polarised light to find the hidden Sun.

Frisch confirmed his observations, which he announced in 1940, by placing food at different distances and in different directions from a hive and seeing how it affected the dance. Much earlier, he had found that bees can distinguish colours, and have a well-developed sense of smell, though their sense of taste is not as advanced as ours. Even earlier, while still in his 20s, he had found that fish can distinguish colours and brightness of light and that their hearing and ability to discriminate between sounds is more acute than ours.

Frisch shared the 1973 Nobel Prize for Medicine or Physiology with two other researchers, Dutchman Nikolaas Tinbergen and Austrian Konrad Lorenz, who had studied the behaviour of birds and fish, helping to understand, among other things, the origins of social organisation. Their findings illuminated the behaviour of mammals, including humans.

Max Delbruck and Salvador Luria: Natural Selection at Work

1943

The invention of the electron microscope in the 1930s transformed biology as profoundly as the original microscopes of the seventeenth century had done (**1673**). Instead of light, beams of electrons were focussed by lenses made from magnetic fields. Objects could be magnified many thousands of times. Again, a whole new level of nature leapt into view. Viruses, believed since **1898** to exist but never seen, could now be imaged and explored; many looked like crystals.

Even more striking were the bacteriophages, viruses that infected only bacteria, first postulated around 1915. French–Canadian biologist Felix d'Herelle named them, even before he had seen one, because they 'ate bacteria'. Under the electron microscope some looked like alien spacecraft or hypodermic needles, able to perch on the outer skin of the bacteria and apparently inject genetic material that forced the bacteria to make many copies of the virus before being killed by it.

Max Delbruck trained in physics but became interested in biology and genetics after emigrating from Germany to the United States. Bacteriophages attracted him because they seemed to be the simplest possible genetic system; physicists often like to strip things down to their bare essentials to understand how they work. Delbruck formed a partnership with American Salvador Luria after a physics conference in 1941, and they formed a 'phage group' to figure out how these 'living-but-not-living' structures worked.

An early finding clarified a key issue in genetics. How do bacteria become resistant to forces like phages or the new antibiotics that can kill them? Do they

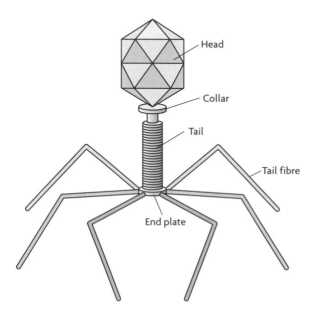

Head

Collar

Tail

Tail fibre

End plate

This extraordinary-looking organism, resembling some alien creature or landing craft, is a bacteriophage, a virus that infects bacteria. The head is filled with DNA (or RNA). The tail acts much like a syringe, injecting the genetic material from the virus into the bacterium. The virus DNA or RNA takes command of the protein-making machinery of the bacteria, which is forced to make many copies of the virus before it dies.

somehow adapt and so survive? Or is it just a matter of luck: those with the right defences at the start will get through and the rest will die.

Delbruck and Luria grew bacteria on a plate of nutritious chemicals, and then infected them with a bacteriophage. The plate, initially cloudy with growing bacteria, became clear in a few hours as they died. But a few hours later, the plate was cloudy again.

It seemed that a few bacteria had been 'fit to survive'; they were immune to the phage attack and so flourished. This was natural selection à la Charles Darwin (**1859**), powered by random mutations that produced the 'survival skill', rather than acquired behaviour being passed on as Jean Lamarck (**1809**) would have proposed.

George Beale and Edward Tatum: 'One Gene, One Enzyme'

Modern biotechnology began with mouldy bread. The making of bread, wine, beer, cheese, yogurt, vinegar and 'smallgoods' are all traditional 'biotechnology'. Louis Pasteur (**1856**) discovered that yeasts and bacteria cause the chemical changes in these foods, collectively known as 'fermentation', so they can get the energy they need to live and grow. Twentieth-century science has added vastly to our knowledge of the detail in all this, so we can create new forms of biotechnology.

1943

Important evidence came from American chemists George Beale and Edward Tatum. They simplified their work by choosing a very basic organism to work with, a red bread mould called *Neurospora*. From work with Thomas Hunt Morgan and his fruit flies (**1911**), Beale knew that even simple characteristics such as eye colour depended on a long series of chemical reactions that were controlled somehow by genes, the units of inheritance. He also knew that the way genes worked could be changed by exposing the fruit flies to X-rays, which caused the genes to mutate (**1926**).

Tatum knew all about growing and studying bread moulds. He and Beale began to expose the mould to X-rays; some of the moulds stopped making certain chemicals they needed to survive, as if some particular chemical reactions had been shut down. By now it was well established that all chemical reactions in living things need one or more enzymes to catalyse (speed up) the reaction. Beale and Tatum argued that the mutations had stopped some enzymes being made. This meant that each gene must be controlling the production of a specific enzyme. One gene, one enzyme.

Generalising a bit further (more research backed this up), all enzymes are proteins, so each gene in the bread mould controls the making of some particular protein, including the proteins from which the mould is built. This seems to apply to all living things. Though the details vary, we could now say that genes exert their influence by controlling how the hundreds of different proteins are made. Mutations interfere with this, perhaps by scrambling the instructions for making the proteins.

Only one question now needed an answer. How do genes actually control how proteins are made? An answer to that was getting close (**1953 Watson and Crick**).

Robert Woodward:
Complex Chemicals from Scratch

1944 Nothing revealed how far chemistry had come since the days of Antoine Lavoisier (**1789**) or Friedrich Wohler (**1828**) than the ability of mid-twentieth-century chemists to construct complex compounds with hundreds of atoms from very simple ones like water, carbon dioxide or ammonia. Chemical synthesis had become a highly sophisticated science. Yet the Nobel Prize citation (Chemistry, 1965) for the top man in the field, Robert Woodward, spoke of his contribution to the 'art' of chemical synthesis.

Perhaps this implied that deciding which of the many possible paths to follow in making something complex from something simple requires imagination more akin to painting or architecture than to laborious manipulation of test tubes and chemicals. But leaps of thought are always important in science—the flashes of insight, the 'Eureka moments'. It is justifying those conjectures by methodically assembling of evidence that takes the hard work in science, as proponents seek to prove they are right (or at least not wrong).

Woodward clearly combined imagination with diligence. American-born of Scottish parents (his given names were Robert Burns), Woodward spent time on analysis as well as synthesis, breaking complex chemicals down to see how they were put together and from what. These were mostly natural products extracted from plants or other organisms, like many medical remedies over the centuries. An early achievement was the first structure for penicillin, extracted from a mould (**1928 Fleming**); others included the poison strychnine and the antibiotics terramycin and aureomycin.

Once the content and architecture of a compound was known, it could, in theory at least, be built up from its simple parts. Woodward triumphed first with quinine in 1944. Trying to make quinine from coal tar had sparked the explosion of organic chemistry in the nineteenth century (**1856**). The list thereafter includes: cholesterol, the fat in human blood; the hormone cortisone; lysergic acid, source of the hallucinogenic LSD; the green plant pigment chlorophyll; and the antibiotic tetracycline. In 1971 he synthesised vitamin B12, a molecule so complex it required 100 separate steps. You could not do that unless you knew exactly where you were going.

Following Woodward, there seemed no reason why any organic chemical, once it had been analysed, could not be constructed, including proteins, the basic building blocks of plants and animals, and DNA, which was to prove to be the carrier of heredity. But there was still some way to go yet.

Edwin Chargaff: Finding the Rules for DNA

During the middle decades of the twentieth century, geneticists continued to close in on the secret of inheritance. The work of Oswald Avery and his colleagues (and earlier Frederick Griffiths) on the 'transforming factor' (**1928**) left little doubt that hereditary information, such as the coded plans for making proteins, was carried by chemicals called nucleic acids.

Researchers knew there were two nucleic acids found in cells. One we call DNA (deoxyribose nucleic acid), the other RNA (ribose nucleic acid). DNA is active in the nucleus of a cell, where it makes up most of the chromosomes (**1869**). RNA seems to operate more in the rest of the cell. If the 'ribose' part of DNA and RNA (mostly a sort of sugar containing phosphorus) was stripped away, what was left was mostly made up of four chemicals (aka bases) made from nitrogen, carbon and oxygen. These were adenine, thymine, guanine and cytosine—more simply A, T, G and C.

Austrian-born Erwin Chargaff had emigrated to the United States in the 1930s. After assiduously analysing DNA from many organisms, he found something so striking that it is now called Chargaff's Rule. Even though the relative amounts of the four chemicals varied from one plant or animal or virus to another, the quantities of A and T were always similar, as were the quantities of C and G. So it looked like A and T were always found together—ditto C and G, . Anyone trying to figure out a structure for DNA or RNA would need to take that into account. Finding a structure was now the goal, since that would show how DNA actually encoded genetic information.

PS: If there were any lingering doubt that nucleic acids, not proteins, were the carriers of genetic information, work by Americans Alfred Hershey and Martha Chase in 1952 would have dispelled it. They attached radioactive atoms to both the proteins and the nucleic acids in the viruses infecting bacteria, aka bacteriophages (**1943 Delbruck and Luria**). Only the radioactivity linked to the nucleic acids (and not that in the proteins) ended up in the infected cell and so was incorporated into new viruses.

> *Science is wonderfully equipped to answer the question 'How?'*
> *but it gets terribly confused when you ask the question 'Why?'.*
> ERWIN CHARGRAFF

Fred Whipple and Jan Oort:
The Truth about Comets

Over the centuries, telescopes have revealed that Earth and the other planets have company as they journey around the Sun. Comets, with their bright heads and wispy tails, have been known since ancient times, and were brought under the same laws as the planets by Edmond Halley (**1705**). Since **1801** (**Piazza**) a growing number of asteroids, too small to be more than dots of light, had been found in the reaches between Mars and Jupiter. Asteroid orbits are nearly circular, but comets travel in elongated paths that take them alternately close to the Sun and far out into much colder regions between the planets.

American astronomer Fred Whipple was the first to make another distinction between asteroids and comets: what they were made of. Asteroids, he said, were essentially rock. Comets were 'dirty snowballs'—dust and rock fragments held together by frozen gases such as water and carbon dioxide.

He based his idea on studies of Comet Encke, which comes into view every three years (**1822**). As a comet nears the Sun, some of its ice melts, then boils, releasing a cloud of dust that catches the light of the Sun and provides the 'hairy star' look. Rock fragments are liberated at the same time, later drawn to Earth to cause meteor showers. This model stood up well to scrutiny at its first big test—the close study of Halley's Comet by Giotto and other spacecraft when it came by in 1986.

There is a problem: if some of a comet's contents boil off every time it goes by the Sun, it will not survive many passages before it has evaporated away to nothing. Even a big comet like Halley's must be less than a hundred thousand years old. On this reasoning, there should be no comets left. As an answer, Dutch astronomer Jan Oort proposed the Oort Cloud, a vast reservoir of comets at the outer edge of our solar system, perhaps a light-year from the Sun. There, many millions or even billions of comets orbit the solar system in the frigid half-light.

From time to time, something disturbs the quietness, perhaps the close approach of another star. Some of the snowballs are bumped out of orbit and start to fall towards the distant Sun. They become comets as we know them, replenishing the stock. They are also missiles, certain at some time to strike the Sun or even one of the planets (**1796 Laplace**). Such a comet broke up in the Earth's atmosphere over Siberia in 1908, flattening and burning 10 000 square kilometres of forest. Larger strikes would do much more damage.

The Battle over the Big Bang

1950

The discovery by Edwin Hubble (**1929**) that the universe is expanding, that the great star systems or galaxies are moving apart, immediately raised questions about its past. If you were to rewind a video showing the growth of the universe, you would see the universe collapse and all the matter and radiation crowd together in an intense fireball of infinite density. Many astronomers now believe that the universe began like this, in a singularity or white hole (**1916**). In 1950 Americans Ralph Alpher and George Gamov pondered what traces such a catastrophic start to the cosmos might have left, other than the fleeing galaxies or their ancestors.

Alpher and Gamov argued that the first things to emerge from the melee as it expanded and cooled would be protons and neutrons. Those would clump together to form the nuclei of the light atoms: hydrogen, 'heavy hydrogen' (deuterium) and helium. Calculations showed that there would not have been time for anything heavier to be 'cooked up'; all the other elements are made inside stars (**1938 Bethe and von Weiszacher**). Analysing the light from stars using spectroscopy (**1814 Fraunhofer**), could show the amount of, say, deuterium, and that would indicate how fast the universe expanded at first and how hot it was.

Not everyone agreed. Since the 1950s English astronomer Fred Hoyle and the Americans Herman Bondi and Thomas Gold had been pushing a very different story. They argued that the universe never had a beginning, that on the largest scale it has always looked much as it does now. As the galaxies moved apart in a 'steady-state' universe, new ones would have been formed in the spaces between. Hoyle, always passionate, publicly disparaged the rival view as the 'big bang'.

Important evidence for this debate soon emerged. Englishman Martin Ryle had spent the war developing radar sets for planes. He then returned to Cambridge University to set up radio astronomy, following up the discovery by the Americans Karl Jansky and Grote Reber (**1932 Jansky**) that radio signals were reaching Earth from space. His radio telescopes were soon finding dozens and later hundreds of 'radio sources', tiny objects in deep space, mostly invisible in ordinary telescopes but giving off radio noise. These seemed to be spread through space like the galaxies; maybe some were galaxies. By 1958 Ryle and his team had some counts of the sources, and that spelt trouble for the Steady-state people.

Assuming that stronger sources were close and weaker ones far away, Ryle found twice as many weaker sources as he expected relative to the number of strong ones. It looked like the weaker sources (far away) were in general closer together (more in a given volume of space) than the stronger sources. The far-away sources were also further back in time (since light takes time to travel), and so represented the universe when it was younger. That fitted the Big Bang picture, which had the galaxies spreading out as the universe aged, but not the Steady-state one.

By 1960 the score was Big Bang 1, Steady-state 0. And there was still more to come (**1965**).

Mass Production, Motor Cars and Microchips

TWENTIETH-CENTURY TECHNOLOGY

As the twentieth century began, the pace of technological change, already high, was accelerating. No aspect of life in industrialised societies escaped its impact. Not only were new machines and devices being invented, but also they were being purchased and put to use in increasing numbers in homes, offices, factories and farms. Human history had never seen such a revolution as was brewing.

Some of the new wonders first saw light in the closing decades of the previous century: the 'horseless carriage', moving pictures, the telephone, X-rays, radio, synthetic materials, recorded sound, the large-scale production of electricity. The 50 years to 1950 would see them everywhere. Other technologies born in this century include aviation, electronics, television, nuclear power, space travel and the immense array of ways to process and disseminate information—arguably the most influential inventions of all.

The meshing of many factors drove this process of change. Productivity increased in all industries, mostly due to new technologies. This generated wealth and liberated resources that fed back into the development of even better goods and services; 'better' in the sense of cheaper, easy to use, safer, more powerful, less polluting or wasteful. However, it was many decades before environmental and social considerations became particularly important.

On the whole, manufacturers provided the things people wanted and were willing to pay for. Consumers responded by beginning to want things they had never dreamed of having. The increased demand lowered prices, raising demand even further and paying for the next round of technological advances.

At the same time, technology was underpinned more and more by science. This trend had begun in the previous century and represented a reversal of the centuries-old relationship. Better understanding of the workings of nature—in energy, in material and in biological systems—made the process of innovation more efficient. Increasingly it was based on scientific understanding, often expressed in mathematics, rather than on traditional trial and error.

Any catalogue of the century's cascading achievements in technology must be incomplete, because there is so much to choose from: the first radio signal across the Atlantic (1901), the first powered flight in a heavier-than-air machine (1903), the Binet IQ test (1905), triode radio valves (1906), neon signs (1910), Henry Ford's 'mass production' of motor cars (1914), commercial radio broadcasting (1920), movies in colour (1922), quick frozen foods (1924), 'talking pictures' (1927), penicillin (1928),

FM radio (1933), fluorescent lights and sodium vapour lamps (1935), radar (1935), television broadcasting (1936), ballpoint pens (1936), nylon (1937), jet engines (1937), photocopiers (1938), helicopters (1939), DDT (1939), 'aqualungs' (1942), LSD (1943), microwave ovens (1945), the first 'atomic bomb' (1945), geodesic domes (1948), instant coffee (1949). That covers only half the century.

By the 1950s many of the technological tracks were firmly laid. Electricity (made mostly by burning coal) and petroleum products were confirmed as the prime sources of power. Aircraft became bigger and faster, employing jet engines. Motor cars were everywhere, and highway networks expanded to accommodate them. Farming became more mechanised and more people left the land. Household 'labour-saving' devices proliferated. Every technology became cheaper and accessible to more people, doing more for less, but often imposing unwanted or unexpected costs. Governments found it necessary to impose tighter regulations to protect human health and the environment.

The most powerful agent of change was 'information technology', or IT. Previous centuries had given us the telegraph and the telephone, calculating machines and typewriters, so IT did not appear from nowhere. But the invention of the transistor in 1947 and its consolidation into the first 'integrated circuits' a decade later began the radical transformation of almost everything we do.

Initially huge, immensely costly and wasteful of power, and complex to use, 'computers' diminished in size and cost as they grew in power and useability. By the 1980s, an immense array of everyday uses had been found for 'microchips', not only in the first personal computers, with their growing store of programs, but also in mobile phones, video cameras, compact discs, smart photocopiers, new diagnostic devices in medicine and office-sized automatic telephone exchanges.

IT found a powerful synergy with new communications technologies using infra-red light sent down threads of glass; the acronym was now ICT (information and communication technology). These made gathering, processing, storing, transmitting and displaying information in all forms so cheap it was virtually free, so fast it seemed instantaneous, so easy anyone could do it. By the end of the century, the drive was for 'convergence' and 'networking', combining many functions into one device and linking such through the internet to multiply their power many thousandfold.

In the closing decades of the century, two new technologies emerged to rival ICT in its potential impact on human life. 'Biotechnology' was built squarely on the new understanding of genetics and the workings of DNA; 'nanotechnology' would exploit what we knew about the behaviour of objects little bigger than atoms. While everyone had run to embrace the use of ICT, 'genetic engineering' and 'nanotech' were having a much harder time gaining public acceptance.

1951–2000

The World Stage

FOLLOWING THE UPHEAVAL AND WIDESPREAD destruction of World War II, Europe endured a new kind of conflict, the 'Cold War'. This split the continent, and much of the planet, into two armed camps, divided by the 'Iron Curtain' and backed by the military might of the 'superpowers', the United States and the Soviet Union. Tension was exacerbated by the awful proliferation of two technological offspring of World War II—nuclear weapons and the missiles able to deliver them over hundreds and thousands of kilometres.

The threat of 'mutually assured destruction' may have restrained the two sides from direct conflict, but pressure burst out elsewhere, such as in the Korean War (1950–53) and the Cuban missile crisis (1962). The impasse ended only with the collapse of the Soviet Union, mostly from internal weakness, around 1990.

Two other major trends affected European nations. One was the steady loss of their overseas empires: colonies in Africa and Asia gained independence from France, Britain, Portugal, Belgium and Holland. Germany's overseas possessions had already been a victim of its defeat in World War I. The Spanish empire, too, was long gone.

At the same time, European states began to seek an organisational unity that had previously come (and only briefly) as a consequence of military adventures such as those of Napoleon and Hitler. Transnational institutions began to grow, starting with the 'customs union' of the Common Market in 1954, which embraced France, Germany, Italy and 'Benelux'.

By the end of the century, a broader and more comprehensive European Union was in place, with 15 members and coordinated policy in many areas; more nations would join early in the new century. Though many strains and points of difference remained, it seemed increasingly unlikely that the states of Europe would ever again go to war.

Germany emerged from its post-war chaos with massive United States aid, becoming again a major economic force. With the fall of the Berlin Wall in 1989, West Germany was reunited with East Germany, previously under the control of the Soviet Union. As 'victorious powers', France and Britain took longer to recover. After decades of autocracy, democracy returned to Spain, Portugal and Greece, as had already happened in Germany and Italy with the defeat of fascism.

International cooperation was already strong in science; it had been for centuries. Due to the growing scale and cost of research facilities, nations pooled resources, leading to the formation of the European Space Agency, the European Southern Observatory, the Joint European Torus (for studies of nuclear fusion), the major particle physics laboratory at CERN near Geneva (birthplace of the Internet), and many other joint programs and facilities.

As the century progressed, much changed. Europe saw, as much of the world did, rising prosperity and living standards, major challenges to the health of the environment, burgeoning technology, especially in communications and transportation, a widespread decline in religious observance and a rise of movements such as feminism and 'gay pride'. Societies were far from harmonious and many people sought alternative lifestyles and systems of belief. In such a climate, science flourished, though it was increasingly subject to public scrutiny and questioning.

ARTS AND IDEAS

The closing decades of the twentieth century were alive with creative activity and new ideas. The publication in 1963 of Rachel Carson's *Silent Spring*, detailing the effects of DDT on wildlife, sparked the environmental or 'green' movement, one of the most powerful forces of the time.

'Consumerism', decried by American Ralph Nader in his 1965 attack on the automobile industry in *Unsafe at Any Speed*, was another major influence. The 'feminist movement' was much stimulated by *The Female Eunuch* (1970), written by Australian Germaine Greer.

Other powerful and often disparate insights came from Americans J. K. Galbraith (*The Affluent Society*), Milton Friedman (*Capitalism and Freedom*) and Jared Diamond (*Guns, Germs and Steel*); German Ernst Schumacher (*Small Is Beautiful*); and *Understanding Media* by Canadian Marshall McCluhan ('the medium is the message').

Literature was equally diverse and profuse. Again, American novelists were prominent: J. D. Salinger (*Catcher in the Rye*), Truman Capote (*In Cold Blood*), Tom Wolfe (*The Bonfire of the Vanities*), Vladimir Nabokov (*Lolita*) and Erica Jong (*Fear of Flying*). To show variety, the list could also include J. R. R. Tolkien (*The Lord of* the Rings), William Golding (*The Lord of the Flies*), Anthony Burgess (*A Clockwork Orange*), Salman Rushdie (*The Satanic Verses*) and J. G. Ballard (*The Empire of the Sun*), all from Britain; Boris Pasternak (*Dr Zhivago*) from Russia; François Sagan (*Bonjour Tristesse*) from France; Günter Grass (*The Tin Drum*) from Germany; Umberto Eco (*The Name of the Rose*) from Italy; and Australians Patrick White (*Voss*) and Thomas Keneally (*Schindler's Ark*).

Prominent playwrights included the Irishmen Dylan Thomas (*Under Milk Wood*), Brendan Behan (*The Hostage*) and Samuel Beckett (*Waiting for Godot*); Englishmen John Osborne (*Look Back in Anger*) and Tom Stoppard (*Rosencrantz and Guildenstern Are Dead*); American Neil Simon (*The Odd Couple*); and Frenchman Eugène Ionesco (*Rhinoceros*).

In music, the 'classical tradition', such as in opera, rested mostly with Englishman Benjamin Britten; Americans John Cage and Phillip Glass were pushing more radical forms such as 'minimalism'. By the end of the century, most leading composers were earning their livings from film scores. In musical theatre, Stephen Sondheim and Andrew Lloyd Webber took over from Rodgers and Hammerstein and Lerner and Loewe.

Jazz continued to build on foundations laid over the previous 50 years, transmuting into new forms. 'Rock and Roll' emerged in the 1950s and changed the face of popular music. Other genres grew vigorously with increasing prosperity and improving technology. The 1960s British pop group The Beatles was a global phenomenon, but only the first of many.

BIG SCIENCE

T he late twentieth century became the era of 'big science'—big in its geographical reach, in resources, in the scale of equipment and investigations, in the number of researchers and the size of teams, in the questions being asked and the impact of the answers. What had begun 500 years earlier as an inquiring spirit among a few philosophers, clerics and physicians was now a global enterprise, involving millions of people and powered as much by governments and businesses as by the curiosity of individuals.

Most nations with any interaction with the traditions of Western Europe were now part of the scientific enterprise, including the more advanced nations of Asia, the Americas and Oceania, such as Japan, Australia and Canada. Developing nations such as China and India were beginning to make a mark, though their time was still to come. European nations were all active, but the centre of gravity of research, marked, say, by the geographical distribution of Nobel Prizes, was now firmly in the United States.

However great and diverse was the enterprise of science during this half-century, some strong uniting strands can be seen: the uncovering of the secrets of the genetic code, leading to an understanding of the deep-rooted mechanism of biological inheritance and the potential for biotechnology; more evidence as to the scale and history of the cosmos that enfolds us here on Earth; more comprehensive models to explain the behaviour of matter at its most fundamental level; clearer delineation of the course of human evolution; more profound insights into the origins of life and into the dynamics of the planet on which we live.

In many areas of science the scale of apparatus and the organisation of research continued to ramp up. The Human Genome Project, for example, was a billion-dollar worldwide enterprise—though driven mostly from the United States—and relied on a new generation of analytical technology. More powerful and sophisticated telescopes were developed to detect visible light, radio waves and other forms of energy from the universe. The biggest astronomical telescope operating in 2000 was twice as large as the record-holder in 1950, and four times larger than any in 1920. By 2000 many instruments had been put into orbit around the Earth to get a clearer view; the best known of these was the Hubble Space Telescope, and instrumented probes launched to explore the solar system. All were backed by increasingly powerful computer systems that could store, analyse and display data.

The same growth had occurred in the machines probing the inner structure of matter. The cost of such equipment was straining the purses of individual nations. That, plus the desire to avoid wasteful duplication, made more and more scientific enterprises multinational. An early example of such cooperation was International Geophysical Year (IGY), actually 18 months spanning 1957–58. This coincided with (and to some extent encouraged) the first launch of satellites into Earth's orbit.

The earliest comparable enterprise had been the international networks of observations of the transits of Venus organised in the eighteenth century by the scientific academies in England and France. Now, such collaborations were a matter for governments. Many nations cooperated in scientific work to better understand the Earth and its many systems. For example, there were major cruises for marine science and oceanography and the first coordinated substantial activities in Antarctica.

Eugene Aserinsky and Nathaniel Kleitman: Sleep and Dreaming

Prior to the 1950s few people thought sleep was worth studying. Sleep was the brain's 'downtime'; unless you were a follower of Freud, dreams were meaningless. The eyes of babies seemed to move behind their closed lids

1952

as they dozed, but nobody thought adults would do the same.

In fact, no one had looked. The first to do so was American Eugene Aserinsky, an impoverished research assistant who did not even have a degree. He worked for Nathaniel Kleitman, a pioneering sleep researcher. In 1952, acting on a hunch, Aserinsky scrounged some equipment so old it constantly broke down, and used his son as his first subject.

Recording 'brain waves', the patterns of electrical activity in the scalp, he found that at times during sleep the brain is as active as when we are awake. Connecting other sensors to record the movement of the eye muscles (he knew of the eye movements of babies), he found that the eyes of sleeping subjects with active brains were in constant rapid movement. He called this REM (for 'rapid eye movement' sleep); the rest of sleep was 'non-REM', with very slow brain waves.

Other discoveries followed. Periods of REM sleep alternated with non-REM sleep in a 90-minute cycle, with the REM periods getting steadily longer. During REM sleep, the bodies of subjects were paralysed; they could not move. Kleitman thought REM sleep might have something to do with dreaming, so he and Aserinsky took to waking subjects, asking if they had been dreaming. About 80 per cent of people in REM sleep reported dreams, and only 20 per cent in non-REM did so. In REM sleep, the dreams were notably vivid and bizarre, with lots of movement and complex plots. Non-REM dreams were more rational and static.

So REM sleep is a third state of consciousness, between waking and deep sleep, generally (not exclusively) linked with dreaming. Presumably our rapid eye movements are 'looking' at the things in our dreams. Our bodies are paralysed so we cannot get out of bed and chase our dreams around the room.

This pioneering research proved that sleep is about more than rest. Half a century later the research continues, though there is not much agreement as to what dreams are 'for'.

The Cosmotron Leads the Way

1952

Since nuclear physics began early in the twentieth century, physicists have sought to break matter apart with missiles of various kinds. Nature provided the first probes, such as the very useful alpha particles that were released by radioactive elements like thorium. With these, Ernest Rutherford and his colleagues had proved that the atom had a nucleus (**1909**), transmuted one element into another for the first time (**1919**) and found the neutron (**1932 Cavendish Laboratory**).

But alpha particles were not powerful enough for more advanced experiments. The answer was the 'particle accelerator', a machine that used high electrical voltages to hurl particles (protons at first) down a tube emptied of air onto a target. With such machines, John Cockcroft and Ernest Walton at the Cavendish Laboratory (**1932**) had broken open many of the light atoms; they had 'split the atom'.

Linear particle accelerators were limited too, and were supplanted by machines more like racetracks than drag-racing strips. Beams of particles could be made very fast and powerful by being sent thousands of times around a circular path, getting a kick along every time they passed 'go'. In the first such machines, known as 'cyclotrons', particles spiralled out from the centre of large metal plates. These were built at the University of Berkeley in the United States under Nobel Prize winner Ernest Lawrence. The cost of scaling these up was immense and generated more new designs. By the early 1950s the 'synchrotron' was the preferred machine; it remains so 50 years later, though machines have grown enormously in size.

In the lead here was the Cosmotron, a synchrotron that opened for business in the United States in 1952. Its 'racetrack' (an evacuated tube in the shape of a ring) was about 10 metres across and delivered bombarding particles to targets with more than a billion units of energy. The first linear accelerators had managed only 100 000 units. By the 1990s machines a thousand times more powerful and kilometres in diameter were in operation. They cost billions of dollars to build and needed teams of hundreds to run them. Nuclear physics had become 'big science'.

Physicists pushed for ever higher energy for two reasons. On the one hand, a more powerful accelerator was like a bigger telescope; it could see finer details and more features within matter. That is how the first 'quarks' were found within protons and neutrons, particles that were previously thought indivisible.

Lots of energy also meant that new particles could be 'made', as positrons and electrons had been created from the energy of cosmic rays (**1928 Dirac**). Using such machines, quarks and other exotic particles were later made real and brought to life in the laboratory (**1996**). The Cosmotron had an early triumph here. In 1955 it converted the energy of its beam into matter, creating pairs of protons and antiprotons, particles that would annihilate each other if they were brought together.

Stanley Miller and Harold Urey: Making the Molecules of Life

In 1953 young American biochemist Stanley Miller filled a glass flask with some basic gases: water vapour, methane, ammonia and hydrogen. Miller's supervisor, the leading chemist Harold Urey, believed that those would

1953

have been the common gases in the atmosphere of the Earth when it was very young. Miller then fired electrical sparks through the mixture for several days. When all was cool, he found that the flask contained more-complex chemicals, including the amino acids glycine and alanine. With about 20 others, those are the basic units of the proteins from which all living things are made and that (as enzymes) make chemical processes work fast enough to sustain living things.

Miller and Urey designed their experiment to advance the debate on the origin of life, and it certainly did. It ranked in importance with the discovery by Friedrich

Wohler in **1828** that a chemical produced by living things (urea) could be made by simply rearranging the atoms in another chemical definitely not associated with life (ammonium cyanate). The Miller–Urey experiment has been re-created and upgraded many times, with ultraviolet light and X-rays added to the mix to provide extra energy and encourage the molecules to combine. Even more complex chemicals have been created, such as the short chains of amino acids called polypeptides.

So it seems reasonable to think that the chemicals needed to construct living things could have been made on Earth by ordinary physical and chemical processes in the earliest days. Of course, that would have been only the starting point. A lot more steps, many of which we do not yet understand, stand between amino acids and the simplest living cell.

James Watson and Francis Crick: The Secret of the Genetic Code

1953

Many would claim as the greatest scientific discovery of the twentieth century the 'cracking of the genetic code' in 1953 by American James Watson and Englishman Francis Crick, working together at the Cavendish Laboratory in Cambridge. Certainly, finding a structure for the 'genetic chemical' DNA was profoundly important; in the wake of this discovery we have deciphered the complete genetic instructions for making whole organisms, including ourselves (**1990**), and generated powerful new biotechnologies.

The discovery had a long prehistory, dating back at least to Johann Miescher and Walter Flemming finding nucleic acids in the nuclei of cells in **1869**. Much later came Edwin Chargaff (**1950**) with his 'rules'. These state that the vital chemical units (the four bases) in DNA are always found grouped together: adenine (A) with thymine (T); cytosine (C) with guanine (G). Others had used X-rays to explore how proteins were put together, finding that many were made like a spiral spring.

Crucial to Watson and Crick winning the 'race' (against leading researchers like Linus Pauling) were X-ray pictures (**1915 Braggs**) taken in 1952 by Maurice Wilkins and Rosalind Franklin, who worked at King's College in London. Such photos took a lot of interpretation, but buried in them were vital clues to the solution Watson and Crick proposed. The pictures indicated that DNA had not one spiral but two, intertwined like the banisters on a spiral staircase. The 'treads' on the stairs were the

> *Science moves with the spirit of an adventure characterised both by youthful arrogance and by the belief that the truth, once found, would be simple as well as pretty.*
>
> JAMES D. WATSON

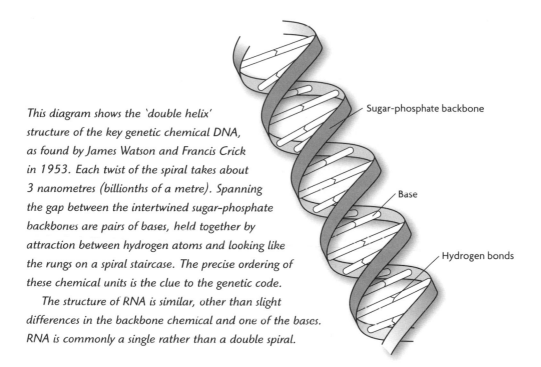

This diagram shows the 'double helix' structure of the key genetic chemical DNA, as found by James Watson and Francis Crick in 1953. Each twist of the spiral takes about 3 nanometres (billionths of a metre). Spanning the gap between the intertwined sugar-phosphate backbones are pairs of bases, held together by attraction between hydrogen atoms and looking like the rungs on a spiral staircase. The precise ordering of these chemical units is the clue to the genetic code.

The structure of RNA is similar, other than slight differences in the backbone chemical and one of the bases. RNA is commonly a single rather than a double spiral.

Sugar-phosphate backbone

Base

Hydrogen bonds

pairs of bases: A teamed with T, and C with G. The treads were joined down the middle by the mutual attraction of hydrogen atoms in each half of the step.

Watson and Crick could now explain how DNA copies itself every time a cell divides, when all the chromosomes with their embedded genes must be duplicated (**1869 Miescher and Flemming**). The hydrogen bonds break, the twin spirals unzip. From the soup of chemicals in the cell, each half of the staircase now collects materials to rebuild its missing side. Each A teams up with a T, each G with a C, the banister is reconstructed and there are now two DNA molecules instead of one.

Laid bare, too, was how the genetic code actually operates. In order to control the making of a particular protein, each coded message (a gene) needs to provide the order in which amino acids should be joined together. The 'letters' of the code were the four bases A, T, G and C. Watson and Crick hypothesised that each 'word' (technically called a 'codon') would represent a particular amino acid in the chain that makes up a protein.

How many letters would each word need? Two wouldn't be enough, as there are only 16 ways to choose and arrange two letters out of four. That would leave insufficient words to represent the 20 known amino acids. But three letters per word made 64 possibilities, leaving some words for other purposes (**1966**).

Watson and Crick showed brilliant insight to discern all this from the complexity of evidence. When the Nobel Prize for Medicine in 1962 was announced, Maurice Wilkins shared it with them. Rosalind Franklin did not; she had died in 1957, aged 37.

William Ewing and Harry Hess:
The Sea Floor in Motion

1953

The almost universal rejection of Alfred Wegener's notion of continental drift (**1912**) stemmed from one serious flaw in the idea. Though he had assembled impressive evidence that the continents had moved, coming together and splitting apart, he could not explain how it had happened. Most geologists thought the idea of the solid continents ploughing through the equally solid bed of the sea was ludicrous. Indeed it was.

The rehabilitation of Wegner's radical idea began in 1953, when American geologist William Ewing published results of his surveys of the bed of the Atlantic Ocean. Using newly developed equipment to measure the depth of the water and so chart the hidden landscape, he found a vast range of mountains running down the middle of the ocean, almost exactly following the shape of the coastlines on either side. Furthermore, the closer you came to this mid-ocean ridge, the younger the rocks became, both the bedrocks and the overlying sediments.

Ewing took this to mean that the floor of the Atlantic was in motion, spreading away from the ridge. He visualised new sea floor being formed from molten rock welling up along the line of the ridge from within the Earth, pushing out in both directions. If Africa and South America were embedded in the sea floor, they would

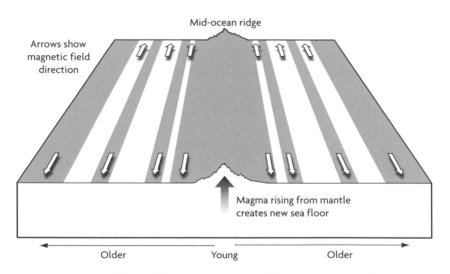

This represents a region of the sea floor several hundred kilometres in extent. The white bands of rocks hold a magnetic field pointing north, as it does now. In the black bands of rocks, the magnetic field points south (reversed polarity). Stripes of similar polarity are of equal width on either side of the central ridge, where the rocks had formed before moving outwards. Rocks at the outer edges are millions of years older than those in the centre. Such sea–floor spreading powers the movement of the continents.

be pushed apart. This was how the continents might have moved. Wegener began his return to favour among geologists.

Even more powerful evidence came in 1962, when American geologist Harry Hess towed a magnetometer over the sea floor. (This was initially done in the Pacific but the Atlantic showed similar results.) His instrument recorded the strength and direction of the magnetism embedded in the rocks. French physicist Pierre Curie (**1898**) had found that when molten rock was cooled below a certain temperature (the Curie Point), it captured any magnetic field it was in and held that ever after. Hess found that the rocks on the sea floor were magnetised in great stripes hundreds of kilometres wide and thousands of kilometres long, the magnetic field in one stripe pointing north and in the next stripe pointing south.

Here was evidence of two fascinating facts. Firstly, the Earth's magnetic field had changed direction many times in the past, with magnetic north becoming south and then reversing again. By dating the rocks with radioactivity (**1921**) we could find when these pole reversals had happened. The most recent one was 500 000 years ago.

Secondly, the pattern of magnetic stripes on one side of a mid-ocean ridge exactly matched the pattern on the other side. This could only mean that those sections of the sea floor, now thousands of kilometres apart, had cooled at the same time and so had once been side by side. Only a spreading sea floor could explain that.

Frederick Sanger: The Architecture of Insulin

Only four people have won the Nobel Prize twice: Marie Curie, Linus Pauling, John Bardeen and Frederick Sanger. Like Bardeen, and unlike Curie and Pauling, Sanger won the same prize twice, for chemistry in his case, in 1958 and 1980.

1955

Educated at Cambridge, English-born Sanger continued to work there for 40 years, perhaps the epitome of the dedicated and methodical researcher. But he was certainly no plodder; rather, he was one of the most influential scientists of his time. War broke out early in his career. While many other scientists went into work of military importance, Sanger was a conscientious objector and stayed in his laboratory.

Sanger admitted to a lifelong fascination with patterns and sequences in chemical substances. He believed that if these could be read, a lot of vital information might be made available. He was not yet 30 when he set himself the task of determining the structure of an important protein, that is, the exact identity and order of the amino acids from which it is constructed. He settled on insulin; it was easily available, not too complex (though it contained many hundreds of atoms) and if the structure could be found, it could be synthesised and perhaps the therapy for diabetes improved in some way.

Sanger had to develop new techniques for his work. To decide which amino acid was which (there were 22 to choose from), he used 'chromatography': the speed with which a solution of an amino acid is soaked up by special absorbent paper depends on its 'molecular weight', the weight of all its atoms combined. The work was laborious. It took him eight years to identify clearly each of the 51 amino acids in insulin and to sort out the order. Insulin had been one of the first hormones ever identified, by Canadians Frederick Banting and George Best in **1922**. Now it was the first whose architecture was completely known.

Many other researchers followed in his path, finding how other important proteins, such as haemoglobin, are built up. Sanger himself moved on. Keeping his interest in sequences, he began to study the order of the parts in molecules of DNA, the genetic chemical. This would lead to the first complete genetic inventory (genome) of any organism and then to the Human Genome Project (**1990**).

Willard Libby:
Ancient Dates from Radioactive Carbon

1958

What is the oldest city on Earth? The answer, first found in 1958, is Jericho, settled some 9000 years ago. How do we know? By radiocarbon dating, the brainchild of American chemist Willard (Frank) Libby. His technique is now used to date all sorts of ancient remains and artefacts, objects up to 70 000 years old.

Like many younger scientists, Libby spent much of his early career doing war work. Only 31 when war broke out in Europe, he was soon working on ways to separate the isotopes of uranium so that an atomic bomb could be made (**1938 Nuclear Fission**). After the war, radioactivity continued to play a major part in his career.

Carbon 14 (C14), first noticed in 1940, is one of the 'isotopes' (**1907 Thomson**) of carbon, chemically the same but with a different atonic weight. It is radioactive (**1899**), unlike the much more common isotope carbon 12 (C12). C14 is constantly made in the atmosphere, as cosmic rays (**1912 Hess**) hit atoms of nitrogen.

Like C12, C14 (as carbon dioxide) is absorbed from the air by plants through photosynthesis, so every living plant will have a small amount of 'radiocarbon' (C14) in its tissues, which is continually replenished while it lives. When a plant (or an animal that has eaten plants containing C14) dies, it no longer absorbs C14, so the quantity of C14 in the organism begins to decay. By measuring the amount of C14 remaining in something once living, say a piece of wooden furniture, ashes from a wood fire, a piece of parchment or old cloth, an Egyptian mummy, Libby could tell how long it was since the plant involved was alive.

Carbon 14 has a half-life (**1903 Rutherford and Soddy**) of 5700 years—after 10 half-lives (57 000 years) only about one millionth of the original radiocarbon remains. This limits how far back we can go with radiocarbon dating—60 000 years at most, even with the most sensitive instruments to measure the level of radiocarbon. Nonetheless, the technology has been of immense value to archaeologists and historians.

Radiocarbon dates have to be 'corrected'. The level of cosmic radiation varies over long periods of time, and so does the production of C14 in the air. Corrections can be found by comparing the radiocarbon date with one found by some other method, say by counting tree rings. The correction can move dates by hundreds of years. The calculation also gives us a way of finding the level of cosmic radiation at any particular time in the past and comparing that with other things in the environment, such as the activity of the Sun.

James van Allen: The Earth's Girdle of Radiation

In October 1957 the then Soviet Union, in the depth of the Cold War, launched Sputnik 1, the first artificial satellite sent into orbit around the Earth. And so 'the Space Age' dawned. Many of the early satellite launches were engineering endeavours, showing what nations could do and using technology developed for military purposes. Few scientific instruments were carried aloft, but as the spacecraft were going to largely unexplored regions, even simple measurements could yield big returns.

Typical of the early spacecraft were the United States *Explorer* satellites; they were about the size of basketballs, with just enough room for a Geiger counter and a radio transmitter to send measurements back to Earth. That initiative came mostly from American physicist James van Allen, who thought that large amounts of radiation, mostly charged fragments of matter such as protons and electrons, should be trapped by the various layers of the Earth's magnetic field and form a 'belt' through which the satellites would pass.

And so it proved. The Van Allen Radiation Belts (there are several) were found to fill a doughnut-shaped region of space a few thousand kilometres above the Earth's surface. The belts held rapidly moving particles (protons and electrons) that arrive as part of cosmic radiation (**1912 Hess**) coming from the Sun and from deeper space beyond. Some of this spills into lower levels of the atmosphere, setting up shows of the aurora, the 'northern and southern lights', when air atoms are induced to glow like the gases in a neon sign.

Charged particles in motion constitute an electric current; currents make magnetic fields. So changes in the radiation belts helped cause the wobbles in the Earth's magnetic field seen 300 years earlier (**1702 Compass**). This shows the ongoing links between events at the surface of the Earth and those in nearby space.

The Secret of Life: How DNA Makes Proteins

The question 'What is life?' has many answers. In 1946 the German Edwin Schrödinger, so influential in setting up quantum physics (**1927**), asserted that life is about information. He proposed that living things use something similar to morse code; they store information about how they are constructed and operate, and pass on the information via their genes. You could say that a plant or an animal is just their genes' way of making more genes (like a chicken is an egg's way of making another egg).

In **1953** James Watson and Francis Crick had shown how the code worked, embedded in the molecules of DNA that make up the genes and the chromosomes that hold them. Schrödinger's hunch was right, though the code uses sets of three symbols rather than the two (dot and dash) in morse code. In 1958 Francis Crick set down a statement so important that it is known as the 'central dogma': genetic information flow in living things is all one-way—it can pass from DNA to proteins but not the other way. DNA makes proteins and can copy itself; proteins do not make DNA.

The process whereby information stored in DNA controls the making of proteins came into focus from around 1959. This was the work of many hands and minds, too many to list them all, though the Americans Sydney Brenner and Matthew Meleson, and the Frenchmen Jacques Monod and François Jacob may be considered first

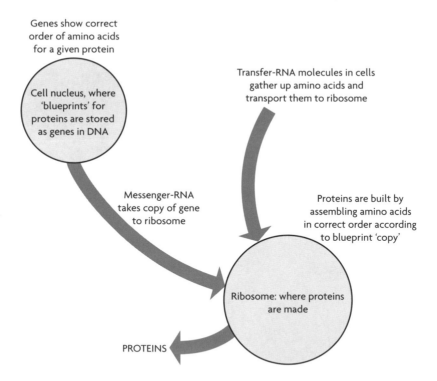

This diagram captures merely the outline of the astonishingly intricate process that transforms information stored as DNA in the nucleus of a cell into complete proteins assembled in the cell's ribosomes. Key roles are played by two forms of RNA: messenger–RNA provides a copy of the 'blueprint' for the particular proteins being assembled and transfer–RNA collects up the needed raw materials in the form of amino acid molecules.

among equals. Attention switched to RNA; this was similar to DNA but found throughout the cell rather than just in the nucleus, so more likely to have an active role in making proteins.

Many ingenious experiments revealed how sophisticated the process of making proteins in living cells is. In its role as information courier, RNA is multiskilled; forms of the chemical play three different parts. Small molecules of transfer-RNA gather up specific amino acids from the 'soup' in the cell, and carry them to 'protein factories' called ribosomes, mostly made up of ribosomal-RNA. The transfer-RNA molecules, with their attached amino acids, line up against what is effectively a copy of the DNA blueprint for the particular proteins being made. That copy has arrived from the nucleus aboard yet another sibling, messenger-RNA.

The end result of this is a string of the right amino acids lined up in the right order. Once these are glued together by enzymes, the string can be peeled off as a complete new protein. The sophistication of the system does not end there. Monod and Jacob found that the DNA has 'on and off buttons' that control which bits of code are copied into messenger RNA and so which proteins are made. All in all, it is a remarkable system to be the product of unguided biological evolution.

Roger Sperry: Two Brains in One Head

In the 1950s doctors treating severe cases of epilepsy, in which large areas of the brain convulsed together, sometimes took the radical step of performing surgery to sever the connection (the corpus collosum) between the right and left hemispheres of the brain. The treatment usually worked; the patients were able to live normally in society, with no one being aware that anything had happened.

1961

Beginning in the 1960s a long series of experiments by the American psychobiologist Roger Sperry (Nobel Prize for Medicine 1981) showed that in fact something was very different. Patients with such a 'split' brain could behave as if they had two independent brains with different capabilities. The subjects used their 'right brain' when recognising patterns or spatial arrangements (the word 'holistic' expresses this); the 'left brain' was better with language and mathematics and things that needed to be done in a certain order ('logical' or 'sequential'). The left brain tended to learn using words, the right brain in other ways, for example, using imagery. Sperry noted that in much of education, verbal forms of learning are promoted over non-verbal forms. Provocatively, he claimed that 'society in general discriminates against the right hemisphere'.

Experiments in this area using both humans and laboratory animals rely on the structure of the nervous system. Generally speaking, the right-hand side of the body, including, for example, hearing in the right ear, is controlled by the left-hand side of the brain, the left side of the body by the right brain. The specialisation goes further in the eyes. When looking straight ahead, what we see in the right-hand side of the visual field goes to the left brain; the left brain sees what shows on the right side of the centre line. Since it is generally the left brain that 'talks', while the right brain is 'mute', a split-brain patient who sees something only in the left visual field cannot describe it in words, though they can draw it and may have already responded to it in other ways, for example, by blushing if it is a 'rude' picture.

In most people, the two halves of the brain are connected and behaviour depends on both hemispheres working together. Yet most people have a preferred way of thinking: 'left brain'—logical/analytical/straightforward, or 'right brain'—visual/imaginative/fanciful. Going beyond these observations becomes controversial, such as making assertions that men and women show different left brain/right brain preferences and that the two sorts of thinking can often interfere with one another, preventing us from doing our best intellectually.

Murray Gell-Mann and George Zweig: The Particles within the Particles

1963   By the early 1960s physics was in crisis over the 'fundamental particles' from which all matter was made. It was like the explosion in the number of known chemical elements in the early nineteenth century. When the number of elements reached 15 or 20 times the four postulated by the ancient Greeks, scientists went looking inside them, hoping to find a small number of tiny bits that could be put together in various ways to make all the rest. Those were found; by the 1930s we knew that atoms are made from protons, neutrons and electrons in various combinations.

Almost at once new fragments of matter began to show up, smaller than atoms, similar to but different from the 'big three' (protons, neutrons and electrons). Soon there were dozens: K particles, V particles, other particles with Greek names, various mesons, not to mention their antiparticles. Some of these flashed in and out of existence in a split second, though they always decayed into something else.

Again the hunt was on for some inner order, some small number of 'really'

> *All of modern physics is governed by that magnificent and thoroughly confusing discipline called quantum mechanics...It has survived all tests and there is no reason to believe that there is any flaw in it...We all know how to use it and how to apply it to problems; and so we have learned to live with the fact that nobody can understand it.*
>
> MURRAY GELL-MANN

fundamental particles that made up the rest. Independently, Americans Murray Gell-Mann and graduate student George Zweig concocted a scheme. This applied to two groups of related particles, all of which 'felt the (strong) force' (**1936**): the baryons (proton, neutrons and their assorted kin) and the group of mesons, obviously related.

According to this new theory, there were three of these *really* fundamental particles; Gell-Mann, who knew his James Joyce, called them quarks, after a line in *Finnegan's Wake*, 'Three quarks for Muster Mark'. Zweig called them 'aces'. The idea was that quarks and their antiquarks would gather three at a time to make the proton, neutron and so on, and two at a time to make up the mesons. Three different types of quarks were distinguished: the up quark, the down quark and the strange quark (the last was needed to explain some particles that behaved 'strangely').

In fact, all the quarks were a bit strange. To explain how three quarks could team up to make a proton with charge of one, the quark had to have less than the charge of the electron, either one-third or two-thirds as much. So a long-held belief had to go—that the electron carried the smallest possible electrical charge. However, the scheme was neat and explained what the physicists had found. It also did what Mendeleyev's Periodic Table had done (**1869**): it predicted particles not known then to exist, to be specific the 'omega minus' particle, which turned up soon after.

Gell-Mann and Zweig had a difference of opinion over one thing. Gell-Mann did not think the quarks actually existed; they were just a useful way of thinking, patterns of behaviour, not real particles. This was like the view some had of Nicolaus Copernicus and his Sun-centred universe (**1543**): it was useful for calculations but the world did not have to be like that. Zweig thought the quarks were real and perhaps could be found. Various schemes were put up to look for them, say in cosmic rays (**1912 Hess**) or in particle accelerators (**1952 Cosmotron**), but none turned up. For a while Gell-Mann appeared to be correct. If quarks existed, we never seemed to be able to prise one free so we could take a proper look at it.

But in the long run, Zweig was right. It was just a question of gathering enough energy to create quarks out of 'nothing' (**1996**).

Mary and Louis Leakey: An 'African Genesis' for Humans

Louis Leakey was born in what is now Kenya. At the time he first became involved in the search for the origins of the human race, most experts thought humans had evolved in Asia. The oldest known human fossils, perhaps half a million years old, had been found in the east, in Java and China, and labelled *Homo erectus* ('upright man') (**1856**). However, Charles Darwin had argued for Africa to be the cradle of humanity, and Leakey set out to test the idea. He was encouraged by the first discoveries by Raymond Dart in 1924 of human-like fossils in South Africa.

1964

Species	When they lived	When first found	Where they lived
'Lucy' (Australopithecus afarensis)	3.8–2.8 my BP	1975 (Ethiopia)	Tropical and temperate Africa
Handy man (Homo habilis)	2.5–1.8 my BP	1964 (Tanzania)	Tropical Africa
Nutcracker man (Paranthropus boisei)	2.3–1.2 my BP	1957 (Tanzania)	Tropical Africa
Java man or upright man (Homo erectus)	1.8–0.2 my BP	1896 (East Indies)	Tropical and temperate Africa, Asia and Europe
Neanderthal man (Homo sapiens neanderthalis)	250–30 ky BP	1854 (Germany)	Europe and Western Asia
Cro-Magnon man (modern humans—Homo sapiens sapiens)	200 ky BP to present	1868 (France)	Everywhere

my BP = million years before present   ky BP = thousand years before present

This table shows how some of the evidence about our ancestry has been assembled over the last 150 years. Generally speaking, the longer ago the pre-humans lived, the more recently their fossils were first found. Unlike our most recent known ancestors (Neanderthal man, Cro-Magnon man), our ancient forebears lived in more remote and often accessible places, such as parts of Africa.

New evidence about our ancestry continues to emerge. The connections between these hominids are still much debated. Not all the more ancient hominids are ancestors of the more recent ones; some may simply have died out.

While in his 20s, he began to hunt for evidence in the Olduvai Gorge, a 100-metre-deep, 50-kilometre-long chasm in modern Tanzania. Later he was joined there by his wife Mary; in time her own work would become at least as important as his. Their son Richard would also build his own reputation in human origins research, and the family business has carried into the third generation.

Only after 25 years of excavation was an important discovery made. In 1959 Mary found the jawbone of an ape-like creature they called Zinjanthropus ('Nutcracker man') in the lowest level of deposits in the gorge, about 1.75 million years old. This and similar creatures (australopithecines meaning 'southern apemen') appeared to be our ancestors but not ancestors of the chimpanzees and gorillas.

If *Homo erectus* (and later *Homo sapiens*) had descended from the australopithecines, where were the 'missing links', the creatures that spanned the gap? In 1964 a key discovery was made: in the same layer of deposits where Zinjanthropus had been found, the Leakeys uncovered bones (initially a jaw and parts of a hand, and later a skull) of an animal they labelled *Homo habilis,* or 'handy man'. Louis Leakey argued that this creature walked upright and had made the stone tools that had been dug up in Olduvai. This was a major advance on the apemen; it had a larger brain and opposable thumbs for making and using tools and weapons. He concluded that this creature should share the genus *Homo* (Latin for 'man') with *Homo erectus* and with us.

Other excavators later found bones of *Homo habilis* and related creatures in other regions of eastern Africa, mostly within the Great Rift Valley, but no similar bones have been found anywhere else in the world. So an 'African genesis' for humans looks very plausible, with the descendants of *Homo habilis*, including *Homo erectus*, migrating from Africa to other parts of the planet.

William Hamilton: The Evolution of Altruism

The Genetic Evolution of Social Behaviour, a paper published in 1964 by the British biologist William Hamilton is, hard as it might be to believe, the most commonly cited paper in scientific history, referred to by more **1964** researchers in their own publications than anything by Albert Einstein. Hamilton has been called the most influential Darwinist since Charles Darwin himself. His paper started a whole new train of thought among biologists, who now find explanations for behaviour in the Darwinian drive for an organism to maximise the survival of its genes. This is now commonly called 'sociobiology'.

The key concept for Hamilton was the notion of kinship. This refers to the other individuals with whom a particular individual shares genes. What matters in evolution is that genes go forward to the next generation, not so much how they get there. So what often appears to be altruistic behaviour—raising your sibling's offspring rather than your own or even sacrificing your life to save relatives—makes perfect sense in terms of evolution, and so will be inbuilt in individuals by natural selection.

Putting it mathematically, which Hamilton was well able to do, an individual should sacrifice him/herself (at least in terms of personal reproduction) to save eight cousins, or four nephews or nieces, or two siblings, or one identical twin, since on average cousins share 12.5 per cent of genes, nephews and nieces 25 per cent, siblings 50 per cent and your identical twin 100 per cent. In all these cases, a total of genes equivalent to your own will go forward to the next generation.

This also explains why altruistic and even self-sacrificing behaviour is very common in colonies of ants and bees, since they are all descended from the one queen and drone and so share most of their genes. This notion can be expanded to cover situations where an individual is not genetically linked to those around them but is dependent on their cooperation or on access to their resources for the survival of his or her own offspring. Highly developed social behaviour can therefore influence not only the survival of the group but also of kinship associations within the group.

Hamilton died of malaria following an expedition to the Congo to gather evidence that the disease HIV/AIDS had arisen from bungled experiments in vaccination involving monkeys. This matter remains controversial.

Cyril Hazard and Maarten Schmidt:
Stars That Were Not Stars

1964

In 1962 British radio astronomer Cyril Hazard headed into the Australian outback to make use of a recently completed radio telescope. He was interested in an object in the heavens called 3C273 because of its place in the Third Cambridge Catalogue put together by pioneering radio astronomer Martin Ryle (**1950 Big Bang**) and his team at the University of Cambridge. The catalogue listed tiny sources of radio static scattered among the stars but not apparently linked to anything visible through a telescope.

At the time, radio astronomy images were inevitably much fuzzier than those taken by ordinary (light) telescopes, so identifying something visible out there that might also be emitting the radio static was a challenge. Hazard's idea was to use the Moon as a sort of shutter. Its position in the sky could be very precisely calculated, so by timing the instant the Moon passed in front of the radio source and cut off the signal he could show precisely where the radio source was.

Photographs taken of the point in the sky where 3C273 lay hidden turned up something very odd—a blob looking like a star, but with a streak of light sticking out one side. Similar objects elsewhere were found to have their own peculiarities. The name quasar for 'quasi-stellar object' seemed right for such things, but what were they?

Dutch-born astronomer Maarten Schmidt, now working in the United States, took the next step. He used the 5-metre telescope at Mount Palomar in California, the biggest in the world at the time, to measure the red shift of 3C273 (**1912 Slipher**). This measurement would indicate how fast it was travelling. But the measurement made no sense. He could not find the distinctive spectral lines where he expected them to be. The only feasible explanation seemed unacceptable. The lines had been red shifted by much more than those of any known star. So 3C273 was not a star in our own Milky Way, but far beyond it, out among the other galaxies.

3C273 was moving at about 15 per cent of the speed of light and so according to the Hubble Constant was about 2 billion light-years away. That set no records; galaxies both faster and further away were already known. But 3C273 was 1000 times brighter than even a big galaxy would appear at that distance. Yet it was a single point of light. What could possibly be releasing so much energy? At that time, nobody was sure (**1999**).

Arno Penzias and Robert Wilson:
The Big Bang Battle (Round Two)

1965

Since around **1950** there had been two rival theories of the origin of the universe, each with its supporters and detractors: Big Bang versus Steady-state; a universe that changed over time compared with one that stayed the same. Both sides accepted the idea of the 'expanding universe' (**1929**)—that the many galaxies like our Milky Way, or at least groups of galaxies, were moving apart.

Both theories had problems. The Big Bang followers believed that the universe had once been exceedingly dense and hot, indeed infinitely so, and that matter and energy had spread out and cooled since that time, condensing into stars and galaxies. But could present laws of physics work in such extreme conditions? For the Steady-staters, the universe could stay looking the same only if new galaxies appeared in the spaces as the existing ones moved apart. But wouldn't such continuous creation violate the laws of conservation of matter and/or energy? (**1907 Einstein**)

The Big Bang theory had already secured some credibility. It fitted the way the galaxies appeared to be spread through space. It could also explain the amounts of the various isotopes of hydrogen and helium that analysis of light from the stars had revealed. This was one of the two 'relics' that American physicists Ralph Alpher and George Gamov had predicted the Big Bang would leave behind.

The other memento would be the remaining glow from the primeval fireball. As the universe continued to grow and cool, the nuclei of hydrogen and other elements would be able to catch and hold electrons, so forming complete atoms for the first time. At that moment, which was about 300 000 years after the beginning, ultraviolet radiation trapped among the particles would suddenly have escaped in an immense flash. Such radiation should still be about, though very much changed. The expansion of the universe since would have stretched and chilled the radiation into a faint wash of short radio waves, the 'cosmic microwave background'. The Americans predicted that it would resemble the pattern of energy given off by a perfectly black object at a temperature of 2.7 degrees above absolute zero.

No one went looking for this relic; at the time it was not thought detectable. In 1965 it was found by accident. Arno Penzias and Robert Wilson worked for the Bell Telephone Company and were trying to locate the source of radio interference on communication channels from satellites. It was a remarkable rerun of the way Karl Jansky had discovered radio waves coming from the Milky Way in **1932**. Try as they might to cut out all interference, Penzias and Wilson still heard a gentle hiss of static coming from all over the sky, as Alpher and Gamov had predicted. Even the temperature of the radiation was right. This was powerful evidence that the universe had once been exceedingly hot and dense.

The supporters of the vision of Steady-state universe (**1950**) had a serious problem explaining the microwave background, since they argued that our universe had always looked the same. So the Big Bang became generally accepted as describing the early days, or even early seconds, in the life of our cosmos. Other researchers would elaborate on the theory (**1980 Guth**), and further studies of the microwave background would fill in some of the details (**1992 COBE**). Most astronomers now accept that our universe had a beginning. They still argue about how (or even if) it will end.

Marshall Nirenberg:
Completing the Genetic Code

1966

Learning to read the genetic code was one of the great scientific achievements of the twentieth century, and in the history of biology. In **1953** James Watson and Francis Crick had discovered how the code must work, with groups of three 'letters' (actually chemicals called bases) representing each of the amino acids. But working out which group of three (called a codon) went with each amino acid took more than another decade of research. The first match was not made until 1961, by the American biologist Marshall Nirenberg.

That set off a race by a number of teams to complete the rest of the matches. Generally the same method was used: a piece of transfer RNA (**1959**) with a known order of bases was let loose in a mix of amino acids to see which one it would pick up. By 1966 all the combinations had been explored, giving the meaning of the code for RNA. The code is the same for DNA except for the substitution of one letter. There are only 20 amino acids and 64 ways to choose and arrange the three letters in the codon, so some amino acids are represented by more than one codon. Some of the unused combinations have other functions, such as acting like punctuation to mark the beginning or end of a message.

Reading the code constituted a powerful breakthrough in biology. Now we could link the DNA in an organism's genes, including ours, with the various specific proteins, and then to the body structures and processes that those proteins built or controlled. A change in a gene, say by mutation, would change the protein it was responsible for in some way and perhaps cause disease or disability. It was now possible to use genetic 'counselling', or gene 'therapy', to correct a malfunctioning gene. Also to come was genetic engineering (**1972**).

Lynn Margulis: Cells that Live in Cells

1967

At first glance, a plant or animal cell is a simple object: it's a microscopic blob of liquid with a wall of some kind around it, and a few other spots of different colour within. Many decades of progress with increasingly powerful microscopes have revealed the hidden complexities of cells (**1838**). We now know that cells have many and complex structures and systems, including systems to store and release energy. In plant and animal cells, the main powerhouses are the mitochondria, which release energy from foods. Green plant cells also contain chloroplasts, which do the reverse: they trap energy (from sunlight) and convert it into sugars and other foods through photosynthesis.

There are many puzzles about these bits of cell, puzzles that in the late 1960s first led American biologist Lynn Margulis to propose a startling theory for their origin. Both mitochondria and chloroplasts, she claimed, were once free-living bacteria-like

THE FIVE KINGDOMS OF LIFE

	Characteristics	Examples/Main divisions
Monera	· Single cells · Cells do not have nuclei · Some do photosynthesis · Many move independently	· Bacteria · Cyanobacteria if they do photosynthesis
Protists	· Mostly single cells (some gather in colonies) · Cells have nuclei · Some do photosynthesis · Many move independently	· Plant-like (with photosynthesis): 'algae', many water weeds, phytoplankton in ocean · Animal-like (without photosynthesis), aka 'protozoa': amoeba, paramecium, zooplankton in ocean
Fungi	· Mostly multicellular · Cells have nuclei · Do not do photosynthesis · Do not move independently	· Mushrooms, moulds, mildews · Yeasts (single celled)
Plants	· Multicellular · Cells have nuclei · Cell walls made of cellulose · Do photosynthesis · Do not move independently · Can be complex in structure	· Without stems (e.g. mosses) · With stems (the rest) · With spores/no sex (e.g. ferns) · With sex but no flowers (pines) · With sex/flowers
Animals	· Multicellular · Cells have nuclei · Do not do photosynthesis · Move independently · Usually complex in structure · Have specialised sense organs	· Coelenterates (with body cavity): jellyfish, anemones · Annelids (segmented worms): earthworms · Molluscs (with soft bodies): snails, shellfish, squid · Arthropods (with jointed legs): insects, spiders, crustaceans · Chordates (with spinal chord): fish, amphibians, reptiles, birds, mammals

organisms that had set up house long ago inside other single-celled creatures (they may have been initially simply eaten) and later became indispensable to each other. Margulis called this 'endosymbiosis' ('endo' means 'within').

Margulis assembled some strong evidence for the idea, and it has been added to since, to the point that her explanation is widely accepted. Both mitochondria and chloroplasts contain their own plans for reproduction; both have DNA and a genetic code that is quite different from that of the cell as a whole, but similar to the code in bacteria. They reproduce by dividing in two when the cell divides and are not, so it seems, constructed under orders from the cell nucleus. The membranes that surround them are also distinctive and resemble bacteria membranes. Chloroplasts are similar in many ways to the free-living cyanobacteria, which can also do photosynthesis. Fossils of these organisms are among the earliest evidence of life on Earth, several billion years ago (**1993**).

Margulis believed she was adding an important new element to Darwin's theory of evolution (**1859**). Darwin himself did not know the sources of the variations between individuals of a species that provide something for natural selection to work on. Later, others suggested mutations (**1901**). Margulis argued that endosymbiosis, whereby a cell gains extra genetic information from the organisms it has engulfed, is a much more important source of variation, especially for simple organisms and the earliest stages of evolution.

Thinking about the origin of cell contents led Margulis to propose a new way of classifying living things; the old simple division into animal and plant 'kingdoms' became inadequate long ago. Her scheme (there are rivals) has three other kingdoms, making five in all. One contains bacteria, which have no nuclei in their cells (technically these are prokaryotes). The second holds protists, single-celled organisms like amoebas that live alone or in colonies. The third contains fungi, from yeasts to mushrooms. Like plants and animals, fungi and protists (once called protozoa) have a nucleus in which DNA is stored (that is, they are all eukaryotes). Both plants and animals have many cells and often complex organisation; a major difference is that plants also carry out photosynthesis.

Jocelyn Bell and Anthony Hewish: Messages from 'Little Green Men'

1967   In 1967 graduate student Jocelyn Bell was monitoring radio signals collected by a new radio telescope that her supervisor, Anthony Hewish, had commissioned at Cambridge University. They were looking for more of the newly discovered quasars (**1964 Hazard and Schmidt**) but something else turned up instead. Bell found a source of radio static that was 'pulsing', growing weaker and stronger many times a second in a very regular fashion. No one was sure what to make of it at first. For a time it was known as LGM1 for 'little green men'; a half-serious suggestion that extraterrestrials were signalling was on the list of possibilities.

That particular explanation did not last very long. The source was clearly a long way away. Bell had found the first 'pulsar', though the star was not pulsing in and out but rather spinning very rapidly. The discovery was not totally unexpected. The violent collapse of a very large star to form a supernova had been predicted to leave behind such a relic, a superdense body called a neutron star (**1936 Zwicky**). As the star collapsed, it would begin to spin faster and faster. It would also develop a very powerful magnetic field and start to beam out radiation like a lighthouse. A pulse of radio waves would be recorded each time the beam swept by an observer. So it seemed the newly found pulsars were the predicted neutron stars.

Pulsars are commonly found in the middle of the clouds of expanding gas left by supernovas. Some give off pulses of visible light as well as radio signals. The spin of the pulsars is so regular that they were soon judged to be the most accurate clocks in the cosmos, but they also slow down over time. The rates of spin and of the slowing down of the spin indicate how recently the supernova happened. Measurements with pulsars gave us the first evidence that 'gravity waves' existed (**1973**).

In 1974 Hewish shared the Nobel Prize for Physics with his Cambridge colleague Martin Ryle (**1950 Big Bang**); this was the first Nobel Prize for work in radio astronomy.

Bell did not share in the prize, even though she had made the critical discovery, and Hewish had been initially sceptical.

Sheldon Glashow, Abdus Salam and Steven Weinberg: The Push for Unity

Over the centuries science has been much concerned with finding unity among events and entities that seem very different. Many scientists, both present and past, are motivated by a belief that there are connections between different phenomena, say between living and non-living things (**1828**). In the nineteenth century, Michael Faraday's experiments (**1831**) and James Clerk Maxwell's theories (**1871**) showed that electricity and magnetism are intertwined. Their work encompassed light as well as other forms of electromagnetic radiation, such as heat, radio waves and X-rays.

1967

Maxwell and Faraday did not know that the twentieth century would bring quantum physics, with lumps of energy (photons) and charge (electrons). Putting these requirements into Maxwell's theory produced quantum electrodynamics (or QED), which explains very well how light and matter interact on the smallest scales.

In the other corner of the global laboratory, another theory was being worked up. This dealt with beta decay (**1899 Radioactivity**), the form of radioactivity that releases electrons. The force that powered this radioactivity seemed much weaker (in fact a million times weaker) than the force of attraction between charges or magnetic fields (the electromagnetic force), so it was called simply the 'weak force'. It also acted only over very short distances—less than the diameter of a single atom—unlike the electromagnetic force, which can be felt over large distances—at least the several kilometres involved in a bolt of lightning.

Yet bright minds could find a way to bring these two forces together. In 1967 two Americans, Steven Weinberg and Sheldon Glashow, and Pakistani-born Abdus Salam, working in Italy, operating mostly independently, announced their new scheme, the 'electroweak' theory.

The three researchers built on the crucial insight from Hideki Yukawa (**1936**) that these subatomic forces involve the exchange of some sort of particles. To explain everything, they found that four particles were needed. One was the well-known photon (**1905**), which travels at the speed of light and therefore has to be without mass. It carries the electromagnetic force. The other three were involved in the weak force. They must have mass, because the very short range of the weak force meant that they would move relatively slowly. Despite the differences, these four particles are a family, with much in common; as with many families, different things are expected of various members (physicists like to call these discrepancies 'symmetry breaking').

The test of a theory lies in what it can predict. The electroweak model maintained that four particles were needed and that the three then unknown would be found, if not free in nature then created out of energy in one of the big new particle accelerators (**1952**). And so they were, a decade or so later, and with the predicted amounts of mass.

So this stage of the ongoing quest for unity in nature was clearly a success. But something bigger was beckoning—a union with the strong force that holds the atomic nucleus together. That was not far away (**1980**).

Herbert Boyer and Stanley Cohen: The Birth of Genetic Engineering

1972

Unlike the technologies of earlier centuries, which were mostly created by ingenious craftworkers, late twentieth-century technology is derived from science. The massive biotechnology industry could never have emerged without the deep understanding of the nature of genes and the genetic code built up over decades of scientific research.

New knowledge does not inevitably lead to technology. There must be a demand for the technology, and someone must take the first steps towards turning the knowledge to a practical end. In the case of genetic engineering, a core element of biotechnology, the key date was probably 1972, and the key names, among many others, the Americans Herbert Boyer and Stanley Cohen.

Following the watershed discovery by James Watson and Francis Crick in **1953**, it was well accepted that the processes of life, such as the fabrication of the proteins that make up and regulate living cells, are controlled by coded instructions carried on long strands of the chemical DNA. Scientists had learned how to read the code, and how the instructions in DNA were followed by cells (**1959**).

By the late 1960s they knew two more things. In bacteria—the simplest living things—not all the DNA is in long strands (the chromosomes); some is in small rings called plasmids. They also knew that the DNA could be cut up into short pieces by special enzymes that always cut at the same point, and other enzymes would join the pieces together again.

Boyer, Cohen and their colleagues extracted some plasmids from bacteria and cut them open. They then inserted pieces of DNA taken from other organisms, to see if the bacteria would follow those new instructions as well as they did the ones they were used to. They did. The bacteria began to make the proteins coded by the spliced-in genes. Thus began 'recombinant DNA technology' or genetic engineering.

Within a decade, one practical consequence of this advance became apparent. Colonies of bacteria were making human insulin (**1922**), following instructions given by human DNA spliced into plasmids. By 1982 government authorities in the United States had approved the use of this insulin, in place of the animal insulin previously used, to treat diabetics. Since then, hundreds, perhaps thousands, of chemicals for diagnosis or therapy have been created by genetic engineering. Much of medical practice has been transformed. By the 1990s the first genetically modified foods were coming onto the market, sparking a continuing controversy.

Knowledge has grown too. By splicing various bits of DNA, such as whole genes, into bacteria, we can see what proteins they are coded for, or even if they do nothing. Early evidence has suggested that many pieces of DNA are 'junk', with no role at all.

This path to knowledge has not been unimpeded. Concern arose early that the most commonly used bacterium for this work (because it was so well understood) was *Escherichia coli* (*E. coli*), found in the human intestine, where it helped helps with digestion. If, say, cancer-causing genes were spliced into *E. coli* to see how they work, they could possibly 'escape' from the laboratory and infect the general population, with disastrous consequences.

So in the mid-1970s researchers obeyed a call by the American Paul Berg for a moratorium on such experiments until proper safety measures were in place, including the use of variants of *E. coli* so weakened that they could not survive outside the test-tube. Once those safeguards had been developed and legislated for, the work resumed and continues to this day.

Joseph Weber: Waves of Gravity

In **1871** the Scottish physicist James Clerk Maxwell predicted that when electric charges move suddenly they generate ripples in the electric and magnetic fields that surround them and these ripples can carry energy across space. In **1888** the German Heinrich Hertz found those ripples, generated by electric sparks and now called radio waves.

1973

The same thing should be true of gravity. When massive objects are accelerated, made to change their speed or direction, ripples should spread through the gravitational field as 'gravity waves' and be detectable somewhere else. But finding them would be a very tough task. Gravity is by far the weakest of the four basic 'forces of nature', though paradoxically it is the one we are always aware of. The 'strong force' (**1936**) and the 'weak force' act only within atoms; the 'electromagnetic force' between charges can both attract and repel and usually cancels out over large distances.

So to make detectable amounts of gravitational radiation would take immense objects moving violently, such as two neutron stars (**1936 Zwicky**) or two black holes (**1916**) colliding, or a giant star collapsing to form a black hole. Even then, the radiation weakens rapidly as it spreads (as light does), and if the object is a long way away, the detectors must be exceedingly sensitive. A gravity wave passing through a solid object should cause it to flex or stretch, but only by as much as the size of a single atom. Other effects—such as the detector's internal heat or any sort of nearby vibration—could cause such changes.

The first 'gravity wave detector' was turned on by the American physicist Joseph Weber in 1973. He used an aluminium bar with state-of-the-art (for the day) devices attached to measure any change in its size or shape. He claimed to have seen gravity waves flexing his bar, but the consensus among other physicists was that his detector was simply not sensitive enough; something else had caused the blips on his meters.

Later generations of detectors used more exotic materials and cooled them to a few degrees above absolute zero to minimise the 'thermal noise'. The most recent machines use laser beams kilometres long to measure the distance between mirrors with the necessary precision. If the theory is right, those distances should change by infinitesimal amounts as a gravity wave passes through.

These first superdetectors went online early in the twenty-first century; so far no one has made a confirmed 'sighting'. They are being set up in widely spaced locations: the scientists will be looking for 'coincidences', when a gravity wave passing through the Earth triggers several such machines almost simultaneously.

Gravity waves have, in fact, been detected, but not directly and not on Earth. In deep space, a pair of neutron stars (**1967**) has been found orbiting each other very fast and very close and apparently radiating gravity waves. That is the best explanation for the fact that the orbiting stars, which are very heavy and are moving fast, are slowing down as they lose energy, by the amount the theory of gravity waves predicts.

Donald Johanson: Lucy and the First Family

1974

When the pop group The Beatles wrote the song *Lucy in the Sky with Diamonds*, anthropology would have been far from their minds. The song was a hymn to LSD, the potent hallucinogenic drug popular with some at the time. But Lucy now has her place in human history, as the name given to a creature that forms a vital link in the chain of human evolution.

American Donald Johanson was a young, little-known fossil hunter when he dug up Lucy, or most of her, in 1974 at a site called Hadar in Ethiopia. The story is that The Beatles song was booming on the record player at the time, so the name came naturally. The site and the region around it lies in the Great Rift Valley of eastern Africa and had yielded a number of important finds. This connects to Olduvai Gorge further south, where Mary and Louis Leakey had found 1.8 million-year-old remains of the tool-using *Homo habilis* a decade before (**1964**). Johanson was to find more complete remains of *Homo habilis* there in 1986.

Compared with some of the nineteenth-century finds, which were mere fragments, the skeleton of Lucy, dated to around 3.8 million years before the present, was remarkably complete, with pieces of skull, a jaw, ribs, vertebrae, most of the arms, pelvic bones and upper and lower legs. From these, Johanson could show that Lucy walked upright, leaving her hands free—she was the first creature in history known to do so.

The region around Hadar is called Afar, and Lucy's formal name is *Australopithecus afarensis*. This name places her in the group of 'apemen', close to but distinct from the genus *Homo*, which includes the human species. Some researchers say she is in our group. Either way, she is a very near relative. In all probability her kind evolved over several million years into *Homo habilis* and may have lived alongside them for a time. *Australopithecus afarensis* was 1–1.5 metres tall, with males larger than females. Further digging uncovered bones of up to 13 other individuals close by, showing that Lucy was a member of the first family of which we have evidence.

The evidence of 'bipedalism' was strongly confirmed the same year: Mary Leakey found an amazing set of fossil footprints preserved in a bed of volcanic ash at Laetoli, not far from Olduvai Gorge. These show that nearly 4 million years ago two adults and a child (and some animals) walked across a layer of fresh ash, which was soon covered up and later turned into rock. The prints leave us in no doubt that some pre-humans walked on two legs at the same time as Lucy was alive. It seems reasonable to assume that the footprints were made by creatures like Lucy.

Mario Molina and Sherwood Rowland: The Ozone Layer in Peril

In the 1930s the DuPont chemical company in the United States began to produce a very useful new group of chemicals, called 'freons' by the makers, and chlorofluorocarbons or CFCs by the rest of the world.

1974

At that time, refrigeration systems used ammonia as the 'working fluid' to transfer heat from inside to outside. Though it performed that task well, ammonia was not ideal. It was poisonous and corrosive, and too hazardous to use in refrigerators in the home.

DuPont developed CFCs to replace ammonia. These compounds of carbon, hydrogen and chlorine were just as good at refrigeration; they were not toxic, not corrosive and they did not burn or support burning. By the 1960s another use had been found for CFCs—driving the contents out of the increasingly ubiquitous 'spray cans' or 'aerosols'.

But some crucial questions about CFCs had not been asked. What did they do to the wider environment once released? Would they be like DDT, hailed in the 1940s as a wonder insecticide that advanced the war on malaria but by the 1960s implicated (for example, in Rachel Carson's book *Silent Spring*) in the deaths of many young birds? Thalidomide, the 'morning sickness' drug taken by pregnant women and that was later proved to do terrible things to their unborn, is another example.

In 1974 American chemists Mario Molina and Sherwood Rowland raised an uncomfortable thought. CFCs were very hardy gases under normal conditions, lingering in the atmosphere for decades. But if they spread upwards 25 kilometres into the stratosphere (**1902 Heaviside and Kennelly**), the molecules could be broken open by intense ultraviolet light (**1802**) from the Sun, releasing free chlorine, a very reactive gas. This would then attack the accumulations of molecules of ozone (the 'ozone layer') that occur there. Ozone, a form of oxygen with three atoms in each molecule, blocks ultraviolet light from reaching the ground where it can harm living things, including ourselves.

The chemical reactions involved went much faster when it was cold. Rowland and Molina predicted that the damage to the ozone layer would be worst near the north and south poles and particularly in spring. The predicted loss of ozone would be enough to pose a threat to health and the environment.

So it proved. Research bases set up in the Antarctic in the 1950s had been routinely measuring the amounts of ultraviolet light reaching the ground, so they had a baseline against which to judge the damage to the ozone layer that began to show up in the 1980s. The scale of the threat was soon obvious, with the polar ozone layer weakening to only half its usual strength in the early spring, and the problem worsening year by year.

The long-term outcome of the story should be good. Alerted by this research, governments and international agencies moved in the late 1980s to limit and then to ban the release of CFCs, under an agreement called the Montréal Protocol. It was the first agreement of its kind to head off a global environmental threat. Since the ozone layer is regenerated by ultraviolet light from the sun, the weakness will heal in time and the problem will go away. But that will take some decades.

There was a bigger picture consequence here, a cautionary tale. We have learned to be a bit more sceptical about 'wonder chemicals', to ask about their 'down-sides' as well as their 'up-sides'. There always are some.

Choh Hoa Li: Our Brains' Own Morphine

1975

The capacity of opium and similar 'opiates' or 'narcotics' to deaden pain and produce a state of relaxation, even euphoria, has been known for millennia. In the early nineteenth century, the action of opium was found to be due to the presence of a special chemical, an alkaloid called morphine (**1818**). But uncertainty continued as to how opiates actually work. One suggestion was that there are certain locations within the brain and nervous system known as receptors, to which these chemicals can 'bind', that is, somehow plug in to influence mood and the perception of pain.

At the time this suggestion was first made, around 1970, it could not be tested. But the necessary technology was soon developed, and by 1983 a team in Sweden and two in the United States independently found the receptors. The immediate question was: why do we have them? The opium poppy is native to the Middle East, yet brains all around the world have the receptors, despite the fact that the human race seems to have developed in Africa, not the Middle East.

A bold answer was suggested. Perhaps morphine works because it mimics a substance that the body already makes, a chemical that binds to the receptor sites, interfering with the transmission of pain signals and making us feel good. Researchers in Scotland found a small chemical made up of a string of five amino acids in the brains of pigs. They called it 'encephalin', which means 'in the head'. It proved to be a weak pain-killer and strongly addictive, so the hunt was on.

A few decades before, American 'neurochemist' Choh Hoa Li had been studying the pituitary gland for another purpose. From the pituitaries of camels, he isolated a substance whose purpose he could not identify. Hearing about the Scottish discovery, he soon realised there was encephalin in his extract. It proved to be a powerful pain-killer, three times more powerful than morphine when injected into the bloodstream, 15 times more powerful when injected into the brain. He called it 'endorphin', which means roughly 'the morphine within'.

So the brain does have its own natural opiates (there are in fact several similar chemicals). These are released by the pituitary gland during times of stress or crisis; they let us ignore pain when our lives or the lives of loved ones are threatened. They generate the 'runner's high'—the sense of euphoria that can be brought on by vigorous and sustained exercise. It is likely that they are responsible for the pain-killing effects of acupuncture, and may even explain hypnosis and the commonly observed 'placebo effect'.

More specifically, the endorphins explain why drugs like morphine, opium, heroin and codeine (all chemically related) can be addictive. These drugs take hold of the receptors for the natural pain-killers, so that less of those are released. When the effect of the ingested or injected opiate wears off, lots of receptors are left empty, leading to a craving for the narcotic and a withdrawal response.

John Corliss: Alvin Finds Oases in the Deep

The discovery of vast undersea mountain ranges in the world's oceans caused by the spreading of the sea-bed (**1953 Ewing and Hess**) was joined a few decades later by another striking find. In 1977 a team of American researchers, led by oceanographer John Corliss, mounted an expedition to explore one of these ridges lying close to the Galapagos Islands, just off the coast of South America. They used an unmanned submarine named Alvin (perhaps called after the cartoon chipmunk popular at the time), equipped with lights, cameras and sampling equipment.

Researchers suspected that along the ridge, where molten rock is welling up from within the Earth's crust to form new sea floor, hot water would also be roaring up, forming undersea hot springs powered by heat from the molten rock below. The water might be at several hundred degrees, kept from boiling by the immense pressure at depth. But the water would be dark, lying too deep (more than

2000 metres down) for sunlight to penetrate and sustain photosynthesis, which is the engine of life elsewhere on the planet. The only source of food in the abyss would be a gentle rain of organic debris from creatures living and dying above. So they expected the rocks to be bare and the waters largely devoid of life.

They could not have been more wrong. They did find the hot 'hydrothermal' vents, which they called 'smokers' because they continuously spewed out mineral particles that clouded the water. Some of the smokers were white, some black. Deposits of these molecules, rich in compounds like iron sulphide, had built up to form chimneys, sometimes many metres high. Around them and even within them, the rocks unexpectedly teemed with life: tube worms more than a metre long, clams, blind crabs and numerous fish, all linked in a complex ecosystem. The mass of living matter per square metre was as high as anywhere on the planet, an oasis in the pitch dark.

This alien world, replicated in the other great oceans, was a stunning discovery and forced a rethink of what life needs to survive. Life must have energy, and sunlight is a source, but there are other possibilities. Down there in the blackness, the energy source proved to be chemical compounds in the hot water, in particular sulphide particles. Highly adapted bacteria, able to survive in the superheated water and to extract from those chemicals the energy they need to live and grow, are the start of the food chain. Many of them inhabit the body cavities of the tube worms, which as a result are provided with food.

We now know that life, even a complete ecosystem, is not necessarily dependent on light. So maybe life does not need light to get started. The deep dark ocean, with its tranquil environment and steady supply of energy from below, may have been where life on our planet first took hold. It is an intriguing idea.

Louis and Walter Alvarez:
The Mark of the Killer Asteroid

1980 The sudden disappearance of many life forms from the surface of the Earth at different times in our planet's history was first recognised 200 years ago by William 'Strata' Smith (**1799**); the idea was developed by his nephew, John Phillips (**1860**). By the late twentieth century, study of the geological record had identified half a dozen such 'extinction events', when more than 50 per cent (and sometimes more than 90 per cent) of all forms of life had disappeared very quickly (in geological terms).

Looking for causes, geologists and biologists had hypothesised massive changes in the Earth's environment, such as huge volcanic eruptions that released gases and dust to change the climate. Strikes on Earth by asteroids or comets were another possibility, especially when it was recognised that a small comet had exploded in the atmosphere above a remote area of Siberia in June 1908, laying waste 10 000 square kilometres of forest and sending a shockwave around the world.

The first strong evidence linking an asteroid with a mass extinction was gathered in 1980 by American physicist Louis Alvarez and his geologist son Walter. Digging near the town of Gubbio in Italy, they found a thin bed of clay that contained much more of the rare metal iridium than is normal. Iridium is much more common in meteorites than on Earth, and that led the Alvarez pair to propose that the enriched clay was evidence of an asteroid strike. Similar clay deposits were later found in other parts of the world, apparently representing the same time, suggesting that the asteroid must have been very big, perhaps 10 kilometres in diameter, to spread debris over so large an area.

In the same layer were other intriguing bits of evidence: fragments of quartz with distinctive fractures showing that they had been hit very hard, and tiny spheres of minerals that could have hardened from molten rock thrown through the air. Most striking of all was the date assigned to the layer—65 million years before the present, linking it in time to the demise of the dinosaurs at the end of the Mesozoic Era and the beginning of the age of mammals. An asteroid so large would have released gargantuan energy, causing massive destruction and even global climate change that dinosaurs and other creatures could not have survived.

To complete the picture, and make the hypothesis even more plausible, surveys in Mexico found a huge crater, now deeply buried under sediments, offshore from the Yucatan Peninsula. It was 180 kilometres across, the sort of hole the 'killer' asteroid would have left, and it, too, was dated 65 million years ago.

Not everyone was convinced, but the Yucatan asteroid strike is a very likely candidate for the destroyer of the dinosaurs. The discovery has sparked a hunt for similar causes of other mass extinctions, and more recently, a search for asteroids roaming near the Earth that might some day collide with it.

Alan Guth: The Universe Inflates

It is no secret that our universe is expanding, that the major objects in it— the vast clusters of galaxies with their many billions of stars—are moving further apart at immense speeds. American astronomer Edwin Hubble announced all this in **1929** and just about everyone accepts it.

1980

To explain why it is so, the Big Bang theory was developed around **1950**. This says that our universe began as a lump of energy and matter both massive and minuscule, with (initially) infinite density and temperature. From this 'singularity' (**1916**) our universe has expanded and cooled, with the galaxies and stars condensing out of the gas and dust to make the cosmos we see today, some 13 billion years later.

This is what is now called the 'Standard Model' of the history of our universe. But it has problems; there are things it cannot explain. One is the 'flatness' problem. In the world we know, parallel lines never meet, so we say our universe is 'flat'. We can imagine living in a different sort of universe where space overall is 'curved'. To take an analogy, meridians of longitude on the curved surface of the Earth are parallel at the equator but meet at the poles. Albert Einstein said that matter bends space (**1915**), so in our universe matter must be so thinly spread on average that space remains flat.

Then there is the 'uniformity' problem. Since Arno Penzias and Robert Wilson discovered the 'cosmic background' in **1965**, scientists have discovered that this 'afterglow' of the Big Bang, which represents the state of affairs about 300 000 years after the universe got going, is very uniform; the cosmos then was almost exactly the same temperature in every direction. But matter is far from uniform: galaxies are spread through space, but with immense voids between them and within them (between the stars).

In addition, there is the 'magnetic monopole' problem. The Big Bang should have made huge numbers of these exotic objects, which are like one end of a bar magnet without the other end (not an easy thing to imagine!). But careful searching has failed to find a single one. Where are they?

The then 34-year-old American physicist Alan Guth was wrestling with the monopole problem when he glimpsed a radical solution. Suppose that very early in its history our universe began to grow faster and faster. Imagine a pea becoming the size of a galaxy in the blink of an eye. After this period of violent 'inflation', the universe would have settled down to its current modest rate of expansion.

This vision solves the monopole problem and many others. Monopoles may have been plentiful in the early days, but the inflation has spread them out over so vast a volume of space that no one has ever seen one. Likewise, matter has been diluted by inflation to the point that it does not bend space overall (just here and there). And a small, uniform universe has inflated to become a very large, non-uniform universe. Tiny irregularities in the early universe have been amplified by the inflation to become the seeds around which the scattered galaxies have formed.

Guth is now the name most commonly associated with the 'inflationary universe' model, though others have helped develop the idea. The energy needed to inflate the universe so far and so fast comes, he says, from something like the release of energy when water vapour condenses into liquid water. This is what makes a thundercloud grow so large. Bizarre as it sounds, cosmic inflation is now widely accepted, and recent studies of the cosmic background have produced still more support (**1992 COBE**).

Stanley Prusiner: The Mystery of the Prions

1982

It was a shock when Dutch researcher Martius Beijerinck first proposed in **1898** the notion of a virus, something much smaller than the smallest cell but still able to cause disease. Viruses are very small and cannot themselves make proteins, as plant and animal cells do. However, they are virulent because they do have the genes that direct the protein synthesis: they reproduce by hijacking the protein-making machinery of the cells they infect.

Surely nothing smaller than a virus could cause infection. Alas, yes. The disease known as scrapie in sheep and goats and the related and much feared 'mad cow disease' (bovine spongiform encephalitis) and Creutzfelt Jakob Disease (CJD) that affects humans all seem to be caused by something much smaller and simpler than a virus. These diseases all cause some form of mental degeneration and appear to be infectious. Around 1974 they caught the attention of American neurobiologist Stanley Prusiner. He set out to find out what causes the diseases to spread.

By 1982 he had isolated the infectious agent. His discovery caused a lot of controversy because the disease-carrying particles, which he called 'prions', seemed so small, even smaller than viruses. Further study showed that they were just bits of protein containing no genetic material at all, not a single gene. So prions could not reproduce, not even as viruses do. They must work by somehow directly causing other proteins to malfunction.

Proteins are long strings of amino acids, but the way they work depends mostly on their structure, on how they are 'folded' into a particular (usually very complex) shape. If the folding of a protein is faulty, it will not do its job. Prions seem to be wrongly folded proteins themselves, which then cause other proteins to fold incorrrectly. The influence of prions seems able to cross species boundaries; mad cow disease appears in cattle fed meat meal made from scrapie-infected sheep.

It took Prusiner 25 years to convince the medical community to accept his ideas. His research could also impact on diseases like Alzheimer's and Parkinson's, which seem similar in some ways to diseases caused by prions, including the build-up of abnormal proteins in the brain.

Luc Montagnier and Robert Gallo: Finding the AIDS Virus

In 1981 a previously unknown disease began to present itself in the United States, especially among homosexual men. The immune systems of victims, which normally protects them against invading disease-causing micro-organisms, seemed severely weakened; they fell ill from pathogens they would normally have shrugged off. There were distinctive symptoms: a rare form of pneumonia, a particular cancer.

1983

By 1982 the disease, now called 'acquired immune deficiency syndrome' (AIDS), was showing up elsewhere. In France health workers observed it among people receiving blood transfusions, such as haemophiliacs. Since blood transfusions were routinely screened for bacteria, it seemed likely that AIDS was caused by a virus.

Researchers at the famed Pasteur Institute in Paris (**1886**) took up the hunt. As with the search for the cause of malaria (**1880**), the key was to identify something in the bodies of victims not found in others. Since AIDS victims regularly lost the critical disease-fighting blood cells called DC4 lymphocytes, the French team, under the leadership of Luc Montagnier, cultivated such cells from AIDS sufferers. They were looking for evidence that a virus was present, in particular a 'retrovirus'. These viruses, which cause many animal diseases, have their genes written on RNA rather than DNA (**1959**). To infect a cell, they need to make a copy of their genes on DNA so that the cell's protein-making machinery will follow it. This copying needs an enzyme called reverse transcriptase. Finding that in a sample would show that a retrovirus was at work.

Within months, the French team had found reverse transcriptase in a culture of DC4 cells that died soon after, so they knew the retrovirus was there. In early 1983 the virus was first seen under the microscope. At this stage, Montagnier turned for help to the American Robert Gallo, who had discovered the only other human retrovirus found to date, one which caused a rare form of leukaemia. Gallo, too, was looking for a virus as the cause of AIDS. By May 1983 the research teams, now numbering many dozens of people across a wide range of expertise, knew enough about the virus, later called HIV (human immunodeficiency virus), to publish their findings.

Over the following months, research in France, the United States and elsewhere strengthened the link between HIV and AIDS. Antibodies to HIV were found in patients with AIDS; this became the basis of blood tests that could show the presence of HIV long before symptoms of AIDS appeared. Since the spread of the virus involved the exchange of body fluids such as blood and semen, preventative measures could be devised. Work began on developing treatments and even vaccines; this work continues to this day with mixed success, with both Montagnier and Gallo prominent, among many others.

Research has also tried to discover the origins of AIDS, which apparently did not to exist in its present form before about 1980. That remains a matter of controversy.

Mitochondrial Eve and the Universal Adam

1987

There are two basic contending ideas for the origins of the human race, *Homo sapiens*. The 'out of Africa' scenario says that all modern humans are descendants of others of our species who migrated from a common homeland in Africa, having evolved there from our ancestor *Homo erectus* (**1856**). The alternative 'multiple origins' theory says that it was *Homo erectus* who migrated, evolving into *Homo sapiens* independently in various locations.

Evidence first published in 1987 indicates that all human beings can trace their ancestry back to a single female who lived in Africa more than a thousand centuries ago. This proposed ancestor is called Mitochondrial Eve (or ME for short). The 'mitochondrial' part of the name means that the evidence comes from comparing the DNA in the mitochondria, the energy sources in our body cells (**1967 Margulis**).

Mitochondrial DNA (and the genes made from it) is handed down only by mothers to daughters, which makes it much easier to trace. A father cannot give his mitochondrial DNA to his children.

The 'Eve' is a misleading title if it suggests that ME was the first woman. This was clearly not the case. She would have been only one of many women alive at the time, but it seems that only her mitochondrial DNA has been transmitted to all of the present generation. The other women have no descendants living today, or at least none who have come down through the matriarchal line, from mother to daughter and so on.

The similarity of our mitochondrial DNA, revealed by increasingly rapid and sophisticated analysis, means that there must have been an ME, even if we can never know who she was. While the mitochondrial DNA is handed down from generation to generation almost unchanged, mutations (**1901 de Vries**) do occur, and the number of differences between individuals constitute a genetic clock, indicating how long ago their common ancestor lived. The rate of mutation is still debated, but widely accepted estimates have ME living about 200 000 years ago. The geographical pattern of the changes points to a common origin in Africa.

Two other points. The Mitochondrial Eve theory does not really rub out the 'multiple origins' view. It could be that groups of *Homo erectus* migrated out of Africa a number of times, but only one of those 'tribes' has any matrilineal descendants alive today. In addition, something similar to ME has been found on the male side. Only men carry the Y chromosome (**1902 Sutton and Boveri**). Comparing the genes found across many individuals points to a most recent common male ancestor, a Y-chromosome Adam, who seems to have lived much more recently than Mitochondrial Eve. The debate seems set to continue for some time.

Stanley Pons and Martin Fleischman: Is Cold Fusion Real?

In 1989 well-known American electrochemists Stanley Pons and Martin Fleischman stunned a scientific gathering with the claim that they had initiated nuclear fusion in a piece of desktop apparatus at everyday temperatures.

1989

The disbelief of the audience was understandable. Nuclear fusion involves the blending together of atoms of an element (or, more precisely, the nuclei of those atoms) to form a heavier element. By this time, it was well accepted that the fusion of hydrogen atoms to form helium was the energy source of the Sun and the other stars (**1920 Eddington**). But hydrogen nuclei are all positively charged and so repel each other. The nuclei must be moving very fast to overcome that repulsion, and that means exceedingly high temperatures of many millions of degrees, such as occur in the hearts of stars. It was inconceivable that such a reaction could take place cold.

As they reported it, Pons and Fleischman had filled a glass vessel with 'heavy water' (water containing the heavy isotope of hydrogen called deuterium, in place of ordinary hydrogen) and immersed an electrode of the uncommon metal palladium. When they passed an electric current through the heavy water, a lot of energy was released, much more than they were supplying. It appeared that atoms of deuterium were fusing together, turning matter into energy.

Such a discovery, if it were true, would not only shake physics; it could also be a solution to the challenge to create energy without pollution. But scepticism was rampant. If fusion was really occurring, radiation should be given off, say in the form of neutrons. But there seemed to be none. Other researchers could not repeat what Pons and Fleischman claimed to have done. The whole thing was denounced as a mistake or incompetence or hype, like Percival Lowell seeing canals on Mars (**1877**), or Linus Pauling's crusade about megadoses of vitamin C (**1928**). Outright fraud was suspected, like the Piltdown Man hoax (**1912**).

There were other parallels—reported discoveries that ultimately did not fit the evidence. In 1966 Russian scientists had claimed they could make 'polywater', water 15 times more viscous than ordinary water, with a higher density and boiling point. Only after 15 years and better experimentation did the evidence go away. Polywater does not exist. Compared with polywater, 'cold fusion' was a nine-day wonder. It was discredited within months, denounced in the press as quackery. Pons and Fleischman were driven by the bad publicity from their jobs at the University of Utah.

Interestingly, the discovery has not yet gone away. More recent attempts to reproduce it have had more success. We should remember that Lord Kelvin once denounced X-rays as a hoax (**1900**); they were not. Louis Pasteur's germ theory (**1862**) was judged absurd; it isn't. Alfred Wegener's 'continental drift' (**1912**) was ridiculed; it is now orthodoxy. Pons and Fleischman may yet be vindicated. Only time and the evidence will tell.

Unravelling the Human Genome

1990

The discovery by James Watson and Francis Crick in **1953** of the nature of the genetic code opened up an exciting possibility. It should now be possible to read the entire genetic message, the 'genome', encrypted inside the cells of an organism. It should be possible to decipher the instructions to make every protein from which the organism is constructed, including all the enzymes to control the chemical reactions that make up life.

Time and the right tools were the key requisites. Even the simplest organisms have dozens of genes. Most have hundreds or thousands, and that makes a very large number of 'base pairs', which are in effect the letters in which the messages are written. Each protein must contain many dozens of amino acids; three letters are needed to identify which ones those are. So there are probably millions of base pairs to be read and placed in the right sequence. Even a single wrong letter means a wrong amino acid and so a faulty protein. Using the laborious methods of the early 1970s, the task would have taken forever.

But bright ideas and new technology arrived in time. The new techniques came from researchers like Englishman Frederick Sanger (**1955**) and Americans Walter Gilbert and Allan Maxam. In 1977 they independently found new and faster ways to 'sequence the bases'. By the mid-1980s, these methods had been automated and methods devised for making thousands of copies of long sequences of DNA so that many sequencing machines could work on them simultaneously.

In parallel with the new technology, the first complete genome was sequenced by Sanger in 1978, listing the 5000 base pairs in a simple virus. By the mid-1990s, researchers knew the 2-million base-pair sequence in the virus causing influenza; then the genome of a free-living organism (baker's yeast), then the human gut bacterium *Escherichia coli*, the workhorse of so much research in genetic engineering (**1972**); then of a multicelled organism (a threadworm). In 2000 the genome of another organism important in the history of genetic research, the fruit fly (**1911**), was laid bare. With 165 million 'letters', it was the biggest genome yet.

The beckoning goal was, of course, the human genome, the genetic message that describes us. The International Human Genome Project was initiated in 1990. Even with 1000 researchers working in 40 countries, it was predicted to take 20 years, and by 1997 only 3 per cent of the genetic sequence had been read. But again new technology and ideas came to the rescue and the work picked up pace. Competition also helped, with the publicly funded project racing against a private-enterprise effort headed by American Craig Ventner and his firm, Celera.

Suddenly in 2001 the job was almost done. Preliminary lists of the three billion base pairs that filled out more than 30 000 human genes were published independently by the two streams of the enterprise: the public and the private. Medical scientists anticipate that knowing the genome will bring new and perhaps revolutionary ways to diagnose and treat disease. Even if progress is slow, we can now say we 'know ourselves' in a more profound way than ever before.

COBE: The Ripples in the Sky

The discovery by Robert Wilson and Arno Penzias in **1965** of a faint 'glow' of microwave energy coming from all over the sky did more than vindicate the Big Bang theory of the origin of the universe (**1950**). The radiation was a snapshot of the state of the universe only a few hundred thousand years after it had formed. It was worth closer examination to see what detail might be embedded in it.

The first measurements of the cosmic background indicated that it was 'isotopic', that is, the same in all directions, showing the same temperature (around 2.7 degrees

1992

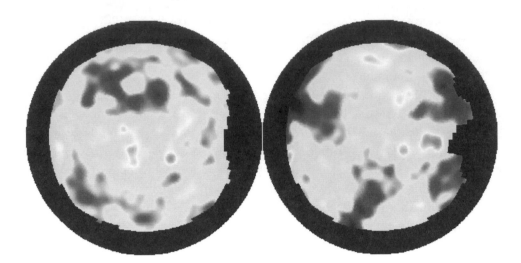

North Galactic Hemisphere   South Galactic Hemisphere

These images of the cosmic microwave background (CMB) radiation reveal the state of the
universe when it was less than a million years old. Made by the Cosmic Background Explorer
(COBE) satellite in the early 1990s, they show the two halves of the night sky, with the light and
dark patches revealing where the temperature of the CMB is slightly higher or lower than the
average (by a few millionths of a degree).

This shows that even when the universe was very young, it was not quite smooth; it had some
slightly denser regions. Astronomers believe that these irregularities or 'ripples' were the grains
around which the galaxies formed over billions of years.

above absolute zero). As measurements got more precise, tiny variations of a few thousandths of a degree in the cosmic radiation were picked up. Some of these turned out to be due to the Doppler Effect (**1842**). They showed that the Earth and the rest of the solar system were moving through space relative to the cosmic background at about 300 kilometres per second. Even smaller variations proved the motion of the Earth around the Sun, the ultimate vindication for Copernicus (**1543**) and Galileo (**1633**).

If these variations in the cosmic background were corrected for, what was left? By 1992 an American-launched space probe called the Cosmic Background Explorer (COBE) had been gathering data for several years. When summed up, the observations showed minute variations in the cosmic background, 'hot spots' and 'cold spots' differing by only 30 millionths of a degree. Tiny as they were, these variations were undoubtedly real, covering a few degrees of sky each and looking like ripples. It appeared that even so long ago, and so soon after the fireball of the Big Bang, the universe was not perfectly uniform—almost, but not quite. There were already regions that were a little hotter and therefore a little denser than others.

The universe is very much less regular nowadays, even on the largest scales. Nearly all of space is empty of matter. The galaxies are not scattered uniformly, but gathered into clusters and superclusters. In some places there are voids 100 million light-years across, apparently with no galaxies at all; in other places there are long bright lines of galaxies. One such structure, hundreds of millions of light-years long, is called the Great Wall.

The pictures from COBE link the long-ago with today. It seems that the tiny fluctuations in the density of the very early universe (represented by the minute ripples) have been made much stronger by gravity over billions of years, with matter being pulled together into the galaxies and clusters visible today. The observations were best explained by the story of an 'inflationary universe' (**1980 Guth**), so that theory has gained credibility. The pictures also supported an even more radical idea, that more than 90 per cent of the universe is not visible (**1936 Zwicky**). It is not made of ordinary matter of the kind that makes up our bodies or our planet. It seems that much of it is undetectable 'dark matter', and even more is made from the mysterious 'dark energy'.

And COBE revealed even more. It provided the most accurate estimates yet of the age of the universe (around 14 billion years), and of the Hubble Constant, which shows how fast the universe is expanding (**1929**). This is an ongoing story. Since COBE, another probe called WMAP has taken even more precise measurements; more advanced instruments are planned. With the 'dark matter' and 'dark energy' riddles still to be solved, the cosmic background has a lot more to tell us yet.

Plate Tectonics: Getting It All Together

The launch into orbit in 1992 of the Lageos satellite, a joint venture between the United States and Italy, was a significant event, doing for the Earth sciences what the advent of microscopes and telescopes in the seventeenth century had done for astronomy and biology. Objects and effects too small or distant to be seen and measured would now be within reach.

1992

For the geologists, the goal was the direct measurement of the movement of the continents. Once Alfred Wegener's initially ridiculed theory of continental drift (**1912**) was rehabilitated by the discovery of sea-floor spreading (**1953 Ewing and Hess**), few doubted that the continents moved, but the movement was so slow as to be undetectable.

Now, laser beams from ground stations reflected off the mirror-studded satellite could measure the distance between two points on the Earth's surface with an accuracy of a few centimetres. So any shift in that distance due to continental drift would show up in a few years. It's now known for sure that on average the continents relocate by a few centimetres every year, about as fast as your fingernails grow.

Continental drift, now proven, has become part of a bigger picture known as plate

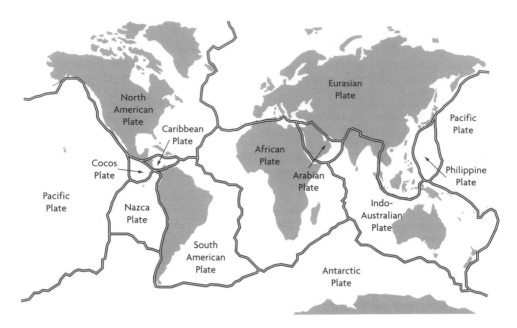

This map shows the major plates into which the outermost layer of the Earth (the crust) is broken. Relative motion between these plates causes the continents to drift, opening and closing oceans, assembling and breaking apart major landmasses.

Along much of the lines of contact between the plates, sea floor is being either created or consumed. Most earthquakes, volcanoes and deep trenches in the sea floor, together with mountain ranges such as the Andes, Rockies, Himalayas and Alps, occur close to plate boundaries.

tectonics. Continents move relative to each other (as do the beds of oceans) because they are carried on different fragments of crust, known as plates. There are about 15 plates floating on the hot, not quite solid mantle that lies beneath the crust. Some plates are huge, like the plate that holds the bed of the Pacific Ocean; others are quite small. They are pushed around by immense forces from below, liberated by the Earth's internal heat and delivered mostly through the mid-ocean ridges where new sea floor is continually forming.

There is now no real doubt about all of this, and explanations of many other phenomena flow easily and inevitably. Take earthquakes, for example. Maps show that earthquakes occur mostly along plate boundaries, especially where fragments of crust push past one another, as around the Pacific. Where one plate is driven under another (a subduction zone), we find volcanoes and deep ocean trenches. Plates in direct collision can raise mountains, such as the Himalayas and the Andes. Plate tectonics is now recognised as a major force shaping the surface of our planet, past, present and future.

Running the movements backwards in our computers, and bringing in other evidence, researchers can estimate how the continents were arranged, and constantly rearranged, in the past. The evidence points to all the continents being joined together some 300 million years ago, when dinosaurs first walked the Earth.

That supercontinent, first imagined by Wegener and known as Pangea (for 'all Earth'), slowly fragmented, first into two—Laurasia to the north and Gondwana to the South—and then into the continents we know today. All this activity left its mark on evolution: the mammals in Gondwana were, and remain, markedly different from those in Laurasia. This fragmentation of the continents has affected climate too: ice ages seem more common when continents are grouped around the North Pole, as they are now.

So plate tectonics has become much more than theory about drifting continents. It is a broad-ranging framework that draws in much more evidence, creating a vision of nature, like atomic theory (**1808**) or Maxwell's equations (**1871**) or the genetic code (**1953**). It represents the unity of understanding for which science strives.

William Schopf: When Did Life Begin?

We would know almost nothing about the history of our planet, or the development of life, if it were not for the 'record in the rocks'. Since the time of Leonardo da Vinci (**1517**), adventurous minds have assumed that

1993

fossils are the remains of organisms once living and perhaps now extinct. Later researchers recognised that the sequence of fossils reflected the evolution of life from simple to complex, from amoebas to humans. It became evident how the proliferation of life had waxed and waned over time (**1860**). With the discovery of radioactive dating (**1921**), we could begin to calculate the ages of the layers of rock and the relics of life within them.

As Charles Darwin had recognised in **1859**, there is little record of the very earliest life; nothing much before the appearance of various shellfish and the like at the start of the Cambrian period, now dated about 550 million years before the present. The oldest rocks seemed to have hardened more than three billion years earlier. So what had life been doing all that time?

As often happens in the history of science, new technology provided the answer. Two techniques in particular were valuable: a way to measure the ratio of the various isotopes of elements in a rock sample, and a way to produce three-dimensional images of tiny structures hidden within the rock. The first could detect traces of carbon and measure the ratio of the two isotopes C12 and C13, which is a good indication that life has been at work. The second could reveal the shapes of once-living cells, if that was what they were.

American 'palaeobiologist' (a biologist of ancient life) William Schopf has been one of the masters of these techniques. Early in 1993 he used these techniques to find evidence of extremely ancient life in rocks in arid Western Australia, close to Marble Bar. Although the rocks are nearly three and a half billion years old, according to the radioactive elements they contain, the fossil organisms in them look remarkably similar to some existing today—the colonies of stromatolites that build reefs in shallow bays along the Western Australian coast. These are cyanobacteria, aka 'blue–green algae', primitive organisms without a cell nucleus but able to carry on photosynthesis, trapping sunlight and carbon dioxide to make their food. It seems that their long-dead relations in the rocks nearby could do the same.

Older fossil-bearing rocks now known in other parts of the world are much the same age or a little younger. So the antiquity of life on Earth seems established, even though it took nearly three billion years for living things to have more than a single cell. There is another conclusion. Even if no older relics of life are found, this discovery means that life began on our planet within a few hundred million years of the surface cooling and hardening. It was as if life was simply waiting for a place to call home.

Galileo at Jupiter: One Mission Among Many

1995 In December 1995 the spacecraft *Galileo* arrived at the planet Jupiter after a six-year, four-billion-kilometre voyage from Earth. The appropriately named mission (**1610**) was the most ambitious to date, as the probe was to go into orbit around the giant planet and explore its system for several years. It would prove a brilliant success, stretching to 35 orbits over eight years, showing how far the 'enterprise of space' had come since the first primitive satellite had gone into orbit around the Earth in 1957, initiating the space age.

If the late twentieth century is when 'big science' emerged, nothing is bigger than the challenge of exploring space. It involves immense distances, huge hardware, hostile environments, massive costs and precision planning stretching over decades. But the returns have been equally great. Our knowledge of our cosmic neighbourhood has been transformed, as well as our understanding of our own planet.

The bulk of the missions were backed by the United States, but Russia, Japan and the nations of Europe also had their names and flags on the sides of space probes, and their scientists waiting for the data to come back by radio. Many things went wrong. Some probes failed in flight or crashed on landing, some went silent, some were lost. But in total, the era of space exploration has been a stunning success for technology, ingenuity, persistence and knowledge. Hardy spacecraft from Earth have carried cameras and multiple scientific instruments on 'fly-bys' of all the planets except distant Pluto, and in orbits around the Moon, the Sun, Venus, Mars, Jupiter and, most recently, Saturn. Our machines have touched down on the Moon, Venus, Mars and Saturn's largest moon, Titan, returning pictures of those harsh worlds.

We have marvelled at an immense variety of scenery, from the red and yellow crust of Jupiter's moon, Io, shaped by dozens of active volcanoes, to the frozen surface of Neptune's companion, Triton, with its geysers of liquid nitrogen; from the Sun-scorched arid rocks of Mercury to the riot of colour that sweeps and swirls through the chemically enriched atmospheres of Jupiter and Saturn.

It's now known that Mars hosts an extinct volcano 20 kilometres high, and a vast valley 100 kilometres wide extends half way round the planet. The three or four rings around Saturn (as seen from Earth) have proven to be hundreds of rings of dazzling intricacy. Membership of the solar system has multiplied, with dozens of tiny moons, invisible in even our largest Earth-bound telescopes, found to be orbiting Jupiter, Saturn, Uranus and Neptune.

While we watched, our automated emissaries flew through the head of Halley's Comet, soft-landed on an asteroid, sampled the solar wind, survived transits of the rings of Saturn, roamed across the surface of the Moon and Mars and mapped the surface of Venus by radar through its almost impenetrable cloud. Two probes have left the solar system altogether, crossing the orbits of Neptune and Pluto and now heading for interstellar space. What would Galileo have thought of it all?

We have learnt so much that any summary must be incomplete. Perhaps most exciting is the growing evidence that Mars, now cold and apparently arid, was once warmer and much wetter, and that Jupiter's ice-shrouded moon Europa has a deeply hidden saltwater ocean. Both of these environments may have been, and may still be, fit for life of some kind.

Finding the First Planet Elsewhere

Over the last 400 years, ever since Galileo (**1610**) and others first turned their primitive telescopes to the night sky, the number of known objects in our solar system has multiplied manyfold: three more planets, dozens of moons, hundreds, even thousands, of comets and asteroids. Scientists suspect there are many more, perhaps millions of lumps of ice and rock orbiting the Sun in the half-dark far beyond the outer planets.

1995

But all of these are still under the Sun's control, presumably all formed from the one vast, rotating cloud of gas and dust, as Laplace first hypothesised (**1796**). Given that there are millions of stars similar to our Sun, we can ask whether the same process has gone on elsewhere, yielding families of planets around distant stars. It sounds reasonable, especially since the Hubble Space Telescope has already found discs of gas and dust around other stars.

But finding a planet elsewhere is a tough task. A planet around even the nearest other star would appear so small in even our largest telescopes, and so swamped by the light of its sun, as to be impossible to detect directly with current technology.

Fortunately, there are other ways, and some of these are beginning to bear fruit, including one discovered by Christian Doppler (**1842**). He predicted that the movement of a source of light, such as a star, will change its colour as it moves; it will be redder if it is going away, and bluer if is approaching. This is how the expansion of the universe was discovered (**1929**). The gravitational pull of a planet on the star it orbits will make the star wobble very slightly back and forth. Light from the star will go a little redder, then a little bluer and then redder again as the planet goes around in its orbit.

It took until the 1990s for light-analysing equipment to become sensitive and precise enough to make this wobble visible. In October 1995 the first planet not orbiting our Sun, a companion to the star 51 Pegasi, was announced. The amount of the wobble and the time it takes allowed researchers to calculate both the mass of the planet and how far it is from its own sun. The planet 51 Pegasi is at least half the mass of Jupiter.

Other discoveries quickly followed and, as technology has improved, the numbers of known 'extrasolar' planets has multiplied. Our solar system is definitely not alone. Within 10 years, more than 100 such planets were known, all much bigger than the Earth and many orbiting very close to their suns ('hot Jupiters'). Other methods should help pick up smaller planets, such as measuring the dimming of the star's light as the planet passes in front. But finding an Earth-sized planet is likely to need a special telescope placed in orbit. The European Space Agency has plans for a mission of this kind, called *Darwin*.

Fermilab Finds the Top Quark

1996 In 1996 the immense particle accelerator located at Fermilab on the outskirts of Chicago found what it was looking for—evidence of the fleeting free existence of a subatomic particle known as the 'top quark', predicted to exist but never before seen. The discovery paralleled the identification of new elements to plug the gaps in Dmitri Mendeleyev's Periodic Table (**1875**) or the discovery of the positron to vindicate Paul Dirac's **1928** prediction of the existence of antimatter.

Particle physics in the late twentieth century represented 'big science' at its biggest. The accelerator hurled protons and antiprotons around a ring 3 kilometres across in diameter at nearly the speed of light. When the beams smashed together, they released energy measured in trillions of 'electron volts': the first accelerators (**1932 Cavendish Laboratory**) could manage only 100 000 such units. From the fireball produced when the protons and antiprotons annihilated each other, new particles could be born, if enough energy was available.

By 1996 the 'Tevatron' at Fermilab was operating above the critical energy level. As the particle beams collided, assembled teams of hundreds of physicists and engineers (another part of 'big science') gathered evidence, not of the top quark directly, since it broke down almost as soon as it was created, but of the debris from its demise. Assembling those fragments, and tracing them back, it was clear that the top quark—indeed many of them—had fleetingly existed.

Predicting the top quark flowed from what the physicists were boldly calling the

MATTER PARTICLES			FORCE PARTICLES
Generation 1	**Generation 2**	**Generation 3**	
BARYONS up quark	strange quark	top quark	photon (electomagnetic force)
down quark	charmed quark	bottom quark	gluon (strong force)
LEPTONS electron	muon	tau	W+, W– (weak force)
neutrino	muon nutrino	tau neutrino	Z_0 (weak force)

This table lists the 16 'fundamental particles', according the currently accepted Standard Model of matter. Twelve of the particles go together in various combinations to make up matter; the other four carry the forces that make the matter particles attract, repel or change.

The matter particles come in three 'generations'. Only the first-generation particles (the up and down quarks—which combine in threes to make protons and neutrons—the electron and the neutrino) are found in everyday matter. The other generations are much heavier and disintegrate or decay quickly. The muon and tau particles behave like heavy electrons; each has its special neutrino (**1932 Pauli**).

The Standard Model does not include gravity, so there is no particle designated to carry that force.

'Standard Model' (**1967 Glashow, Salam and Weinberg**), a set of relationships that could explain the bewildering array of 'fundamental particles' known to exist. The Standard Model incorporated the notion of quarks, three in number, first put forward in **1963** as the building blocks of protons and neutrons in particular. As the theory developed, it seemed that six such particles were needed to explain the full gamut of observations, but only two quarks (the 'up' and the 'down') were found inside everyday matter. The rest, all unstable, may have existed in the hugely hot and dense days of the early universe; now scientists have to make them in huge machines like those at Fermilab.

As the energy of these machines was ratcheted up, the quarks were created one by one: the 'strange' quark, the 'charmed' quark, the 'bottom' quark. As the heaviest, the 'top' quark came last. Physicists seem confident that there are no more to be found, but they have not stopping building and hunting. The next target is the 'Higgs particle', which the Standard Model needs to explain why all the other particles have any mass. The idea is that mass is the result of other particles exchanging Higgs particles, just as the force between electric charges comes from the exchange of photons.

To find Higgs particles—and other things—the world's biggest particle physics laboratory at CERN outside Geneva in Switzerland is currently (2005) readying a gargantuan accelerator, housed in an underground ring 27 kilometres in circumference. It is a billion-dollar project. That's how important it is to physics to find the Higgs particle: it will make the Standard Model just about complete.

James Thomson: The Promise of Stem Cells

1998

The use of 'stem cells' derived from human embryos for research or therapy is controversial today, with the opinions of the public and of governments divided on the matter. The debate is very new, since it was only in 1998 that cultures of such cells were first created.

The American James Thomson was first to do it. He led a team of biologists that established five independent lines of stem cells based on human embryos not required for an in-vitro fertilisation program. These were not the first embryonic stem cells, only the first human ones. Previously they had been derived from smaller animals such as mice. Thomson himself had obtained them from Rhesus monkeys.

Stem cells are taken from a fertilised cell less than a week after conception, at a time when the cell mass is a hollow ball called a blastocyst. A mass of cells on one side of the inner surface of the ball will become the embryo; the stem cells are found there. Since no organisation or differentiation of the cells has yet taken place, they all have the potential, depending on circumstances, to develop into any sort of cells found in the body, in the nervous system, the gut, the blood, muscle, bone, cartilage or any organ. Once set up in a culture, the cells continue to divide and increase in number, essentially for ever, but not to change their form. They retain their unique capacity to become any sort of cell.

Many researchers see great promise in embryonic stem cells, both in increasing our understanding of how the human body works at the level of the cell, such as why diseased cells die, and also in developing ways to treat diseases such as diabetes, Alzheimer's, Parkinson's and cancer. Stem cells of a kind exist in adults, such as those that continually replenish blood cells or neurons or skin. But they are already specialised to some extent, so that skin-forming adult stem cells cannot replace blood cells. Neither would help repair a diseased organ.

So the best results are likely to come from using embryonic stem cells, but a great deal of research remains to be done.

A Violent Universe: Black Holes and Bursters

1999

If you wander outdoors on a starry night and gaze upwards, all seems peaceful. You have no sense of living in a universe permeated by violence. Yet cataclysmic events are constantly occurring out there. Some of these are visible to anyone, such as the supernovas that occurred in **1572** and **1604** (**Inconstant Heavens**). Others need the sort of detecting and measuring instruments developed only recently.

For example, there are now instruments that measure the flood of X-rays coming from certain parts of the sky. These 'X-ray hot spots' are almost certainly the result of matter being violently heated in the intense gravity around a black hole, the kind Karl Schwarzschild predicted (**1916**).

Scientists also believe that huge black holes many millions of times the mass of the Sun lurk within galaxies, including our own. Stars orbit the cores of such galaxies so fast that some immense mass must be driving them. These black holes seem peaceful now, but in galaxies far away when the universe was much younger they devoured gas and dust, and even ripped apart whole stars, lighting up as the dazzling lightshows now visible as quasars (**1964 Hazard and Schmidt**).

Another family of violent objects, the gamma ray bursters (GRBs), was found by accident. In 1973 satellites in orbit to detect the clandestine testing of nuclear weapons recorded a burst of gamma radiation, such as a nuclear test would produce but coming from deep space. Something odd was going on. It was not until 1991 that a special gamma ray telescope was put into orbit to find more such outbursts. It recorded them arriving about once a day, spread evenly across the sky. This showed that they were generated far beyond our galaxy, otherwise more would have been found in the region of the sky where the centre of our galaxy lies (**1918**).

Later still, a network was set up so that optical telescopes could be quickly pointed to photograph the part of the sky where a GRB had been reported. Such a chance came in 1999. The telescopes found a bright point of light. This quickly faded, but there was time to measure a red shift (**1912 Slipher**). This indicated that the source of the gamma rays was a stupendous nine billion light-years away, more than half way to the edge of the observable universe. To be detected at all at such a distance, the energy released must have been equivalent to our Sun being suddenly and totally converted to energy. Experts are still puzzling at how that can happen.

Had that GRB, one of the biggest ever seen, been within in our galaxy, say 2000 light-years away, its light would have made the night sky twice as bright as day. Much closer and it would have fried us. But nine billion light-years away means nine billion years back in time. Perhaps these outbursts, like quasars, happened only long ago. If they do not happen now, they will not happen anywhere near us.

2001

and Beyond

F ive centuries of investigation and insight, drawing on a tradition and practice
that grew up in Europe, have revealed much of the previously hidden (or even
unsuspected) workings of the natural world. Knowledge of what scientists
have discovered has been widely disseminated throughout society. Any high-school
student knows about things that would have astounded da Vinci or Galileo.

What lies ahead? What will future decades and centuries of science reveal? The
answers, of course, are unknown, but we know some of the questions that will be
asked. The major strands of inquiry that we have discerned in the previous 500 years
will continue to advance. On the biggest scale, we will understand more fully how
the universe came to be, its history and its future (**1950 Big Bang**). We will learn
about the nature of the 'dark matter' and 'dark energy' that seem to make up most of
the discernible universe (**1992 COBE**).

Studies of our near neighbours, such as Mars, and Jupiter's moon Europa (**1995
Galileo at Jupiter**), may reveal traces of life, present or past, elsewhere than on Earth.
We will find Earth-sized planets around other stars (**1995 First Planet**), since there
is little doubt that these exist, and take pictures of them. Studies of their
atmospheres should reveal whether or not they bear life, long before we are able to
get near them.

At the other end of the size scale, the final touches will be put onto the Standard

Model of matter and forces (**1996**). We will find the Higgs particle, if it exists. But if previous experience is any guide, we will then start looking for an even more comprehensive explanation of why matter behaves as it does. This will include drawing gravity into the Standard Model to produce a 'Theory of Everything'.

The Human Genome Project (**1990**), now almost complete, will spawn new endeavours. One is already under way: 'proteomics' (as opposed to 'genomics') seeks to identify and describe each of the tens of thousands of different proteins whose manufacture is controlled by the genome (**1959**), and which make up and regulate human beings and other organisms. The practical benefits of such knowledge should be immense.

We should be able to fill in the missing steps in the story of our own evolution (**1964 Leakey**) and form a more complete understanding of the total pattern of life on our planet (**1967 Margulis**). We will come to understand more completely how complex organs such as our brains (**1961**, **1975**) actually work. We will surely at least come close to creating living things from non-living chemicals (**1953 Miller and Urey**).

In the earth sciences, we will continue to draw on the power of plate tectonics (**1992**) to explain, and even perhaps foretell, how our planet behaves or will behave in the future. A capacity to predict earthquakes, volcanic eruptions and tsunamis is a tantalising vision, though it is at least several decades away, and may never be achieved.

In these and related fields, such as meteorology, researchers will be empowered by the rapidly strengthening technology of information processing, which allows ever more complex natural systems to be 'simulated' inside supercomputers. Predictions of weather and climate change (**1922 Bjerknes**) will become more reliable and longer range. Chemists will be able to more completely analyse useful molecules and design better ones. Physicists will create a vision of the earliest moments of the universe, or its last.

New knowledge will come with new ways of looking at the world. Devices to collect and analyse information, of a power and sophistication never before seen, are coming down the track: telescopes with light-gathering power one million times greater than Galileo's (**1610**); orbiting and earthbound observatories tuned to every sort of signal the universe generates; microscopes that chart single atoms; and bursts of laser light short enough to catch chemical reactions in the act.

Many such technologies will be housed in great international facilities, as we have already begun to see. 'Big science' looks set to become still bigger. The research teams that use these facilities will also be large; again that change is already evident. This suggests that no single scientist will dominate the scene, as did Galileo, Newton, Lavoisier, Pasteur, Darwin or Einstein. Nor will any individual researcher range over as broad a spectrum of phenomena as did da Vinci, Hooke, Cavendish or Michell. Multidisciplinary teams will continue to be essential for most challenges, but individual players will each have their own small role.

It has been suggested that we will soon see the 'end of science', that all the serious questions have been asked and answered. This, of course, has been said before (**1900 Kelvin**), and is likely to be proven wrong again. Five hundred years of scientific discovery has shown that every advance in instrumentation reveals new, often unsuspected, phenomena; that every question answered asks several more; that previously well-supported theories must often yield, a little or a lot, to new evidence.

And even if we ever do know everything about how the world works, there will always be a profound purpose for science, even beyond its powerful stimulus to the imagination and the gratification it provides to endless human curiosity. We need science to help us see the consequences of particular courses of action, such as the ones we are currently following.

Major issues such as global warming, the best use of resources, genetic engineering and nanotechnology are more than challenges to scientific understanding. They commonly require urgent collective and political action. To make the best informed choices for the present and future, we need the best available information and insights. Science is a vital tool for the survival of humans and other species, and for enhancing the quality of human life. That is why it will continue.

Further Reading

BOOKS

Dictionary of the History of Science, W.F. Bynum, E.J. Brown and Roy Porte (eds), Macmillian Reference Paperbacks, UK, 1981.

A comprehensive overview of the history of science and related fields, with contributions by many of the leading experts.

Science in History (four volumes), J.D. Bernal, Pelican Books, UK, 1969 (third ed.).

A classic study by a leading physicist of science interacting with society and politics.

A Short History of Nearly Everything, Bill Bryson, Broadway Books, USA, 2003.

Science, Past and Present, F. Sherwood, Taylor Mercury Books, UK, 1945.

Another classic, with extensive quotes from historic papers. Written by a leading chemistry educator.

ONLINE RESOURCES

reference.allrefer.com/encyclopedia/Science_and_Technology/Biographies.html
www-history.mcs.st-and.ac.uk/history/BiogIndex.html
scienceworld.wolfram.com/biography/

The above three sites have short biographies of many hundreds of workers in science and technology.

www.psigate.ac.uk/newsite/science_timeline.html provides an excellent timeline of discoveries in physical sciences.

http://www.strangescience.net/timeline.htm provides a timeline of biology and fossils.

www.physics.ohio-state.edu/~wilkins/science/sctmln.html
http://www.encyclopedia4u.com/l/list-of-themed-timelines.html

These two sites have many more specialised timelines.

The Galileo Project at http://galileo.rice.edu/lib/catalog.html is a fascinating collection of over 600 detailed biographies on members of the scientific community during the sixteenth and seventeenth centuries.

www.chemheritage.org/explore/explore.html is typical of what is available in particular areas of science (in this case, chemistry).

100.1911encyclopedia.org houses the 1911 edition of the *Encyclopedia Brittanica*, perhaps the most comprehensive ever.

Wikipedia at http://en.wikipedia.org/wiki/Main_Page, the monumental collaborative online reference, is invaluable.

nobelprize.org provides comprehensive information about winners of the Nobel Prize.

quotationspage.com is a good source of quotations.

Index